Think to New Worlds

The Cultural History
of Charles Fort
and His Followers

Think to
New Worlds

Joshua Blu Buhs

The University of Chicago Press

Chicago and London

The University of Chicago Press, Chicago 60637
The University of Chicago Press, Ltd., London
© 2024 by The University of Chicago
Published 2024
Printed in the United States of America

33 32 31 30 29 28 27 26 25 24   1 2 3 4 5

ISBN-13: 978-0-226-83148-0 (cloth)
ISBN-13: 978-0-226-83149-7 (e-book)
DOI: https://doi.org/10.7208/chicago/9780226831497.001.0001

Library of Congress Cataloging-in-Publication Data

Names: Buhs, Joshua Blu, author.
Title: Think to new worlds : the cultural history of Charles Fort and
    his followers / Joshua Blu Buhs.
Description: Chicago : The University of Chicago Press, 2024. | Includes
    bibliographical references and index.
Identifiers: LCCN 2023049035 | ISBN 9780226831480 (cloth) |
    ISBN 9780226831497 (ebook)
Subjects: LCSH: Fort, Charles, 1874–1932—Influence. | Fort, Charles,
    1874–1932. | Fortean Society. | Science writers—United States. | Science
    fiction, American—20th century—History and criticism. | Science in
    literature. | Authors, American—20th century—Biography. | BISAC:
    LITERARY CRITICISM / Science Fiction & Fantasy | SCIENCE / History
Classification: LCC PS3511.O63 Z56 2024 | DDC 813/.52—dc23/eng/20231127
LC record available at https://lccn.loc.gov/2023049035

♾ This paper meets the requirements of ANSI/NISO Z39.48-1992
(Permanence of Paper).

To Kim, to Quin:
tattooed on my soul, scrimshawed on my bones

No sense in no sense innocence of what not and what of delight. In no sense innocence in no sense and what in delight and not, in no sense innocence in no sense no sense what, in no sense and delight, and in no sense delight and not in no sense and delight and not, no sense in no sense innocence and delight.
Alright.

GERTRUDE STEIN, "Are There Arithmetics," 1927

Contents

1

First Must We Think
to New Worlds

How shall I be a mirror to this modernity?
When lo! in a rush, dragging
A blunt boat on the yielding river—
WILLIAM CARLOS WILLIAMS, "The Wanderer," 1917

Charles Fort, a man of steady, uncomplicated habits, watched, be-mused, as his body refused the motions. Evenings had long been given to walks with his wife Annie through their Bronx neighbor-hood, usually to the movies. These came less frequently now, the pace slower, the destination without joy, since sitting for long peri-ods hurt—the fleshy padding that had protected his bones was melt-ing away. The hard cheeses he'd enjoyed noshing, the homebrew he imbibed—it was 1932, alcohol sales prohibited—these he could not stomach. Nor did he find pleasure in smoking, or in sleep. He was startled by the gaunt face that greeted him in the mirror. Shaving presented new challenges, hollows inaccessible to the blade.

Go to the doctor, Annie said. But he refused. Fort had been collecting weird notices in newspapers, had made a special study of medical follies that confirmed in him a long-standing skepticism of all authority, religious, civil, scientific, medical. A scribbler for more than three decades, he considered compiling his clippings into a book but, already struggling to complete a different work, didn't think there'd be days enough for *Medi-Vaudeville*. Stubbornness would have to be his last word on the subject. Fort tried to turn the situation into a joke: "So many friends now know about me and no doctor that it is a challenge to me to get well." The joke was feeble, and he could not rise to the challenge. Winter went, spring came, and he found himself confined to bed.[1]

Fortean Modernity

Broke, struggling, Fort was in no position to see that, through the three books he'd published over the past decade or so and the fourth he'd turn out on his deathbed, he would have an influential afterlife, his name transformed into an adjective for an entire category of existence. Once he'd been friends with literary giants, Theodore Dreiser, Booth Tarkington, Benjamin DeCasseres. Once the genre of writing he'd invented had astounded readers and inspired imitators. Religious studies scholar Jeffrey Kripal calls it "science mysticism," a mythology of the inexplicable. "The history of the paranormal," he writes, "is a series of footnotes to Charles Fort."[2] Fort scanned newspapers, scientific journals, and magazines for earnest reports of bizarre happenings. He made notes on rains of blood and frogs, on dogs that talked and vampires, on things that appeared and disappeared without obvious reason, on strange beasts and visions in the sky, on unaccountable fires and impossible powers. He strung these together in books resembling Robert L. Ripley's *Believe It or Not* comics and books, which were popular at about the same time. But while Ripley's freak-show theatrics were always undergirded by respect for science—the events he depicted were either true or they were not, and science would decide—Fort defied science's pronouncements. His was a radical skepticism that refused to accept anything as absolutely true or absolutely false.

Once, not long before his illness forced him to bed, Fort had been the center of a society, praised as the father of a new philosophy. The iconoclastic writer and adman Tiffany Thayer had dreamed up the idea with Dreiser and the newspaper publisher J. David Stern. They had recruited friends and fans, made a big to-do. But the society had fallen apart, and Fort's world had been restricted until it was hardly larger than his apartment. Fort had endured such contractions before, lows that counterbalanced the highs. He'd been considered a kook, a nut, an atavism. Fort, noted one reviewer, "flings away a world of science for a medieval alchemy of ghosts, occult forces and spiritualistic controls."[3] He was out of step with the modern temper, even to his biographers, one of whom called him a prophet, another of whom subtitled his book "The Man Who Invented the Supernatural." And this illness, he knew, would be the end.

Although gifted with a powerful imagination, it would have been difficult for Fort to foresee that, in time, he would again be the center of a society, beloved of the cynical and the playful, embraced as well by writers, artists, and visionaries. His strange books would not be his only legacy. Forteans, participants in this contentious, ragged movement would call themselves. Fort and his creations are relatively well known, Fort the subject of two full biographies and a couple of longer biographical investigations, his ideas analyzed by scholars and enthusiasts, his books still in print.[4] Less understood are his followers; they've been ignored or dismissed as etiolated imitators.[5] This is unfortunate. Grappling with Fort's own contradictions, the lessons he taught, the patterns of thought he crafted, his toggling between naive wonder and pinched cynicism, they forged a unique response to modernity, influential and important, a Fortean modernity that has not been studied by historians but deserves to be. Contrary to the reviewer who accused Fort of trading science for medieval alchemy, Forteans did not (usually) see Fort or themselves as throwbacks, escaping the contemporary world for the safety of the past. They were forward-looking, fully modern in conception and in fact. Forteans had their largest effect on the practices of science fiction, aesthetic modernism, and UFOlogy, pursuits seemingly peripheral to the mainstream. But it was in these cultural arenas, places both highbrow and low, that essential questions were asked—and answered—about the limits of imagination, the structure of reality, and the workings of power.[6] Forteans were central to these investigations. Understanding Forteans offers a new perspective on twentieth-century cultural history, from an oblique but important angle.

"Modernity" is a complicated, much-debated term. Generally

speaking, it refers to a set of massive, interlocking changes that rocked the world starting in the late eighteenth century and lasted, at least, to the middle of the twentieth: the birth of the world we know, continuous with what came before but in many respects fundamentally different—new. Fort was moved by the irruption. The creation of liberal states, and their near destruction by war, by political and economic crises. The breakdown of religious authority, and the revelation of what seemed a new pantheon that humanity had unleashed and now served: technology, science, capitalism. The invention of new methods to investigate these subjects: history, sociology, economics. Psychologists mapped the mind, the subconscious, the unconscious, fragments of the self, explored humanity's interior states, just as other scientists elucidated the intricate mechanisms of nature. Artists imagined radical new forms to understand this new world, describe it, contest it.[7] New prophets arose, and from them came paeans and jeremiads.[8]

The historian Michael Saler notes that for many analysts of modernity, seeking to comprehend its heterogeneity, confusions, and reach, there has been a touchstone, a constant inside the diversity, what has come to be called "disenchantment." Disenchantment is the idea that all those changes, economic, political, scientific, and technological, contrived to empty the world of wonder, to sever connections between humanity and ultimate meanings, higher planes.[9] "God is dead," said the German philosopher Friedrich Nietzsche.[10] Science showed nature to be ruled by forces soulless, indifferent, but inescapable, said the sociologist Max Weber.[11] And theirs became the conventional wisdom. Sociologists proved to their satisfaction that the world was on a path toward secularism, spiritual yearnings sloughed off; historians, that humanity had been thrust from the warm embrace of a world live with animism, thrumming with the pulse of God, into one cool and cruel.[12] Underneath the welter, the fundamental dislocation of modernity was negative. This understanding of the period was pervasive, persuasive into the twentieth century. And it was wrong. Over the last twenty-five years, a new generation of historians, sociologists, and philosophers has challenged the idea that modern people lost their sense of wonder, surrendered their relationship to the ultimate.[13] On the contrary, they have found that modernity released the marvelous, expanded the possible ways in which humanity came into the presence of the awesome. The death of God and the rise of science as the preeminent process for creating truth opened new provinces, provided new materials for imagining, inventing, experiencing enchantment.[14]

Response to the prospect of disenchantment and the possibilities of new forms of enchantment was complex and contradictory.[15] Modernity, Saler writes, was defined by "unresolved contradictions and oppositions, or antinomies."[16] But order can be imposed. There were indeed those who understood modernity as the shedding of religion, who accepted, implicitly or with great fanfare, the requirement of dressing in the garb of scientific reasoning. Others thought science might bring spiritual experience under its aegis, might prove, for example, that spiritualist mediums really did communicate with the dead. Perhaps ghosts were real, the afterlife discernible.[17] Modernity was not disenchanted, concludes Kripal, just differently religious.[18]

Some predicted that modernity would carry humanity into a new era of undreamed-of achievement, a time of much amazement: the future enchanted.[19] Others agreed on the deal's terms but refused to pay. These antimodernists, with whom Fort's critics associated him, tried to bring enchantment to their lives by praising the past or costuming themselves in the trappings of non-Western cultures that presumably retained their sense of awe and wonder.[20] There were those who marveled over science and technological innovation, the delicate mechanisms and strange laws science revealed, the ingenuity displayed by researchers in elucidating facts about the world, the clever machinery of engineers.[21] And there were those who played with what Saler calls the ironic imagination.[22] These acted as if the world were enchanted, allowing themselves to touch the marvelous, but recognized that they were only playing. Thus, audiences flocked to see magicians perform tricks, felt a sense of amazement well up in their chests, but knew all along they were being deceived, witnessing legerdemain, not real magic.[23]

The sociologist Simon Locke identifies four ways in which moderns mixed enchantment and disenchantment, basing his ideas in the work of Max Weber.[24] In 1917, as the world neared the end of its first great war and embarked on a recovery that would lead to a second a mere two decades on, Weber told students at a German university that theirs would be a life of disenchantment—especially those who chose a scientific career.[25] Science replaced spiritual explanations with mechanical ones, he said, and its progress left scientific work without ultimate meaning.[26] Each generation of scientists watched their ideas be disproven by those who succeeded them. At that exact moment, for example, Albert Einstein and other physicists were disproving the ideas of Isaac Newton, which were the very model of science, or, if not disproving exactly, then showing them as provincial. In a fundamental way, scientific research was meaningless. Citizens of modernity, scientific and lay, had to face this new condition bravely. The modern world would circumscribe their lives. Combined with the rising importance of bureaucracy and the growth of state power, science was constructing about each person, Weber said in another context, an "iron cage of rationality."[27]

Locke argues that Weber was not as disconsoling as he might seem. Weber knew there were ways in which the world was still experienced as enchanted but died before he could fully develop them. So Locke took up the task, almost a century on. Science, he noted, provides resources for moderns to push back against science and construct new forms of enchantment. Scientists make mistakes. Science's reach often exceeds its grasp—the facts uncovered on lab benches spin out into larger narratives. At the time Weber was writing, for instance, scientists were uncovering biochemical reactions in humans, and from these studies came claims that humans lacked free will. Only bags of flesh, people were puppets controlled by the mixing of chemicals.[28] Yet by definition, Locke notes, science also opens realms of the unknown, the fantastic and nonfactual. As physicists in the early 1900s probed the structure of the atom, they created space for moderns to speculate about what lay just beyond the scientific theorizing. What strange realms existed beyond humanity's ken? Even as research disproved the idea that space was filled with an insensible matter called ether—space was, rather, a void—scientists left behind the notion of ether as something that could be played with or sewn into an alternative system of thought, one that could feel more enchanting.[29]

The resources provided by science—its mistakes, its overreach, its generation of new realms—were used to build, by Locke's count, four responses to modernity and its supposed disenchantment. These responses contested science and left room for enchantment, but they were thoroughly modern, borrowing from science as needed, facing forward, engaged in the forging of a new, vibrant world. One was fundamentalism, whose adherents see the world as controlled by a singular being using spiritual means (Christian fundamentalism being the best-known example). Often seen as the very opposite of modernity, a retrenchment, fundamentalism is in fact recent, research has revealed, a modern innovation in religion.[30] A second response, that of "New Age" moderns, also posits a world manipulated by spiritual forces, not in this case a single author of reality but multiple agents—angels and demons and fairies as well as discarnate forces.[31] A third response is conspiracism. Conspiracists, like fundamentalists, see the world as controlled by a lone being or perhaps a small cabal, but these rulers—often though not always hidden—use earthly powers to conduct their business: politics and economics, propaganda and intrigue. Conspiracism has a bad reputation, associated as it is with fringe movements, paranoia, and, very often, violence, white supremacy, anti-Semitism, and other revanchist fantasies.[32] But these connotations, while correct, are limited, for there are indeed conspiracies, documented and real.[33]

Locke named his fourth category, Forteanism, after Charles Fort. Unconcerned with spiritual matters, Fort focused, like conspiracists, on mundane forces, albeit some that were not yet recognized or understood. He played with the idea of a singular ruler and invented a philosophy that had only a single ultimate law, but his writings continually invoked many centers of power, distinguishing him from conspiracists. His emphasis on the mundane was not unlike that of science. But he rejected science's role as the sole arbiter of truth, while also refusing a return to past verities. He saw history as progressive, moving toward a time when the marvelous and the so-called scientific were both accepted as components of reality. Forteanism was the philosophy that came after science.

Although Fort was more or less a hermit, his ideas had an active life throughout the twentieth century. The society founded in his name collapsed, was reborn, almost perished again, but eventually came to form the backbone of Forteanism and Fortean modernity. At the heart of the Fortean Society was Tiffany Thayer, a novelist, adman, and provocateur. The story of Forteanism is entangled with Thayer's personal life and idiosyncrasies and shaped by his organization, so although he seems a strange subject, explaining Fortean modernity requires attending to his peccadilloes. (Forteanism is, after all, about the odd.) The story extends, as well, into other cultural forums. Science fiction fans and authors were particularly intrigued by Fort's writings, and many fashioned themselves into Forteans of various types. The anomalies he compiled, the ideas he derived from them, his reconceptualization of the cosmos—these were inspirations for those who tried to imagine the future. Fort's skepticism of science could be taken at face value or reconfigured to support new sciences. Some avant-garde writers also found Fort valuable to their practices. His writing style was exciting—inventive, epigrammatic, spiraling—and his philosophy could be used to explore the functions of mind and imagination. Finally, UFOlogists saw Fort as their direct forerunner, proving the phenomenon that so excited them even before it was given a name. UFOs were a popular topic of the mid-twentieth century, in both senses of the word *popular*: taken up by everyday people and of interest to many.

Science fiction writers, experimental authors, and flying saucer enthusiasts overlapped in complicated ways, sometimes sympathetically, often contentiously. But the groups were united in their use of Fort, who was seen as addressing three topics central to each of their disciplines. First, Fort extended the imagination. He provided means for exploring outer space and inner space without leaving one's armchair. Second, his writings could be used to probe the hidden structures that organized reality. Numerous science fiction authors took the anomalies he collected as clues that, if followed, might lead to the engine of history. They imagined a time when humans might control that motor via scientific or even magical methods, and thus create the future rather than having it thrust upon the world. Some avant-gardists saw in Fort a means of building a space durable enough to preserve individual liberty against the encroaching forces of science, the military, and the state. Others saw him as a model for exploring the organization of the mind, the yet-undiscovered forces that impelled it and thus the actions of humanity. Far from fringe, Fort's ideas were useful for investigating reality's core. Third, Fort helped science fiction authors, modernist artists, and UFOlogists understand how power operated in the modern world. How was the universe authored—by a single being or many? According to spiritual forces or mundane ones? Did earth belong to aliens? asked some science fiction authors. Was there a secular god? asked the novelist William Gaddis, whose ineffability could be limned using Fort's ideas. Were there Space Brothers and shadow people? asked UFOlogists.

Fortean modernity was never strictly Fortean. Rather, those inspired by Fort profligately mixed other responses to the prospect of disenchantment with their interpretation of Fort and the world. The answers to questions about the flow of power in the modern world tended toward the conspiratorial. Forteans could be dispirited by modernity's hollowing out of spiritual experiences. They could be energized by the intricacies of science, the possibility that scientists might one day explain—and thus make real—ghosts and communications with the dead. Forteans could be ironic, playing with ideas without believing in them. They could be dogmatically opposed to science, bureaucracy, the whole of the modern world, could pine for an earlier, better time. There were fundamentalist Forteans; through the 1930s, for example, the pseudonymous newspaper columnist Victor Burr used Fort as a battering ram against hubristic science, incorporating elements of Fort's ideas into his Catholicism.[34] There were New Age

Forteans, too. Fort himself flirted with New Age thinking, imagining that reality might be controlled by a pantheon of gods. Forteans dabbled in astrology, believed in parapsychology. Others saw themselves as anarchists, as socialists, as capitalists, as fascists. What made them all Forteans was a shared appreciation for Fort and a willingness to apply his approach to the same weighty matters: What was the role of imagination in the modern world? What organized reality? Who was in control? The story of Fortean modernity can be understood by examining these four groups—the Fortean Society, science fictioneers, avant-garde artists, UFOlogists—and exploring how they answered those questions. It is a complicated tale and an important one. The Fortean focus on the anomalous, unacknowledged, and unallowed informed midcentury politics, underwrote significant and long-lasting forms of mass culture, inspired imposing works of art, and gave substance to novel varieties of spirituality.

This book takes on each of these groups and their use of Fort. The remainder of this chapter traces Fort's life. Chapter 2 examines the history of the Fortean Society, which despite unceasing tension with Forteans remained the core of the Fortean community. Chapters 3, 4, and 5 focus, respectively, on science fictional, experimentally artistic, and UFOlogical Forteans and their investigations of the modern imagination, social structures, and power. And a concluding chapter sketches the future of Forteanism after the collapse of the society and the rise of postmodernism, not in a systematic way but to sharpen understanding of what came before and as a prolegomenon to any future study.

Marksville, 1947

Fish fell.

Fell from the sky.

A Fortean event.

"It would be dogmatic to deny it flatly," said Bergen Evans, "but no trained observer has ever been on hand when such an event happened."[35]

Alexander Bajkov and his wife Virginia ate breakfast at a restaurant in Marksville, Louisiana. Thursday, 23 October 1947, between 7 and 8 a.m. Fish fell from the sky, the waitress said.[36]

"Detritus of the old belief in spontaneous generation," Evans had suggested a year earlier, or "fossilized 'evidence' for the waters which the Bible says are 'above the firmament.'"[37] An English professor at Northwestern University, Evans considered himself a skeptic, by which he meant a modern man, secular, untainted by folk or religious beliefs. Science alone described reality. And there was no scientific proof that fish rained from the sky, or frogs, or any of a hundred different objects reported to have done so. The persistence of such stories only demonstrated that the rest of the world had not yet joined him in the modern moment. "We are nearer the past than we know, and spooks and demons play leapfrog with dreams of plastics and television in our minds."[38]

Tires exploded fish guts across roadways. Dogs and cats gorged.[39] "The fish that fell in Marksville were absolutely fresh, and were fit for human consumption," Alexander Bajkov said, having left the restaurant with Virginia to witness the prodigy.[40]

Government scientists and ichthyologists noted how frequently fish falls appeared in the historical record. Probably waterspouts, concluded Waldo Lee McAtee in 1917. The idea is hard to credit, he admitted. "However, so many wonderful things occur in nature that negation of any observation is dangerous; it is better to preserve a judicial attitude and regard all (authentic) information that comes to hand as so much evidence, some of it supporting one side, some the other, of a given problem." McAtee tidily arranged the hundreds of reports he'd gathered into twenty-two categories, red rains to spores, jelly to frogs, and published them in *U.S. Monthly Weather Review*, October 1917.[41] "Whirlwinds can pick up objects, whirl them to considerable heights, often up into the thunderstorm clouds themselves, and transport them some distance. . . . The objects fall when the spirals disperse," concluded the US Bureau of Commercial Fisheries, fourteen years after the Bajkovs' interrupted breakfast.[42]

Evans doubted the explanation, even when offered in good faith by scientists and government officials.[43]

So did Charles Fort.

Whirlwinds could not explain the reports he'd read, whatever McAtee and those who followed him said. "How is it that when, in tremendous numbers, little frogs fall from the sky they are unmixed with other objects and substances?" Fort demanded in 1924.[44] Vortices could not be so selective in what they picked up and deposited. Had he been alive in 1947, he might've asked why there were no aquatic weeds, frogs, snails on the streets of Marksville.

Fort suggested—seriously? jokingly?—a variation on belief in waters above the firmament: a disinterred fossil, perhaps, but articulated in unexpected ways, stripped of religious connotation. "My own preference is for the idea that they came from some other world and not through an intensely cold void, such as is popularly supposed to surround this earth, but by way of a warm air current which explorers from this earth may some day traverse; that they came from some world which is not millions of miles away."[45]

Born in Russia, Alexander Bajkov fought in World War I, with the Russians against the Bolsheviks, and later with the British Royal Field Artillery.[46] He went to college, became a professor, standing out for his originality, his eccentricity. In the 1920s he emigrated to Canada to continue his scientific research without the burdensome bureaucracy of Old World universities. He specialized in the biology of fish.[47] He was in Louisiana on a research trip.

That Thursday in 1947 was foggy, the air calm. Winds did not exceed 8 miles per hour, with no whirlwinds reported. The fish fell in pulses over an hour. Three employees of the Marksville Bank were hit; the director woke to hundreds of fish in his yard. Over a strip a thousand feet long, eighty or so feet wide, the fish lay one per square yard or so.[48] About ten thousand altogether.

A lot of fish.

Along Main and Monroe Streets, Bajkov recognized the fishes, all species common to surrounding bodies of freshwater: large-mouth bass, goggle-eye, sunfish, minnows, hickory shad. Two to nine inches long. He slipped some into a jar of formalin.

"Certainly occurrences of this nature are rare," Bajkov wrote in *Sci-*

ence, but that did not make them false. There was more to the subject than folklore or superstition. He'd read the government reports, knew the ichthyological literature. "Why can't fish be lifted with water and carried by the whirlwind?" Only Wednesday, he and a colleague had seen several "small tornadoes" in the vicinity.[49]

It is the nature of Fortean events to resist interpretation. The reports from Marksville could be dismissed as folklore or attributed to misplaced religious belief. The fish falls could be described in terms of scientifically known processes or made the excuse for baroque speculations about the nature of the cosmos. But no single explanation was definitive.

Almost a quarter century on, the biologist Ivan Sanderson named the strange objects that rained from heaven *fafrotskies*: (things that) *fall from the sky*.[50] Fish, frogs, stones, flesh, blood. In his opinion, they were common enough to warrant a specific term. Missiles of startling audacity.

McAtee: "I am only a collector in the forbidden fields and have no theories to uphold or great works to build. I simply collect what appeals to me, record it, and forget about it. It may be of use to someone sometime and that is enough for me."[51]

Fort: "So leaves of trees, carried up there in whirlwinds, staying there years, ages, perhaps only a few months, but then falling to this earth at an unseasonable time for dead leaves—fishes carried up there, some of them dying and drying, some of them living in volumes of water that are in abundance up there, or that fall sometimes in the deluges that we call 'cloudbursts.'"[52]

"Up there." That was the "Super-Sargasso Sea."[53]

Down here: an aquarium.

Something, some demonic presence, plied the waters of the Super-Sargasso Sea, cast its lines, abducted humans. This was not science, nor was it religion or folklore; rather, an amalgamation of them all, a system of thought—ensconced in a quirky philosophy—that would attract followers, cut a swath through the modern, secular world that was neither congruent nor opposed to Evans's ideas but orthogonal to them: a different kind of modernity, dark, Luciferian, fragmented, simultaneously serious and ridiculous.[54]

"I think we're fished for. It may be that we're highly esteemed by super-epicures somewhere."[55]

Strange Gods

Albany, New York, 6 August 1874. Charles Hoy Fort was born into a prosperous family and difficult circumstances. His mother, Agnes Hoy, died before he was five, leaving Toddy, as he was called, and his two younger brothers to the widowed Charles Nelson Fort. The paternal Fort was strict, physically abusive, bringing tears to Toddy's eyes, blood to his nose—a tyrannical figure who cowed his sons into compliance but not respect or love. Impish from an early age, Fort developed an independent streak, perhaps in reaction to his father's despotism, an intransigence that matured into a skepticism toward all forms of authority. He rejected religion—and what was taught at school. A compulsion to collect overcame him when he was young, another trait that would organize his adulthood, collecting and contumacy. Fort dropped out of high school, moved to Brooklyn, where, as he had at home, he worked as a journalist. In 1893, he used a small inheritance to travel, covering thirty thousand miles in three years. "All this to accumulate an experience and knowledge of life." In 1896, illness forced his return to Brooklyn. He got reacquainted with Annie Filing, whom he'd known in Albany. She nursed him back to health. They married in October. The couple struggled to eke out a living, Annie becoming a laundress, Fort a dishwasher.[56]

As Fort's physical purview shrank, metaphysical and imaginative realms expanded before him. It was about this time that he met his gods.[57] These gods—he named four in correspondence—drove him: they were orthogenetic, he said, working toward an unseen goal, transforming the rebellious son of an Albany grocer into the "enfant terrible of science." Decomposition. Amorpha. Syntheticus. Equalization. The pantheon may have been a private joke, but if so, the lie became a kind of truth.[58] Decomposition, he said later, was the god immediately in charge of the world, "this beautifully rotten existence of ours."⁵⁹ What better symbol for a life of toil and unfulfilled dreams?

Amid these trials, he began to scratch out short stories, breaking into the magazine market in 1905, mostly with closely observed stories about tenement life in New York. Theodore Dreiser, then an editor with the publishing house Street & Smith, thought Fort's "were the best humorous short stories that I have ever seen produced in America."⁶⁰ He became a patron, buying some of Fort's stories for the publisher's flagship magazine *Smith's*, offering critiques, cajoling and encouraging. He arranged the publication of Fort's novel, *The Outcast Manufacturers*, in 1909. A few years later, another pair of inheritances stabilized the Forts' financial situation, freeing both Charles and Annie from the need to work. Fort attributed this turn of fortune to the arrival of the goddess Amorpha, who prized orderliness. The new situation provoked a crisis. Fort fancied himself a bohemian, but now Annie demanded creature comforts: bigger apartments, bathrooms, more respectable neighbors—bourgeois amenities that Fort scorned.

He wanted, on some level, the kind of life Dreiser was living, outsize and outrageous. Dreiser's writing exposed and excoriated the social and religious systems that immiserated the poor, the immigrants. His novel *Sister Carrie* had him proclaimed a literary genius, his virtues trumpeted by the influential critic H. L. Mencken. A succession of novels followed, and essays and plays and poems. Dreiser moved through literary circles, took up residence in Greenwich Village, adopted radical views.⁶¹ In 1915, his novel *The Genius*, which fictionalized the dissolution of his marriage, his many infidelities, incited bluenoses, who had it banned. Fort told Dreiser he longed for that kind of notoriety. In a world ruled by Decomposition, one should luxuriate in decadence, he thought.

The form Fort's writing had taken since the arrival of Amorpha, he knew, was unlikely to cause such a frenzy. It became increasingly abstract.⁶² He'd been unable to write a second novel and, now liberated from the need to sell short stories, he spent his days in the library, endlessly digressing through the study of "all the arts and sciences," wearing out "eyesight and pencils and breeches" as he took notes on everything from evolution to calculus.[63] These he put to use supporting a theory he'd developed: "That all things are one; that all phenomena are governed by the same law."[64] Or, as he put it in the book that eventually emerged from this note-taking:

> But it is our expression that there are no positive differences: that all things are like a mouse and a bug in the heart of a cheese. Mouse and a bug: no two things could seem more unlike. They're there a week, or they stay there a month: both are then only transmutations of cheese. I think we're all bugs and mice, and are only different expressions of an all-inclusive cheese.[65]

Fort had become a monist, his theory exemplifying what philosophers call existence monism: only the universe exists; all the common objects of life are mere parts of this whole. In a fundamental way, they are not real.[66] He played with the idea in one story, a scientific romance about a human blood cell that realizes it is part of a vast, interconnected system, just as the human in which it lives is part of a cosmic order. The heterodox lymphocyte is shouted down by conservative cells. Fort titled the story "A Radical Corpuscle." He saw himself as a radical, too, aligned with other nonconformists of the era who offered similar philosophies.[67]

Over time, Fort's readings expanded from what was known to what was not, from the lawed to the lawless. He kept uncovering anomalies, reports of raining frogs, mysterious disappearances, baffling objects in the sky. These fascinated him, and he started taking notes on them too, worrying their meaning. Setting aside fiction, Fort now concerned himself with the nature of existence. "I see this all as a travail," he said, "of emerging more or less a metaphysician from a story writer."[68] He arranged the tens of thousands of notes hoarded about his apartment according to some thirteen hundred categories. Compelled "orthogenetically," he turned in November 1915 to synthesizing his notes, his thoughts on the anomalous.

Under the thrall of his third god, Syntheticus, Fort spent the next six months working on a new book. The manuscript is lost, but his biographers have pieced together the argument. Fort suggested that existence was organized by an undiscovered force, what he called X, that created all things: "you, me, all animals, plants, the earth and its fullness, its beauty and variety and strangeness," as Dreiser said.[69] Sent by Martians, for reasons unknown, X was often invisible, its operation mistaken for natural law or free will. But by studying the anomalous, Fort could detect X's presence, hypothesize the existence of the mysterious ray. In a hundred thousand words, Fort reinterpreted human history in light of this force. He was pushing toward a new genre of writing, as revolutionary as Dreiser's, though less profane.

From "X," he moved on to "Y," financial security allowing for productivity—days spent at the library, evenings devoted to shuffling notes, writing, before Annie served a simple, hearty meal, the night capped with a stroll, the movies, a late snack of cheese and beer. "Y," Fort said, was a complement to "X." Although reconstruction of this manuscript is even less sure, the remnants more fragmentary, it seems that Y was connected to a place Fort called Y-land, somewhere near the North Pole. Attempts by explorers to travel though the vast and harsh Arctic, he said, would bring about a new era. Humanity and the citizens of Y-land would merge, their coupling neutralizing the force of X. What would follow would not be free will, though, but rather a state of nirvana.

Fort's theorizing had an immediate effect on Dreiser. Like Fort, Dreiser did not believe in free will—the characters in his books are motivated by base emotions, biochemical reactions—but for all his materialism, he still longed to discover a transcendental realm, a source of ultimate meaning, an enriching one unlike the mean religion of his youth.[70] Fort's orthogenetic gods helped him solve this conundrum, unite the mundane and the spiritual:

> I once believed, for instance, that nature was a blind, stumbling force or combination of forces which knew not what or whither. I drew that conclusion largely from the fumbling nonintelligence

(relatively speaking) of men and all sentient creatures. Of late years I have inclined to think just the reverse, i.e., that nature is merely dark to us because of her tremendous subtlety and our own very limited powers of comprehension.[71]

There was a force, Dreiser concluded—"Ontogenetic Orthogenesis" or "autogenetic orthogenesis"—that drove reality toward some higher, if unknown, good.[72] Dreiser illustrated these themes in a pair of plays, *The Dream* and *Phantasmagoria*, both of which present existence as the dream of the gods, thus synthesizing his realism and transcendence, humanity bereft of free will yet still in the palm of divinity.[73] Convinced that "X" was "one of the greatest books I have ever read," Dreiser assumed the burden of getting it published.[74] He shopped "X," then "X" and "Y," to editors, publishers, even a movie studio, all to no avail.[75]

Meanwhile, Fort fretted. He vacillated between concern that his gods had abandoned him and worry that they, or some related cosmic force, were preventing the publication of his manuscripts. He fretted, but, for the moment, he was not ready to quit. In December 1916, he decided on his next project, "a study of—occult things you know—things that have been called souls or spirits."[76] Its composition, though, proved difficult. At the end of each day, after hours at the library, he could not bring himself to write. He wasn't sure how to apply "the kind of brains I have" to the subject. "Mine is a coarse and more cynical mind than those that have heretofore examined such phenomena, also it has some other qualities and a different attitude toward what is called the scientific method."[77] He thought of this book as "Z."

Over the next several years, Fort transformed the manuscript into *The Book of the Damned*. Beyond breaking the alphabetic naming convention, he presented his material in a new fashion. The bulk of the book comprises the strange reports he'd gathered, particularly stories of things that fell from the sky, interwoven with attacks on Euclidean geometry, Darwinian evolution, Newtonian mechanics, geology. Though superficially confusing, a mass of disparate mysteries, there is an underlying argument, Fort's eccentric theorizing not absent, just less prominent. *The Book of the Damned* advanced Fort's monism. His quarrel with science centers on its failure to recognize the universe's wholeness. Scientists mutilate reality, cut it into pieces, some called truth, others damned as false. Christianity, in days past, had used the same process to create its own regime of Truth—a Religious Dominant now superseded by an equally incomplete Scientific Dominant. "Demons and angels and inertias and reactions are all mythological characters; But that, in their eras of dominance, they were almost as firmly believed in as if they had been proved."[78] In rescuing the excluded, damned facts, Fort aimed to demonstrate the continuity of existence, to complete the puzzle by providing the missing pieces. He hoped to initiate a new Dominant, one that recognized the universal continuum.

This new era would not be a featureless state, all forces offset; rather, it was to be in dynamic equilibrium, a constant flux. The dynamo powering the turmoil was the instability of the categories "true" and "false." True things—what he called "heavenly" objects—were forever being declared untrue, while the untrue—the damned—were regularly resurrected as true.[79] Existence, as it was experienced by humanity, was intermediate between the heavenly and the damned, roiled by constant transformation. "We are not realists," he writes. "We are not idealists. We are intermediatists—that nothing is real, but that nothing is unreal."[80] The intermediate realm, life itself, is purgatory. Fort was, in this sense, mapping the universe's metaphysical departments, as though he were a modern-day Dante. He told Dreiser, when he sent him a copy of the manuscript, "The Book of the Damned . . . is a religion."[81]

In erecting this strange theology, Fort teetered on the edge of pure crankery, inventing a new system, as he had done in "X" and "Y," as did authors of books on Atlantis or the earth's hollow core. But *The Book of the Damned* was playful, slippery, so unlike the earnest tomes common to the genre. If there is no difference between the real and the unreal, anything is possible, and just as likely impossible. "So there it is. I've given up fiction, you see," he told Dreiser. "Or in a way I haven't. I am convinced that everything is fiction; so here I am in the same old line."[82] Perhaps there was a Super-Sargasso

Sea in the upper atmosphere into which were carried objects from earth—frogs, fish, leaves—and from which they later rained. Perhaps the universe was a living thing, rains of blood its bleeding. Perhaps in 1903 the earth, in its orbit about the sun, passed through the remains of a world destroyed in an interplanetary dispute, the particles falling as rains of dust and redness. Perhaps humanity was controlled. "I think we're property," Fort wrote.[83] Or, perhaps not; so skeptical he could not accept even his own authority, he had given up theorizing. "We have expressions: we don't call them explanations: we've discarded explanations with beliefs."[84]

Dreiser was impressed. "Wonderful, colorful, inspiring," he congratulated. "Like a peak or open tower window commanding vast realms."[85] Around the time he received Fort's manuscript, Dreiser had taken up with a new publisher, Boni & Liveright, which was intent on limbering up the ossified business of publishing, introducing new methods, taking risks on experimental writers, American and European. According to Dreiser, he forced the publisher to put out *The Book of the Damned* on the threat of his leaving. Probably that was not the only reason; Dreiser so frequently vowed to leave for another house if he didn't get his way that his fulminations must have registered as bluff.[86] Whatever the cause, Boni & Liveright added *The Book of the Damned* to its fall list, while Fort tinkered with the manuscript into May; it stayed on the list even when a printer's strike roiled the industry. Originally scheduled for an October release, the book did not appear until the first day of December 1919, in time for the Christmas rush.[87] Fort's first book in a decade was priced at a reasonable one dollar. Sales were brisk, a second printing coming in January. "One of the most amazing books ever issued," said a reviewer for the *Philadelphia Inquirer*.[88]

Destroyed Works

"Step forward, ladies and gentlemen, and wonder—not at marvels, nor a house, a city, a kingdom, or a world of marvels—but a whole universe of the most prodigious marvels."[89] That was the opening to a review of *The Book of the Damned* in the *Arkansas Gazette*. For a few months in early 1920, Fort and his book, with its eyebrow-raising title, were a minor sensation. It was an incongruous start to what would be a difficult decade for Fort, the last full one he lived. *The Book of the Damned* was reviewed in numerous publications, from the prestigious to the middling to the local. The reviews were mostly—though not entirely—positive, or at least offered befuddled appreciation.

"As I read it I became more and more conscious of the fact that I was in the presence of a genius who, if he has hit a bull's eye in his overwhelming deductions, will easily jostle Euclid, Columbus and Darwin off their pedestals," wrote the maverick journalist Benjamin DeCasseres.[90] The playwright and novelist Booth Tarkington, whose novel *The Magnificent Ambersons* won the 1919 Pulitzer Prize, read the book while sick with the flu, having picked it up by mistake—studying criminology during his convalescence, Tarkington was misled by the title—and was blown away.[91] "Who in the name of frenzy is Charles Fort?" he asked. "He's 'colossal'—a magnificent nut, with Poe and Blake and Cagliostro and St. John trailing way behind him. And with a gorgeous madman's humor!"[92] The most effulgent praise came from Ben Hecht in the *Chicago Daily News*. He too asked, "Who is Charles Fort?" before offering his own reply:

> Charles Fort is an inspired clown who, to the accompaniment of a gigantic snare drum, has bounded into the arena of science and let fly at the pontifical seats of wisdom with slapstick and bladder. He has plucked the false whiskers off the planets. He has reinvented a god. He has exposed the immemorial hoax that bears the name of sanity.[93]

Hecht, a journalist and reviewer, fancied himself an alienist, a "lunacy expert," and he saw in Fort someone who refused to be confined by rationality but embraced the wisdom of the fool.[94] Hecht promised readers willing to brave *The Book of the Damned* that they would be rewarded with the liberation of madness. "For every five people who read this book four will go insane."

There were bad reviews, too. "The thing leaves me puzzled," the writer and critic H. L. Mencken told Dreiser.[95] Mencken was an advocate of science (to a point) and stood against superstition, religion, and other forms of what he considered loose thinking; the book was bunk. *Life* magazine thought the *Book of the Damned* fitting only for a Hamlet touched by Ophelia's madness—both indecisive and insane.[96] *The Nation* concluded that Fort's book was "dedicated to the propaganda of systematic idiocy."[97] From a certain perspective, even the favorable reviews were underwhelming. Tarkington admired Fort's "bulliest dementia," not his monism. Most reviews focused on Fort's collection of amazing facts, his outrageous hypotheses, but not how these fit into his view of universal flux. The *New York Herald* blamed Fort's style for readers' confusion: his "prose is so consciously artificial that it actually restrains the reader from admitting the author's premises and considering his arguments in due fairness."[98]

Into the summer, Fort remained stalwart against these dismissals. His fortitude, though, was brittle. Fort's fourth god, Equalization, was the greatest of them all.[99] Equalization assured that any approbation would be balanced by disdain. Thus, Dreiser was punished with the censorship of *The Genius* because he had helped Fort. Equalization, unmentioned in *The Book of the Damned*, was the essence of Fort's argument, the force that motivated—or perhaps only a convenient fiction describing—the endless recycling of the heavenly and the damned. "Religions come, and religions go. . . . Sciences rise by displacing superstitions, only to be found out later as delusions."[100] Fort's equalization came when he learned that the New York Public Library classified *The Book of the Damned* as "Eccentric Literature" and grouped it with books by cranks.[101] Fort could not stand an equalization so cruel. In late October, he took out an ad in the *New York Tribune*, in which he noted that "even with all the heresies," *The Book of the Damned* had received a respectful hearing from the likes of *Popular Astronomy*. His book "represent[ed] enormous labor" and deserved better treatment.[102]

Fort fell into a foul temper. He "burned all [his] notes, 40,000

Theodore Dreiser, left, and Charles Fort at Dreiser's retreat Iroki, in Mt. Kisco, New York, October 1931. (Courtesy University of Pennsylvania. Dreiser papers, ms. coll. 30, vol. 438, item 153.)

of them."[103] Likely, it was during this bout of depression that "X," "Y," and whatever remained of "Z" were dispatched. All of Fort's labors seemed destined for the dustbin, his pleasures gone. One day, when he was out on a mad, having told Annie he'd never return, she "threw away all his old junk," including the magazines in which his short stories had appeared.[104] Not for the first time, or the last, Fort was on the verge of forgoing writing. Embarrassed, he could not bring himself to return to the library for more research. "Forces are moving me to London," he wrote to Dreiser. He didn't name the forces but, in that letter, still referred to the presence of strange, orthogenetic gods in his life.[105]

London restored Fort's confidence. He and Annie established a new routine. Fort resumed research, now at the British Museum. In time, they would return to New York, and Fort to its public library, but the forces were restive, and so through the 1920s, the couple cycled back and forth across the Atlantic. Overcoming writer's block, Fort put together another book, *New Lands*, which vamped on the same themes as *The Book of the Damned* while advancing the argument—the expression—that space was tiny, the distance be-

tween planets trivial, that in the future humanity would take interplanetary day trips. The sky falls he chronicled were the equivalent
of detritus from North America washing onto European shores in
the years before Columbus's first voyage. Boni & Liveright published *New Lands*, with a foreword by Tarkington. But the book was
mostly ignored. After its failure, the publisher rejected Fort's other
manuscripts.[106] One reviewer said *New Lands* reminded him of the
kind of crackpottery kooks sent as letters to the editor.[107]

Unable to build on his previous success, Fort descended toward
such crankery. He contemplated witchcraft and people possessed of
mysterious powers, thought perhaps his own apartment was haunted
by these forces. Evenings he argued with soapboxers at Hyde Park.
Deprived of other outlets, Fort became an inveterate writer of letters
to newspapers, riffing on his heterodox ideas, pleading for additional
information on anomalies. "I have many data," he told readers of the
Philadelphia Public Ledger, for example,

> which indicates that there may be warm air currents in outer
> space. My suggestion is that between this earth and other worlds
> there may be definite currents to which living things in other
> worlds respond migratorily. If living things can come to this earth
> from other worlds, we have the material for visions such as have
> not excited imaginations upon this earth since the year 1492.[108]

Toward the end of the decade, he invented a new game, Super-
Checkers, which required hundreds of pieces, twice as many squares,
the impulse that drove him to collect and arrange his thousands of
notes put to a different purpose. He and Annie bought a pair of
parrots.

His eyesight, never strong, faltered. "I cannot stand living in
blindness," he said.[109] He would no longer be able to read, to write,
to play Super-Checkers. He would no longer be able to see Annie.
Fort slipped into a funk. The orthogenetic gods that had structured
his life were no more. After he left for London the first time, he

never mentioned them again. Fort understood himself differently now, not as driven by unseen forces, rather as a bundle of identities roughly held together by skin and brain. In 1926 he sighed, "It's pretty hard to keep track of my selves."[110] Four years later he guessed, "I have about seventeen selves."[111]

Fort made these admissions to DeCasseres, the two writers forging a mutual admiration dyad in the mid-1920s. Fort had been reading the essayist for some time before DeCasseres introduced himself (Fort may have even borrowed the idea of strange gods from him) and was impressed by his style, his sensibility.[112] He delighted in DeCasseres's interpretation of the *Titanic*'s sinking as a joint operation of God and Satan to sustain in humanity a fear of the divine.[113] Fort invited DeCasseres to join a gag group, an "expedition of Neo-Puritans," the name playing on his notion that the current historical moment paralleled 1492, the discovery and colonization of "new lands" imminent. Fort recognized that, like himself, DeCasseres was composed of many selves, but he most admired his apocalyptic-prophet persona, and so invited him to be the expedition's theologian. A famous tippler, DeCasseres joshed that he'd rather be a bootlegger, to which Fort replied, any theologian worth his salt could justify being both rumrunner and man of God.[114]

DeCasseres returned Fort's compliments, similarly seeing Fort as Fort most hoped to be seen. DeCasseres had read Nietzsche and Spinoza (a distant relative), and took from those philosophers the notion that all reality is flux: there are no foundational truths. Like Fort, DeCasseres considered science a convenient fiction:

> A superstition is said to be a false belief. Well, so are all the postulates of science. I am no enemy of science—their fairy-tales are to me the most fascinating in existence; the fairy-tale of the electron, the fairy-tale of the atom, the fairy-tale of the germ, the fairy-tale of history, the fairy-tale of the phonograph, of the radio, of the telephone, of the motion-picture, of the great murder-engines of war, etc.[115]

The differences between the two were not in substance but in approach, in how they made sense of the world: Fort focused on raising the damned; DeCasseres was fascinated by decomposition, the dissolution of past verities. "The flat surface of our ancient heavens is crumbling over the world like a rotten ceiling," he wrote in 1912.[116] DeCasseres recognized Fort's place in this order, calling him "satanic."[117]

It had been one of Fort's fondest desires, since Dreiser's difficulties with *The Genius*, to be considered demonic. "I write of the attractions of planets, and the affinities of atoms. These are lusts," he said in 1916. But he could "not convey evil notions or astronomic and chemic obscenities."[118] DeCasseres had seen through Fort's chasteness to the imp underneath. In particular, DeCasseres meant—and Fort understood—that Fort was Luciferian. Romantic writers in the nineteenth century had revalued Satan as a positive force, unleashing sexual energies, driving social and scientific progress; Lucifer was the name of this version of Satan.[119] The Luciferian created from the wreckage of civilization, built heaven inside the hell of this world.[120] That was Fort's project, though it was often obscured in his writing. He dissected science and arranged something new from the pieces. The satanic was Fort's most prized self.

He knew, though, that being "hellish" was not his prime quality.[121] That honor belonged to a more disciplined self. "'I' am so numerous that 'I' am sure that 'you' won't hold it against the rest of us when 'I' tell 'you' that one of my 'selves' is a Puritan," Fort told DeCasseres. "This fellow is a tyrant, of course."[122] *Puritan* was a term that Mencken had transformed into an epithet, a word made to stand for everything repressive and mean-minded in American culture.[123] Puritans censored Dreiser. Puritans valued literature not for its aesthetic qualities but for its moral values. Puritanism was incarnated, Mencken thought, in the novels of Harold Bell Wright, a former pastor whose mawkish, maudlin scribblings had made him one of the most popular authors of the early twentieth century.[124] Fort feared that he was heir to this moralizing tradition. "I am Harold Bell Wright," he conceded bashfully, "and no Rabelais, and have a primitive faith in Sea Serpents, and the truth, beauty, and the goodness of them."[125] That he saw in himself a Puritan gave a different tenor to his fake plan for interplanetary travel: it was not just a reference to humanity being on the verge of great discoveries but a description of his work habits and an admission of weakness.

In 1928, Fort gathered his many selves and with Anna and the parrots returned to New York, settling in the Bronx. Although he had several manuscripts in hand and knew Dreiser had the pull to see them into print, Fort did not announce his arrival to his friend. Instead, he found support from another advocate. Tiffany Thayer, a newly famous writer, having published *Thirteen Men*, an innovative mystery that became a bestseller, offered his services in getting Fort's work before the public. Fort welcomed the younger man's adulation, but with some suspicion. "He first read me when he was about twenty years old, and thinks he owes me a lot for it."[126] Thayer repaid the imagined debt by arranging to have one of Fort's manuscripts put out by his own publisher, Claude Kendall. Kendall's list was eccentric, comprising occult works, decadent grotesqueries, attacks on American culture. Fort's strange philosophical meditations fit well.

After extensive discussion with Thayer and Kendall's partner Aaron Sussman, Fort's book was titled *Lo!* It continued his two-decade project. "We shall pick up an existence by its frogs," he wrote:

> Wise men had tried other ways. They have tried to understand our state of being, by grasping at its stars, or its arts, or its economics. But, if there is an underlying oneness of all things, it does not matter where we begin, whether with stars, or laws of supply and demand, or frogs, or Napoleon Bonaparte. One measures a circle, beginning anywhere.[127]

Having staked this monistic position, Fort concerned himself, once more, with impossible rains, with strange appearances and disappearances that he attributed to a force he termed teleportation. He implied the solar system was small, the earth stationary, the stars points in a surrounding shell. He attacked astronomers—from a groundless position, he conceded, because in tension with his insistence upon monism was a withering skepticism. "I believe nothing of my own that I have ever written," he wrote. "I cannot accept that the products of minds are subject-matter for beliefs."[128] The contradiction was held together by the book's structure—inferential, skittering, refusing to settle on any interpretation—and by Fort's prose, the most confident and approachable of his career.

Thayer wrote an introduction. The publisher commissioned illustrations by Alexander King, who'd drawn covers for Mencken's magazine *The Smart Set*. There would be publicity, something purposefully bumptious. Once the book was assured publication, Fort contacted Dreiser, who was pulled into the scheming. Dreiser pressed copies of Fort's other books into the hands of literary friends, among them H. G. Wells, to whom he sent Fort's first book, and *Lo!* when it was finished.[129] He tried to get excerpts run in *Cosmopolitan*.[130] The plotting would bring a happy ending to a tumultuous decade. Thayer and Dreiser planned to make Fort famous.

The Fortean Society

The urge to venerate Fort was present from the beginning of his career. There's a sense in which Dreiser, Fort's patron, "became his first disciple," as Fort's biographer Jim Steinmeyer noted. "When I sensed the imaginative power" of "X," Dreiser said, "I was in a worshipping state of mind."[131] Ben Hecht, in Chicago, had also declared himself a supplicant, usurping Dreiser's priority. "I am the first disciple of Charles Fort," he wrote in his review of *The Book of the Damned*:

> I . . . rush to surrender my homage. Whatever the purpose of Charles Fort, he has delighted me beyond all men who have written books in this world. Mountebank or Messiah, it matters not. Henceforth I am a Fortean. If it has pleased Charles Fort to perpetrate a Gargantuan jest upon unsuspecting readers, all the better.

If he has in all seriousness heralded forth the innermost truths of his soul, well and good. I offer him this testament. I believe.[132]

Hecht would soon expand his literary frontiers from journalism, poetry, plays, and short stories to novel writing. In 1924's *Humpty Dumpty*, the hero, Kent Savaron, a man possessing unconventional genius, shows off his books, proof of his heterodox discernment. "The scientists are in that other box. My favorite is Charles Fort."[133] Hecht's delight with Fort's iconoclasm persisted into the 1940s.

And there were other disciples. Miriam Allen DeFord checked out *The Book of the Damned* from the library and brought it home to her husband, Maynard Shipley. (Shipley was interested in criminology, and DeFord had likely made the same mistake as Tarkington, misinterpreting the book's title.) The two were enraptured, "reading the book aloud to each other, unable to put it down."[134] They struck up a correspondence with Fort, and DeFord even traveled to Chico, California, from the couple's home in the Bay Area to follow up on a reported rain of rocks. Fort confirmed their experiences. They had seen strange things, objects that moved on their own, windows once rusted closed suddenly flying open.[135] Socialists devoted to materialism, they rejected spiritual explanations, just as Fort did. Their only hesitation was Fort's antiscience posture; they accepted science as true. Shipley was a vociferous defender of evolution against fundamentalist Christian attempts to make its teaching illegal. Conceding that Fort's wild theorizing could be "grotesque," they still found him a "strange genius."[136]

Even Mencken saw Fort as a messianic figure. Certainly, Fort peddled "highfalutin balderdash." Certainly, that his maunderings had been "accepted as gospel by a considerable body of presumably sane men" offered "melancholy evidence that the human race has a long way to go before it will deserve to be called rational." Every scientific discipline was shadowed by a "grotesque Doppelgänger." Think only of medical quackeries, Mencken said. Think of physics. Once it had made visceral sense; now quantum theory and relativity transformed physics into "a New Theology, with a mathematical

God." Fort's *Lo!* was yet another ridiculous shadow of real science. But despite all these caveats, and they were many and substantial, Fort was perversely admirable. "He knew how to make even the most extravagant nonsense palatable." Mencken thought it a "pity" that Fort's career was cut short before he could finish his strange *Summa*. "There was in him more than one hint of the special talent of St. Thomas Aquinas."[137]

The 1920s saw a couple of attempts to channel this esteem. After reading *The Book of the Damned*, the newspaper publisher J. David Stern proposed that there should be a society devoted to Fort and his data. The exact details of the first society's origin, institution, and structure, and even if it was ever more than a mere notion, are lost to time. "The great trouble," Fort said, "is that the majority of persons who are attracted are the ones that we do not want; Spiritualists, Fundamentalists." Their objection to science was not Fort's. He wanted to push past both the Religious Dominant and the Scientific Dominant, while those he was attracting, it seemed to him, wanted a return to the Religious Dominant. "I think I make it plain in the books that I am not out to restore Moses," he told a friend.[138]

Fort, though, was not ready to concede defeat. He gathered a network of correspondents, "people in about twenty of the States, in Canada, South Africa, and Australia."[139] These like-minded folks clipped stories of the odd, the uncanny, and passed them on to Fort in London. He doubted that the motley group could be organized but couldn't resist the compulsion to try. His ideas, however lightly he expressed them in his books, were matters of grave importance: his jokes hid weighty matters. "Some day an expedition may sail out from this earth and return with news such as has not opened new vistas and stimulated imagination and enterprise since the year 1492. But if we ever sail to new worlds, first must we think to new worlds. We must have hosts of data to think upon."[140] His proposed Interplanetary Society, not a mere joke, was a winking extension of ideas he considered momentous, urging humanity in a new direction, past religion, past science, to a brand new era.

As with Stern's plan, though, Fort's idea of a globe-spanning lattice of correspondents was never realized. The failure can probably be explained by lapsing interest among his correspondents coupled with his own melancholic disposition. He'd peppered newspapers with requests for details about anomalous reports that supported his theories and pleading for participation in his society, but the letters were too often printed under mocking headlines—"Well, Well! See a Shower of Froggies Lately? Fate of World Hangs in Discovery."[141] His writing was not going well, his eyesight faltering, and he had Super-Checkers to occupy his time. "Something has isolated me, and mostly it has been because I have put in my time as a writer of treatments and subjects with which the world would have nothing to do," he thought.[142]

A few years later, while Thayer worked out an advertising blitz for *Lo!* Stern passed his idea for a Fortean Society to him; Thayer saw its potential.[143] "The publishers, this time, want to start me off with fireworks, and this time have me seen," Fort had said.[144] Here were fireworks. Here was a way to make Fort seen. Thayer convinced Aaron Sussman to help him. Sussman also understood the commercial power of the idea. He was an old adman himself. So renowned was his ability to sell, there circulated the story that he'd once caught a mouse in a trap loaded with only a picture of cheese.[145] Thayer and Sussman drew up letterhead and started recruiting members in November 1930. In December, Stern's *Philadelphia Record* published a story on the society's imminence. *Publishers' Weekly* and the *New York World* followed.[146] By January 1931, the month *Lo!* was published, Thayer and Sussman were sending out press releases announcing new members. "So rapidly are we growing," Thayer crowed, "that while we eat our lunch, figuratively, new members walk in."[147] They signed up friends and followers, among them the novelist John Cowper Powys.

None of us was "quite clear as to what the gospel was," said the bon vivant Alexander Woollcott, one of the Fortean Society's founding members.[148] Composition, purpose, organization—these were obscure. Dreiser told Powys that the society was intended to make scientists take Fort seriously—as a thinker, not a crank—and to provide him with more data.[149] The historian Harry Elmer Barnes joined, however, because Fort was supposed to be in "straitened circumstances" and needed help marketing his works. He offered no more support than to advertise Fort's books as "a contribution to curious literature and chaste levity."[150] The poet Edgar Lee Masters didn't even realize he was a member—he'd had a look at *The Book of the Damned* but was "not much interested."[151] The admen Sussman and Thayer did not bother to clarify, opting instead to overwhelm confusion with celebrity, explanation with braggadocio.

Fort was not happy when he received a letter on Thayer's Fortean Society stationery. "Freaks," he thought, and refused to have anything to do with the matter, brushing aside Thayer's request that he act as honorary president.[152] He was uncomfortable having the society named after him. Although he would agree to a couple of interviews after *Lo!*'s publication, he was not keen on exposure. It risked revealing to the world the mundane selves he tried to hide beneath his satirical prose. Fort said one way to tempt him to join would be to call the group the Society of the Damned—a joke on the members that simultaneously cast him as satanic and emphasized his data—or, riffing on the group he imagined in the mid-1920s, the Stellar Exploration Club, which would take the focus from him and highlight his cosmological theorizing.[153]

Despite Fort's misgivings, despite confusions about its meaning, a Fortean moment was aborning. *Lo!* went through three printings in six weeks. There was even a British edition, put out by Victor Gollancz. *Lo!* received notice in publications from college newspapers to the *New York Times*—where it got a glowing endorsement from Maynard Shipley—and *The Nation*, hundreds of reviews in all. H. G. Wells, however, tossed his copy of *Lo!* in the trash after reading *The Book of the Damned*. Fort "writes like a drunkard," he said, and his attacks on science lacked merit.[154] Wells had published his own book that month, *The Science of Life*, which celebrated biology and biologists. A few reviews echoed these misgivings. But the vast bulk of them were positive, exceptionally so. "Charles Fort has engaged in investigations which make Einstein's seem piddling," said Burton Rascoe in the *New York Herald Tribune*. *Lo!* was the "'De Revolutionibus' and the 'Principia' of a new era."[155]

DeCasseres wrote a long article on Fort in the magazine *The Thinker* that appeared in April 1931. While not part of the formalities surrounding the Fortean Society, the piece still served to introduce the man to a wider public. DeCasseres started from a proposition he had long championed, that the universe was too vast, too weird to be comprehended by the human mind. The only proper relationship with the magnificence of reality was ironic, advancing ideas, beliefs, philosophies, even as one knew these were incomplete, wrongheaded. Each person should be at liberty to devise or embrace whatever theories they preferred. Fort, he said, provided one of the most amazing ways to look at this quizzical universe. If one were to choose, freely, how to understand life, why not opt for an entertaining way: why not embrace the Fortean fantasy, as he called it?

Powys, too, was sensitive to this Fortean moment. All through November and December of 1930 he seemed to float in a Fortean reverie. He read *The Book of the Damned*. He sent a letter to Dreiser for use in advertising *Lo!*[156] He joined the Fortean Society.[157] He spent an evening discussing Fort with Dreiser and Edgar Lee Masters.[158] He read newspapers with a Fortean eye—in the *Times*, for example, he noted a story about a red rain that was explained by an African sandstorm.[159] Of course, Powys thought, that was exactly the kind of faulty explanation Fort would have predicted the press to offer. In December, he read that a poison fog had descended on Belgium and parts of London. It was explained as "a Dissipated Comet" emitting a toxic gas that sickened hundreds, killed scores. "Is it?" he asked his diary. "Is it? Up in one place, down in another. Panic in the Meuse valley Brussels. Panic! What can it be? a Poisonous Toy from what? Sensational statements. A narrow dark cloud 1897; 1902 the same thing occurred. See Mr. Charles Fort . . . what is this? Eh? Ha?"[160]

However reluctant Fort was to enter the hubbub, the laws of motion were against him. On the evening of 26 January 1931, Fort made his way through the cold, snowy streets of New York to what he thought would be an intimate dinner with Dreiser and Stern in Stern's opulent suite at the Savoy-Plaza Hotel. He was surprised to be greeted not only by those two friends but by Thayer. And Sussman. And Claude Kendall. And book reviewers J. Donald Adams and Burton Rascoe. And journalist H. Allen Smith. And more. They were holding the first annual meeting of the Fortean Society.[161] This was "a live man undergoing the process of being erected into a cult in his own presence," said one of the journalists on hand.[162] Before dinner, there were "several hours of interviews and the informal presentation of Fort's 'third assault upon dogmatic science.'"[163] After was a general discussion of the society's purpose, the meeting lasting late into the evening.

Dreiser, as president of the society, and Thayer, as secretary, acknowledged Fort's preference to create a Stellar Exploration Club with the goal of flying a rocket to Mars, but declared that idea was "so much more confining than the true scope of Fort's thought, so immaterial and useless in comparison to the real importance of what he has given the world, that the Society rejects that as part of its program." Rather, they finally agreed, the Fortean Society's aims would be institutional, propagandizing. There would be a governing body and dues. Fort's books would be disseminated, his files enlarged and willed to the society. Members would "harass and bedevil the established seats of learning until they admit their incompetence or are shown ridiculous for their silence," while alternative viewpoints would be promoted—Fort's certainly, but also those that most assiduously critiqued orthodoxy and fostered the "Fortean imagination." Fort had lost control of Forteanism, as the society took on a clarity of purpose it had previously lacked, if only for the night. Dreiser and Thayer agreed: "The Fortean Society is not so much interested in foisting the views of Charles Fort upon the world as it is in exploiting the mental attitude Mr. Fort brings to his work."[164]

H. Allen Smith watched these proceedings with a sense of unease. He was an acolyte of H. L. Mencken, having sloughed off his childhood Catholicism and the stultifying Victorianism of small-town Illinois. "I hold myself to be among the fortunate few of the world who have not a single superstition," he bragged, having crafted for himself the persona of a plainspoken, slightly naive—but unflustered, commonsensical—debunker.[165] Books had been the engine of his escape, educating him after he dropped out of high school. He lionized authors—Dreiser, DeCasseres—whom he saw as geniuses. They appreciated Fort—why? After that evening, Smith interviewed Fort at his apartment.[166] He read *Lo!* but perceived only a reworking of spiritualist ideas. The experience catalyzed a conversion. "I lost some of my blind enthusiasm for book authors."[167] Fort and his disciples were ridiculous, no smarter than the rest of humanity; probably, in fact, more benighted. Their *soi-disant* skepticism—unlike his, which was rooted in common sense and appreciation for science—was indistinguishable from the Victorian superstitions he had outgrown.

The Loch Ness Monster

A monster emerged.

Emerged from the aqueous depths.

A threat from below.[168]

"Sssnnnwhuffffll? / Hnwhuffl hhnnwfl hnfl hfl?" said the beast, as the poet Edwin Morgan imagined it four decades on.[169] Morgan: meaning, sea chief. Fortean Society member 2508.[170] "Zgra kra gka fok! / Grof grawff gahf?"

Scotland, May 1933: sea monster seen. I. O. Evans, sensible English civil servant by day, off-hours agitator for a new social system, nudist, and writer, snooped around Loch Ness.[171] In boyhood, his imagination had been set to flight by Jules Verne.

When Verne wrote, the matter seemed settled: of course, the sea held secret monsters.[172]

Look at Norse mythology.

Look to A. C. Oudemans, Dutch zoologist, his 1892 survey *The Great Sea Serpent.*

Waldo Lee McAtee, chronicler of fish rains, knew: a member of the Fortean Society now, he collected sea monster reports from the nineteenth century, contributed them to the Fortean Society.[173]

Look to Rupert Gould. A retired British navy man, Gould had circumnavigated the loch on a motorbike, hoping for a sighting. Already he'd written two anthologies of scientific anomalies, and one sustained argument: *The Case for the Sea Serpent.*[174] Convinced that anomalies could be resolved by the proper application of scientific thought, he sought to make sense of the Scottish mystery using the same tools he'd brought to investigating sea serpents.

Gould had been a member of a British club, the Sette of Odd Volumes, before being driven out by scandal. His erstwhile best friend there, E. G. Boulenger, director of London Zoo's aquarium, called the Scottish reports "a striking example of mass hallucination."[175]

Journalist H. Allen Smith, seeing the stories come over the wires, thought immediately of Fort, dead these nineteen months. Smith wrote about the creature in January 1934, hanging his article on a Fortean hook. Fort had written a fair amount on sea serpents, he noted, as well as other mysterious monsters: the Jersey Devil, the Blonde Beast of Patagonia, a host of bizarre creatures skulking about Australia. Lest the reader think Smith had joined Gould and Fort in embracing such foolishness, he ended the article with the evaluation of William K. Gregory from the American Museum of Natural History. The scientist's credentials received an entire paragraph, followed by a single, devastating quote from him: "Barnum was right."[176]

The debate swirled, thoroughly mixing with Forteanism.

I. O. Evans became a member of the Fortean Society. Having suffered a crisis of faith and joined the Church of England, he used Fort to keep his mind limber, trying on ideas he did not accept.[177] But the Loch Ness monster was real, he said, no doubt: a flexible-spined creature resembling a plesiosaur, though probably a mammal.[178]

Rupert Gould wrote a book on the Loch Ness monster, insisting that it was real. He refused, though, Thayer's overtures to join the Fortean Society in the mid-1930s, found distasteful Fort's contrariness, and Thayer's.[179]

The monster was a primitive relic, said one member of Thayer's organization.[180] Ella Young, Irish poet, Yeats's compeer, moved to Berkeley, to Carmel, sent reports to Thayer of Loch Ness monsters and Oakland mermaids.[181]

In 1948, naturalist Ivan Sanderson (coiner of the word *fafrotskies*) tried to reverse elite opinion in the *Saturday Evening Post*: "Don't Scoff at Sea Monsters," his headline read.[182] "Pretty mamby-pamby stuff," scoffed Tiffany Thayer. He was a real radical and didn't even believe in Sanderson.[183] Sanderson shrugged off Thayer, became a society member anyway, kept writing, systematizing the various reports of monsters, those of the sea and those of the land. Cryptozoology, he called his field of study. The word was taken up by the Belgian biologist Bernard Heuvelmans, who went on to write a series of books on cryptozoological creatures, most famously *In the Wake of the Sea-Serpents*.

In 1954, a fishing vessel's sonar seemed to detect the Loch Ness monster.[184] John J. Graham—compiler of ten thousand notes on scientific anomalies, designer of never-built rocket systems, fan of Fort but refusing to join the Fortean Society—demanded the British Museum of Natural History explain the signal.[185] Explain, as well, he demanded, the strange biological remains that washed ashore in Dungeness; and the abominable snowman; and how the coelacanth had been rediscovered off the coast of Africa.[186]

A publicity stunt, thought the science fiction author Eric Frank Russell, second-in-charge for most of the Fortean Society's lifespan.[187]

Legend.

Fact.

Humbug.

Truth and beauty.

Necessary dream for modernity, said novelist John Steinbeck, who'd a passing infatuation with Fort in the early 1940s.[188]

Sea monsters belonged to what cultural historian Christopher Partridge calls "occulture": the paganism, New Agery, esotericism, Theosophy, and mysticism central to twentieth-century life.[189] Popular culture mined this occultural stream, animated the world with all sorts of magical objects. Jeffrey Kripal suggests that occulture is the dress worn by something real, a genuine paranormal realm that, in the modern world, can only be experienced through popular culture.[190] One need not go so far—one need not insist that there is a mystical world—to accept that modernity was always enchanted. That wonder, awe, uncanniness, amazement, enthrallment—that all of these were fundamental parts of modernity. Sea monsters, from this perspective, swam not in the waters of the world but in the stream of magical beings, creatures not marginal but essential to the modern world.

Steinbeck: "For the ocean, deep and black in the depths, is like the low dark levels of our minds in which the dream symbols incubate and sometimes rise up to sight like the Old Man of the Sea. And even if the symbol vision be horrible, it is there and it is ours. An ocean without its unnamed monsters would be like a completely dreamless sleep."[191]

The Death of Charles Fort

Alexander Woollcott was, by the time of his association with the Fortean Society, a well-known reviewer with a much-feared and much-admired wit.[192] He missed the Fortean Society's meeting, as he was visiting Asia, but sang the praises of *Lo!*, Fort, and the society when he returned, devoting most of his June 1931 column in *McCall's* to the subject. Never one to be confined by facts, Woollcott portrayed himself and all the absent members as attendees, dealing with each other—negotiating feuds, temperamental and philosophical differences—and the press. "Strange bedfellows," he said, explaining the various Forteans as driven by their own, personal furies—Dreiser beguiled by superstition, Tarkington by Fort's literary style, Hecht by contrarianism and cynicism. Science was accepted in this modern age as religion was in the past, and both eras were improved by their heretics. Thayer, he suspected, wasn't a convert at all, rather "one wishing to see any fight from close range."[193]

Woollcott said he'd been a Fortean for five, six years, ever since Tarkington introduced him to Fort, but it seems that reading *Lo!* (subtitled, in his mind, "and Behold") induced in Woollcott a new way of thinking. His column dwelled on several mysteries Fort discussed, most notoriously the story of Dorothy Arnold, a famous socialite who disappeared in 1910. (Fort coyly suggested she may have been transmogrified into a swan, a particular striking example of the bird having appeared in Central Park on the day Arnold vanished.)[194] A few months after the *McCall's* column, Woollcott produced the first of a new genre of writing, what he called "folklore in the making."[195] His *New Yorker* column told of a woman who'd dreamed, three times, of a hearse driver with an ugly face. When visiting New York later on, she saw the same face on an elevator operator; she bolted, saving herself, as that car crashed, killing all the riders.[196] Woollcott would go on to write about other such tales, later collecting them in *When Rome Burns*, which some folklorists regard as the first book on urban legends.[197] In all of his reporting on folklore in the making, though, Woollcott never mentioned Fort. In fact, Woollcott never again publicly mentioned Fort. Even as the Fortean Society announced itself to the world, his attention was pulled elsewhere. He had read all fifteen hundred pages of Wells's *The Science of Life* in five days' sailing on the Yellow Sea, fascinated by the biology he'd missed in his patchwork education, pleased to find himself understanding everything except the section on genetics.[198] He handed out copies of the massive tome to friends and, in February 1931, reviewed *The Science of Life* on his radio show—it was his favorite book of the year.[199]

Nor was Woollcott's silence unique. Hecht, an in-demand screenwriter who spent much of his time in Hollywood, did not mention Fort again for almost a decade. Tarkington circulated between Indiana and Maine, briefly mentioning Fort in a March 1931 interview but otherwise silent on him.[200] *Lo!* had flopped for its British publisher, Victor Gollancz, so he passed on Fort's final book.[201] In July 1932, H. Allen Smith wrote an article about the reading habits of famous authors.[202] Among those he interviewed were Woollcott, DeCasseres, Hecht, and Harry Elmer Barnes, all strongly associated with Fort or the Fortean Society, as well as Carl Van Doren, who was rumored to be a Fortean, and Clarence Darrow and Gene Fowler, whom Thayer claimed were Forteans.[203] Not a single one of them listed Fort's titles among their favorite books. The Fortean moment was closing. While Woollcott, in his column for *McCall's*, said that the fête in January had been the first of an annual series, he later admitted it had been a one-off.[204]

Other stunts were planned but never came to fruition. In March 1931, Sussman organized an event with the bookseller Moss and Kamin. The press release said Fort would be at the George Washington Hotel, signing books and speaking. Puckishly, the title of his lecture was to be "Fortism and Its Dangers." The occasion was billed as a "Neo-Fortean Gathering," a nod to Fort's hurry to end the Fortean Society.[205] "The Forteans are very old-fashioned," he told De-Casseres. "They're three months old, and that is long enough for any cult. We are the Neo-Forteans." Tarkington, Dreiser, and the illustrator Alexander King were all scheduled to appear. But there was no such gathering. Later, Sussman and Dreiser considered ways to drum up publicity for Fort's last book. An argument, they suggested, carefully arranged, with Dreiser and the journalist Reed Harris on one side, conservatives on the other, debating Fort's merits. But Fort nixed the idea. Dreiser then gave an interview to Harris, recounting his history with Fort, explaining Fort's ideas, but once more Fort demurred, and the article never ran.[206]

With the failure of the Fortean Society, Fort devoted his full attention to his next book. The rapidity with which he completed the manuscript—it was published not quite a year and a half after *Lo!*—suggests he salvaged pieces of the other manuscripts he'd brought with him from England. Fort drove himself relentlessly, working through dinner, skipping the movies; staying home was easier anyway, since walking took so much effort now. He was sick, his body failing him. This book dealt with ideas he'd been contemplating since at least the mid-1920s, not cosmic forces but sublunary ones: pictures falling from walls, mysterious diseases, vampirism, bodies bursting into flame, saintly miracles, psychic powers. Fort implied—expressed—that these happenings, strange to contemplate but common in the record of humanity, could all be accounted for as a kind of witchcraft, or what he called talents. The forces were mundane but multiple. Some people possessed special faculties that allowed them to accomplish the seemingly impossible, though in an uncoordinated fashion; the talents were uncultivated, not yet systematized in the way religion had once been and science now was. Hence the book's title, *Wild Talents*. Fort still riffed on monism, poked fun at science—especially the latest fashion, Einstein and quantum mechanics. But the core of the book concerned witchcraft and his prediction that in the future such talents would become a common feature of everyday life:

> Girls at the front—and they are discussing their usual not very profound subjects. The alarm—the enemy is advancing. Command to the poltergeist girls to concentrate—and under their chairs they stick their wads of chewing gum. A regiment bursts into flames and the soldiers are torches. Horses snort smoke from the combustion of their entrails. Reinforcements are smashed under cliffs that are teleported from the Rocky Mountains. The snatch of Niagara Falls—it pours upon the battlefield. The little poltergeist girls reach for their wads of chewing gum.[207]

Wild Talents, like Sussman's and Dreiser's plans, failed to grab the attention of the public. There was a sprinkling of positive reviews—an acquaintance of Tarkington thought the book exciting; a reviewer for the *Detroit Free Press* said, "It stimulates one's serious thought, and at the same time provides one with the startling thrills of a ride on a roller coaster."[208] But most reviews were unkind—only *New Lands* suffered a colder reception. "A connoisseur of newspaper clippings . . . and rather dull he was, too."[209] "Drivel."[210] "Just what is wrong with an intelligent public that will listen to or read such cracked wisdom, punctuated by wisecracks in bad taste, or with publishers who write absurd blurbs to absurd books, cannot be diagnosed in a sentence or a paragraph. Whether the bottom has dropped out of thinking as out of the stock market, is one question; why, is another. Except to satisfy curiosity, there is no need to consider 'Wild Talents.'"[211]

Mencken savaged *Wild Talents* and the Forteans in the *American Mercury*.[212] While admitting that Fort's method was subtle and not without humor, he dismissed his conclusions as rigamarole. Elsewhere he wrote, "I looked upon Fort as a quack of the most obvious sort."[213] From Fort, Mencken's review moved on to other targets: the *littérateurs* who toasted him, and their penchant for displaying credulity in so many arenas of life. They embraced "the imbecilities of Communism," quackeries of all manner. "The remarkable thing about . . . the vogue of the late Mr. Fort," he said, was not his very commonplace ideas, but that his theories "should be received with all gravity in a presumably civilized country and accepted as gospel by a considerable body of presumably sane men." Without naming names, Mencken aimed at one *littérateur* in particular, Dreiser, his previous enthusiasm for America's great author gone. The "eminent novelist," in his eyes, had become "footling and bewildered." Mencken wanted nothing more to do with him, wanted to make clear his separation from Fortean foolishness. In an article, H. Allen Smith had mistakenly numbered Mencken among those who had joined the Fortean Society, which Mencken said "was a libel of virulence sufficient to shock humanity." He told Smith, "As a Christian I forgive the man who wrote the story and the news editor who passed it. But both will suffer in hell."[214]

While he worked on *Wild Talents*, Fort's health continued its precipitous decline. Overcoming his distrust of officialdom, he visited a doctor. His biographer Jim Steinmeyer thinks he probably had leukemia.[215] "I am a god to the cells that compose me," Fort wrote in *Wild Talents*, a callback to his first public expression of monism, "A Radical Corpuscle."[216] One measures a circle beginning anywhere, and Fort's life had almost completed its circuit. The cells were revolting, as had that heretical white blood cell, turning against their god. His disease was caused by "a conscious, mysterious parasite which had seized" him, he told Dreiser's mistress, Helen Richardson. "There was no escape."[217] As Claude Kendall's house put together *Wild Talents*, and Fort idly contemplated the shape of another manuscript, *Medi-Vaudeville*, Annie waited upon him, turning him in bed, feeding him meat broth.[218]

On 2 May 1932, Annie called the ambulance, and Fort was taken to Royal Hospital. The next day, Sussman brought Fort an advance copy of *Wild Talents*; it would be officially published 5 May. The night of Sussman's visit, Fort slipped in and out of delirium. He died just before midnight, aged fifty-seven, contumacious to the very end: "Drive them out!" were his last words.[219]

The day after *Wild Talents* was given to an apathetic world, a Friday, seven people gathered in an "obscure" parlor in the Bronx to celebrate the life of Charles Fort before his mortal remains were sent to rest in the family's Albany plot. Dreiser gave the oration. "He taught me to think for myself, and I curse the days I spent in college being taught by old professors with their hideous dogmatism," Dreiser said. He was joined by his mistress Helen; J. David Stern and his wife Juliet; Fort's wife Annie; his brother Raymond; and the journalist H. Allen Smith, who wrote up the memorial for the United Press wire service.[220] The *New York Herald-Tribune*, *New York Times*, *Time* magazine, and the Associated Press also noted Fort's passing. "Charles Fort is dead," said another newspaper writer, "and the world is a less exciting place than it was."[221]

Sad, suffering from bronchitis, Dreiser sailed for Galveston a few days after the funeral, then traveled by car to Los Angeles, troubled by strange dreams.[222] His relationship with Fort had been "intimate . . . just as though he lived with me in my home." They'd met in person only a handful of times over the course of a quarter century. But "he was never really out of my mind. Always he was one who seemed to be talking with my own voice, with my own moods."[223] In Los Angeles, he connected with the journalist George Douglas, whom he'd first met more than a decade before. They planned to write a summation of Dreiser's philosophy, bridging the gap between materialism and mysticism. But Douglas died unexpectedly, and Dreiser never completed the project.[224]

Thayer had left New York even before the funeral, also for California. His books had made him famous, dime-store novels stuffed with sex, murder, perversities, sex, outrageous scenarios, and sex, all leavened with an unabashed intellectualism.[225] So synonymous had his name become with readable smut that he believed "Tiffany

Tiffany Thayer, with Julia Sherman, 1935. (Los Angeles Times Photographic Archive, Library Special Collections, Charles E. Young Research Library, UCLA. Creative Commons Attribution 4.0 International License.)

Thayer" on the cover would be enough to sell an expensive book of erotica in the midst of the Depression.[226] His first three novels had sold, combined, almost four hundred thousand copies in just three years.[227] And so Hollywood had come calling.[228] Within a year, a pair of novels had been adapted for the screen and he was under contract with Paramount, which made a third novel into a movie and purchased an original story.[229] He joined the party circuit, crafted for himself an outrageous persona: he wrote only in pencil, without regard to his surroundings—as long as no one near him chewed gum. Sometimes, as he composed, he draped over his shoulders his twelve-foot-long boa constrictor, named Sir Hissen-Farben-Dunghi.[230] He was arrested for drunk driving.[231]

Annie was left alone in her apartment, with the parrots but, for the first time in almost thirty-eight years, without Charles. They'd weathered poverty, crisscrossed the Atlantic, lived a simple life al-most always in each other's presence. She was Fort's rock, and he hers, and now she was unmoored, vulnerable. Fort's family tried to usurp her inheritance. It took years for Thayer to connect with her again, guiltily sending her alcohol to calm her nerves at the end of each day. "I make more money than I have any right to and there is nothing on earth I had rather do with it than assure you a night-cap when you want it."[232] Sussman, too, seems not to have visited. But she was sometimes comforted by the sound of Fort's voice echoing through her home, bangs she was certain he had made—the kind of uncanny phenomena he'd written of in *Wild Talents*. She was sure he was reaching out to her from beyond the grave, a spiritual realm he'd refused to accept in life, but that she under-stood as real.[233]

A few years after Fort's death, Dreiser realized that his friend was still unhonored. "Most would read Fort's books with repugnance and fear. Others would cast them aside with a smile and call them childish fairy tales." But Fort would not be forgotten. There were those who "would shudder with delight, recognizing the poetry, the truth, insight and the marvelous intelligence of Fort's conception." His oeuvre would become "classics," if only to the delectation of a few.[234] These thoughts continued the obsequies he'd offered to those gathered in that obscure parlor. "Future generations will pay him tribute," he'd prophesied then.[235]

John Cowper Powys and the Ichthyosaurus Ego

In October 1920, Dreiser and Helen, and John Cowper Powys and his brother Llewelyn, gathered in San Francisco. (It was here that Dreiser first met Douglas.) Helen suffered from stomach pains, as did Powys. Llewelyn had tuberculosis. A poet friend of Dreiser recommended the group visit the maverick doctor Albert Abrams, inventor of the Oscillo-cast. By producing vibrations he called Electronic Reactions of Abrams, or ERA, the contraption could supposedly diagnose and treat any disorder. After Helen was subjected to the machine, Dreiser reported, "All her symptoms have disappeared and she is as healthy as ever."[236]

The American Medical Association was not so sanguine about Abrams's methods, nor were scientists. The magazine *Scientific American* put together a team to investigate the doctor; among the examiners was Maynard Shipley. Initially, Shipley was open to the possibility that Abrams had discovered a new therapeutic tool, even excited about the revolution the Oscillocast might bring to medicine. After scrutinizing Abrams's practice, however, he concluded that the doctor was deluding not only clients but probably himself.[237] Mencken, an anti-Semite, also visited Abrams and deemed him "the usual Jew doctor," unable to do what he claimed.[238]

Dreiser's and Mencken's differing assessments were indicative of a developing break. Mencken told Burton Rascoe—future Fortean, and Dreiser's first biographer—that he and Dreiser "remained on good terms so long as I was palpably his inferior—a mere beater of drums for him. But when I began to work out notions of my own it quickly appeared that we were much unlike. Dreiser is a great artist, but a very ignorant and credulous man. He believes, for example, in the Ouija board. My skepticism, and, above all, my contempt for the peasant, eventually offended him. We are still, of course, very friendly, but his heavy sentimentality and his naive yearning to be a martyr make it impossible for me to take him seriously."[239] Mencken disliked Dreiser's later literary effusions for the most part, lambasting *An American Tragedy*.

From San Francisco, Powys returned to his career as an itinerant lecturer on matters literary—he was much sought after, praised for his wide-ranging intellect and eloquence. "Leaving the hall after his lectures," remembered the author Henry Miller, "I often felt as if he had put a spell upon me. A wondrous spell it was, too."[240] If Powys felt better after his encounter with Abrams, the treatment did not stick. He was racked with pain for the rest of his life, his diaries revealing him as a bundle of neuroses, concerned with stomach aches and his elimination schedule. Powys lived in genteel poverty at the time, supplementing his income by writing for a publishing company with whom his lover worked, though he was resentful because he too was anti-Semitic and the publisher was Jewish. In 1929, just before he discovered Fort, his financial situation settled, his book *Wolf Solent* a success, followed by other novels of some renown.[241]

At about this time, as well, his philosophy crystallized. Powys valued both pluralism, with its emphasis on extreme individuality—"isolating the self from the 'herd' by the development of a 'crystal core of inviolability'"—and "a breaking down of the boundaries between the real and the imaginary, between the civilized 'self' and the 'other'" that came with monism. His book *In Defence of Sensuality*, published in September 1930, just before he discovered Fort and read *The Book of the Damned* and *Lo!*, synthesized these impulses. The solution to modernity's disenchantment was a sea monster—the ichthyosaur, a prehistoric reptile of the oceans. Humans needed to develop an ichthyosaur ego that would "hold in creative suspension two opposing ideas: the solitary apartness of the reptile, at the same time, the immersion in the natural world that characterizes primitive beings. His 'ichthyosaurus ego' would retain its individuality while being part of the magical world of the unconsciousness."[242]

Powys's ichthyosaurus ego predisposed him to Fort's heterodoxy. From one perspective, he said, Fort's writing "seems to afford a wonderful liberation to my mind," dissolving the chains that had him "enslaved by the 'Dominant.'" He called Forteans the New Protestants, likening them to those who broke from the Catholic church, who insisted that the most important source of religious authority was the Bible and the individual's relationship to it. With Fort's guidance, "a person learns to think for himself." In this sense, Fort fostered individuality. From another perspective, Powys valued how Fort dissolved the individual into the universal flux: "One is left after reading 'Book of the Damned' with that open-mind towards the mystery of life which allows for all manner of strange and even 'improper' occurrences."[243]

Long after the majority of the founders of the first Fortean Society had forgotten Fort, had dismissed the society—and after Thayer had written Powys off as bewhiskered, his fiction maundering—Powys found himself musing on Fort in his diary.[244] His late novel *The Inmates* had a Fortean outlook, concerned as it was with those consigned to a mental institution (damned, as it were) who nonetheless understood truth more clearly than did officials of church and state.[245] Theirs was the wisdom of fools. They were free, as he thought Fort and his acolytes were free, of "those sublimated herd-dogmas of science."[246]

2

A Budget of Paradoxes

River no sun falls through,
Doubt will flicker a dim
Cavern in which the blue
And silver fishes swim.
Doubt is the light and shade
Over the water, the clear
Image that will not fade.
Doubt is the atmosphere.

MARION STROBEL, "Doubt," 1931

In September 1937, Tiffany Thayer mailed out, to erstwhile founders of the Fortean Society, to members of the press, friends, influential persons, the first issue of his new magazine, *The Fortean*. Possessed of the adman's acumen, Thayer sought to provoke controversy. Its lead sentences: "I won't be quiet—not if this is the last collection of words I set down on paper—not if this is the last breath I draw. Amelia Earhart and Fred Noonan were murdered by Dogmatic Science." He explained that scientists had falsely professed to understand earth's geography, misleading Earhart and copilot Noonan, sending them to their deaths. Thayer meant the magazine's fifteen pages to restart the Fortean Society, to launch a satirical attack on "the smug Complacency of Authority."[1] He imagined *The Fortean* in the tradition of *Puck*, the fin de siècle magazine once helmed by H. L. Wilson, whom he called "the chiefest literary talent this country has produced since Mark Twain."[2] Thayer's ambition was boundless. "Give me a broadcasting station and a battery of color presses," he wrote in another context, "and I'll show you a miracle in ten years."[3]

When Wilson read the magazine, though, he was puzzled—he respected science, saw no call to attack it. Nor were other founders of the Fortean Society sympathetic. Tarkington was confused.[4] Woollcott ignored the amateurish publication altogether.[5] Thayer himself was, underneath the cocksurety, uncertain too—he had "decidedly undefined purposes," as Dreiser said when Thayer broached the idea of resuming the society.[6] Thayer didn't even think to copyright the magazine.[7] There would be a second issue in November, but the third did not appear until January 1940, and the fourth a year and a half after that. Although Thayer would provoke the ire of "smug Authority," it was not until the end of World War II that the Fortean Society really caught on. Despite the confused responses, a dewy optimism also followed publication of *The Fortean*'s first issue. The *Palm Beach Post-Times*, unaware of contentiousness among the founders, saw the magazine as the unified voice of literary paragons. The reviewer, his attention likely caught by Thayer's opening claim, announced the magazine's arrival in verse:

> I never heard of Charlie Fort—
> How dumb a person I must be—
> For judging by his partisans,
> His was a great mentality.[8]

Through the course of its history, the Fortean Society was the backbone of Fortean modernity. It fulfilled the promises Thayer made that snowy evening in 1931, promoting Fort's ideas and extending them, especially into the political arena. Held together by the force of Thayer's personality, the society reflected his idiosyncrasies but, despite claims to the contrary, wasn't quite a one-man show. It gathered under its aegis a wild diversity of cultural mavericks and malcontents, among them science fiction authors, avant-garde artists, and UFOlogists of various stripes. Like Thayer, they used Fort to

extend the reach of imagination; they claimed to uncover the organization of the world; they offered their own explanations of who
controlled the world and how. Often, other Forteans' ideas differed
from Thayer's, and their holders contended against him, strained to
push the society in new directions. Such disputes shaped the society
and defined the contours of Fortean modernity.

Little Dog Lost

In 1916, aged fourteen, Thayer quit high school and ran away to
Chicago to live with his mother Sybil Farrar.[9] His parents were ac-
tors, had met on tour, divorced when their only child was five.[10]
Thayer apprenticed as a commercial artist, dabbled in journalism,
sold books.[11] He acted, too, as his father's son, Elmer Ellsworth Jr.[12]
In 1921, he married Minnie Roe while on tour.[13] Thayer's persona
was circumscribed by heredity. One grandfather had been a sober
religious man, the other a ne'er-do-well; industry and rascality would
mark his life. While he failed to find a profession in Chicago, he did
discover a philosophy: religion, science, social mores—all were mere
conventions. There was no good, no evil, no beautiful, no ugly, except as preferences.[14] The enlightened man behaved according to his
innate tendencies, to hell with social niceties.[15]

Around 1922, Thayer read *The Book of the Damned* and installed
Fort in his literary pantheon, alongside Hecht, Mencken, Nietzsche,
drawn to Fort's metaphysical clowning. Thayer fancied himself "a
jester . . . whimsical, irresponsible, agnostic, careless, analytical, im-
provident and very, very silly."[16] Shortly after *New Lands* was published, he wrote to Fort, the two bonding over disreputable pasts—
Thayer's admission that he was an actor met by Fort's that he'd been
a journalist.[17] The correspondence was short-lived, however; Thayer
moved to New York around this time, his marriage likely ending.[18]
In Manhattan, he became an advertiser, promoting "mash to make
hens lay, strong men's courses, beauty clay, and a wild variety of
other products."[19] Nights he spent writing; he'd started churning
out short stories in Chicago and continued to do so through the de-

cade. In 1930, he published *Thirteen Men*, an homage to his literary heroes. The first chapter has Frank Miller kill thirty-nine people; each of the twelve that follow deals with one of the jurors before a concluding chapter returns to Miller at trial. He'd been shaped by his reading, Miller lectures; he contrasts the conventional jurors with subversive books he'd read, among them Fort's. So Fortean was *Thirteen Men* that Fort's friend, the newspaper publisher J. David Stern, suspected Fort wrote it.[20] Shortly thereafter, Thayer created the Fortean Society to promote *Lo!*, then left for greener pastures.

Having launched himself into Hollywood, Thayer was disappointed to find not an oasis but a mirror confronting him with his own limitations. He married again: Tanagra Duncan, a dancer.[21] He opened a theater.[22] He started his own publishing house and opened a bookstore—Tiffany Thayer and Other Good Books—for his stepdaughter to run.[23] But the marriage failed, as did his acting career. Paramount, the movie studio for which he worked, forced him to turn out "hackneyed" plots.[24] His father died. Thayer subsumed his dissatisfaction into a series of novels, most of which recrafted autobiography as "phantasmagoria."[25] His kiss-off to Hollywood—and one of his last novels—was 1938's *Little Dog Lost*, which tells the story of a movie mogul uncertain of his true calling or his identity, variously called, throughout the book, Frank, Stanly, John, and Franklin. Disgruntled by Tinsel Town's vacuity, Frank-Stanly-John-Franklin abandons marriage and daughter, becomes implicated in a kidnapping, government intrigue, family melodrama, a love affair. None of the events move him, though; he thinks himself untouched by conventional emotion. Only at the end does his veneer break when, watching a boy search for a lost dog, he realizes that his egotism has been a defense against empathy. The true existential prerogative, Frank-Stanly-John-Franklin concludes, is to "live dangerously"— not to scoff at others but to follow one's ideals, whatever the cost.[26]

Thayer himself left Hollywood, his wife, and his stepdaughter in 1936, returning to New York and advertising. He joined the J. Walter Thompson agency, where, among other projects, he worked as a continuity writer on the Chase & Sanborn Coffee account.[27] Thayer was obsessed with an idea that H. L. Wilson had floated in his novel *The Wrong Twin*: "a good loose trade," a job a man could practice anywhere without being tied down.[28] He seemed to think advertising fit the bill. Evenings were still devoted to writing, long-hand, as well as fencing and painting. Reportedly, he slept only four hours a night. His literary output, however, declined precipitously (after *Little Dog Lost* there would be only an edited volume, a few rewritten classics), as he became lost in the composition of a single novel that took him twenty years to produce—and in running the Fortean Society. Thayer had announced the society's imminent resumption in the spring of 1935; he enrolled friends, acquaintances, and the idly curious as members or "regional directors."[29]

Thayer expected the magazine's first issue to center on Dreiser, but Dreiser was fed up with Thayer's shenanigans and wanted nothing to do with the project.[30] Dreiser was trying to write a biography of Fort, but Thayer had absconded with Fort's notes in 1932 and refused to return them when Dreiser, on Annie's behalf, asked. Nor would he do so after Annie died, in August 1937, and willed the notes to Dreiser. Dreiser demanded in 1935 that his name not be associated with the Fortean Society at all.[31] Most of the other founders were equally dismissive. The plan to restart the society and publish a magazine was a waste of time, Thayer's friend, the Chicago bookseller Ben Abramson, told him, refusing to join the society even as he was named a "regional correspondent." Thayer, he admonished, was forgoing a lucrative writing career for pure crankery.[32]

The poor reception did nothing to diminish Thayer's ardor. H. Allen Smith had called him an owl collector, and that remained so.[33] Thayer signed up the maverick astrologer Alfred H. Barley and his wife Annie, extended membership to Raymond Cass, who claimed to hear the voices of the dead in radio static, and Albert Page, a mailman who derived an abstruse theology from an unusual interpretation of atomic theory.[34] He promoted B. J. S. Cahill's "Butterfly Map" of the world and collected the works of Isaac Newton Vail, which argued that the earth was surrounded by a ring of water from which fell organic matter.[35] Thayer nursed his own heterodox ideas as well, ideas that stretched the imagination. He speculated that the earth was growing, alternating between a cube and a sphere—a play on the notion of squaring the circle—which accounted for earthquakes, evidence of ice ages, the fitting together of land masses, the dissemination of ideas across distant cultures, the myth of the flood, Atlantis, flat earth theories, and why seven is sacred.[36] If Earhart had understood, she wouldn't have relied on scientists, and wouldn't be dead.[37]

Thayer also reached out to the agricultural scientist and author T. Swann Harding after reading a pair of his books.[38] Both were skeptics building on the nineteenth-century free-thought movement, which sought to curb the cultural power of mainstream Christianity, expanding that critique to encompass science as well—but they differed as to how vehemently to oppose the scientific project. Harding, an acolyte of Mencken, respected scientific benchwork; he believed that scientists discovered facts but that these facts were then embedded in stories, myths, and partisan tales only vaguely connected to truth.[39] Like Mencken, he especially distrusted modern physics, and he worried over the manipulation of science by the press, by advertising.[40] Thayer doubted that there was such a thing as a scientific fact (although he was slippery on the subject) but was nonetheless excited by Harding's ideas.[41] He tried, to no avail, to get an article Harding had written for Dreiser's magazine *American Spectator* turned into a book.[42] He could've published at least excerpts in *The Fortean* but told Harding he wanted his magazine to avoid the "advertising, political, economic angle" in favor of "natural wonders" and the errors of scientists.[43]

Running the society, though, was difficult. Thayer does not seem to have budgeted adequate time for the endeavor, or he may not have anticipated the effort it would take. He repurposed Fortean Society letterhead he'd printed in southern California, sloppily crossing out the old address.[44] Each issue of *The Fortean* ended with a page or so of its namesake's gnomic notes—Thayer's girlfriend took on the challenge of deciphering and transcribing Fort's scribblings. The department lasted the entirety of the society's existence and confused readers for as long.[45] And there were other demands on his time: his mother needed tending in Illinois, his stepdaughter help with the bookstore.[46] Issues 1 and 2 had come out a month apart, in September and October of 1937. After that, he tried to switch to an annual format, but he was unable to meet even that schedule for the third issue, instead tossing off a summative article for the magazine *Ken*.[47] His Fortean focus contracted to a single point: publishing an omnibus edition of Fort's books. Once more, he underestimated the challenges. Neither he nor Aaron Sussman had copies of all of Fort's books and, with Kendall's recent (and mysterious) death, he could not turn to the publishing house for them.[48] They had to dun friends and advertise to gather all four. The scrambling paid off: in early 1941, Henry Holt published *The Books of Charles Fort*. Thayer wrote the introduction and, masochistically—given the demands on his time—indexed the thousand pages himself.[49]

By then, Thayer had grown palpably more cynical; his introduction for *Lo!* had been playful, the one for the omnibus so hectoring that, at the request of the publisher, Sussman intervened and made him soften the rhetoric.[50] *The Fortean* required no such compromise. The magazine did report on natural wonders, including stories Harding culled from the papers, and it celebrated those Thayer designated as Founders—Woollcott in issue 1, Tarkington in 2, Rascoe in 5, Powys in 7, Wilson in 8—but these features were increasingly drowned out by Thayer's cantankerous criticisms of science (he called out scientists by name), religion, the government, and schools. Despite what he told Harding, he did not avoid the political, economic, advertising angle. In the second issue he wrote, melodramatically:

> The soul-withering horror that grips my bowels and makes me writhe in futile frenzy is the nefarious and noxious means by which the Glory of Modern Science and all its works is maintained. Popular education has been forced upon mankind in the guise of a boon. THEY taught us to read.[51]

In the next issue he extended the Fortean ambit to media criticism. Newspapers were not neutral conveyors of information, mere sources for reports of the strange, as Fort sometimes seemed to imply; rather, the press was in cahoots with scientists, part of a vast system imposing "mental discipline."[52] For the rest of *The Fortean*'s run, Thayer concerned himself with "what mankind is fed by the eyes and the ears to warp and twist and retard the normal growth and development of its mentality."[53] These were the people and institutions that structured the world and organized perception.

Thayer came by his bilious paranoia honestly—it was part of his makeup. But the world's inexorable descent into a second global war in a generation made it impossible for him to keep his views in check. "It was only the phenomena of war propaganda which forced us into political lists," he told the poet Ezra Pound after the war.[54] As he saw it, the Fortean Society was a bulwark against creeping tyranny. Scientists, astronomers, doctors, teachers were "the first conditioners of infant mentalities, and they create the thought habits (genuflections) which permit all the abuses which the political and economic reformers are trying to rectify."[55] By undermining their authority, he hoped to liberate humanity. Given the global situation, though, and his own thoroughgoing pacifism, he felt compelled to move beyond scientists and their ilk, focusing his hostility on those with even more power, politicians and the financial elite. He uncovered not only the world's hidden structure but the way power moved within it, reshaped it. Thayer suspected that under the pretty gobbledygook offered by the press—the war as a defense of democracy, the liberal order, and everything that was good—something more nefarious was at work.

He made clear his ideas in issue 6—or, rather, let his cynicism run wild. Like the narrator of *Little Dog Lost*, Thayer had an empathetic core. He was genuinely devastated by war, its grinding up and spitting out of bodies.[56] He'd seen what DeCasseres called the "planetary cannibalism" of the War to End All Wars, had no taste for more.[57] But, as in *Little Dog Lost*, he cloaked his empathy in causticity. Thus, the issue opened with a two-page editorial titled "Circus Day Is Over" that denounced the war as a hoax sacrificing millions of lives to sustain those in power and feed the insatiable maw of fat-cat bankers. A mere month after the attack on Pearl Harbor, he wrote that the Axis and Allies had "planned this gigantic hoax upon their peoples in all personal friendliness."[58] The next issue, which did not appear until June 1943, again opened with an editorial, "The Socratic Method, or, Pass the Hemlock," a three-page list of leading questions that implied the war was the culmination of a conspiracy: Does anybody believe ballots are counted? Is the US fascist? How could one escape propaganda when newspapers owned radio stations? Why did the Bank of Italy rename itself Bank of America?

Thayer expected to get a rise but was unprepared for the intensity of the reaction. Woollcott, who was propagandizing for the war, quit the society after issue 6, and enjoined Tarkington and Hecht to follow.[59] In April 1942, J. David Stern reported Thayer to the FBI for sedition.[60] The bureau opened a file and seized some of the issues being sent internationally, scaring Thayer so much that for a while he stopped using Fortean Society letterhead and insisted that Fortean correspondence be sent to his work address, not the post office box he'd rented for society business.[61] It must've been a self-destructive impulse that compelled him to publish issue 7. This time Tarkington, also producing wartime propaganda but much more genial than Woollcott, quit.[62] Sussman did too. "I am disgusted," he wrote. "What you are doing, unfortunately, under the guise of publicizing [Fort], is making his name a synonym for the dirtiest kind of subversive business."[63] Sussman's remonstration abashed Thayer—his only friend among the founders, one of his few friends at all. "I trust that if our critical attitude toward current political events disappeared at once from the Magazine that you would reconsider your decision," he responded.[64] Thayer retrenched. Issue 8 came out in December 1943, with six pages devoted to an excerpt from a book criticizing the Indian caste system by Kanhayala Gauba; future issues were filled with additional long excerpts. He frittered away space with other eccentrics.

The Fortean Society limped along through the war years, its magazine's schedule so erratic that the FBI thought Thayer had given up.[65] Rather, he was trying to broaden the group's base of support.

For a time, he courted astronomers, though he thought their science "too old, respectable and firmly entrenched" to be truly Fortean.[66] They offered no new stimulus to the imagination, no new revelations about the world's organization or the powers that controlled it. There were half-hearted attempts to link Fort with parapsychologists who studied mediumship, the existence of a spiritual realm, the exercise of wild talents such as ESP. It is too much to say that the society was moribund, that Thayer wasn't trying; there were joys, new reports to investigate, new sciences and governmental activities to criticize, new heresies to promote. But the war years were lean, and Thayer's project spiraled toward obscurity. Shipley passed in 1934. Wilson, Woollcott, Tarkington, DeCasseres, and Dreiser all died between 1939 and 1946.

Wild Plum

Thayer: "Politicians tell us that 'the price of liberty is eternal vigilance' and go on to prove that vigilance can be no more eternal than voting for them will make it, but despite the abuse of that truism it is highly pertinent here. No single despotic power Congress has turned over to the President is so great a menace to the vestige of 'liberty' remaining to us as the 'rapidly spreading' use of the 'lie detector.' . . . This is no abstract 'issue' which Forteans may regard objectively. It is a matter of immediate, vital, personal concern to every one of us. It is of particular Fortean interest because the power being invoked to gain absolute supremacy over mass mentality is awe for and confidence in Scientific 'achievement.' . . . If we permit its adoption as standard equipment in our police departments and law courts, we shall be—in essence—kneeling before a myth, first cousin to that other omnipotent wraith, the truth-loving Holy Ghost."[67]

28 March 1944, Bismarck, ND

Charles Schwartz, the state fire marshal, heard of disturbing events at Wild Plum School, out in Stark County.

Mrs. Pauline Rebel was the teacher at a one-room schoolhouse there, taught eight kids outside of town.

They'd all watched as a pail of coal stirred to life, pieces bouncing around the room, knocking into walls. One hit Jack Steiner on the noggin.

The bucket turned out; the coal ignited. Window blinds smoldered; a bookcase caught fire.

School officials rushed to the site, saw the coal alive, moving as though "reacting to a mysterious force." A dictionary moved on its own.[68]

29 March 1944, Richardton, ND

Schwartz got out there the next day, strapped the teacher and students to a lie detector. Not one failed.

He sent the dictionary, the coal, the pail, too, out to the college, the university, even the FBI. Nothing unusual about them, they all said.

Schwartz talked with the students—five belonged to the Steiner clan, a farming family, papa and mama from Russia.[69]

He talked to their dad, George, who suspected tomfoolery until a piece of coal jumped in his hand, which made him think otherwise. He spoke with Helen and Ismara, George's eldest, heard their stories.[70]

Mrs. Rebel was Russian, too, hardly older than her students.[71] She wondered if the events were connected to threatening letters she'd found. Or a hooded man.[72]

The school was closed.

News of the weird events at Wild Plum appeared in newspapers around the country, mid-April, sent over the wires by the Associated Press.

Time magazine reported on the "witchery" in its 24 April issue. The magazine explained the events with a baroque bit of scientific speculation: the accidental organization of molecules into a particular arrangement—purely by chance—caused the coal to jump.

* * * * *

"No event since the boom in the sky over Brooklyn has been reported by so many members or covered so thoroughly by them," Thayer said.[73]

Here was proof of some new, unseen force, witchcraft, a poltergeist, wild talents. "To put it mildly," Schwartz said, "we are puzzled."[74]

Wild Plum had been under siege since January.

* * * * *

Walter Dunkelberger, science fiction fan, MFS (Member of the Fortean Society) was the society's Johnny-on-the-Spot.

He applied himself to the story with diligence, clipping reports in newspapers, even corresponding with state officials.

He compiled an eighteen-page "dossier l'affaire Prunelle Sauvage," as Thayer called it, sent copies of all he had to New York City.[75]

Thayer was always on the lookout for failures by the police, loved tales of strangers with identical fingerprints.

He didn't write up Wild Plum until that summer, by which time there had been some incredible developments, a solution that wasn't.

The Fortean Society caught the authorities with their pants down.

18 April 1944, Dickinson, ND

"The case of the leaping lignite at lonely Wild Plum rural school in southern Stark County is solved," reported the Associated Press.[76]

Late the previous evening, Schwartz and a lawman had questioned the students again, this time in front of their parents.

Four of them said they'd created the prank—with the help of "several other students."

Mrs. Rebel was near-sighted; gullible, too. So, the kids threw coal. They planted lit matches. Jiggled the pail with rulers.

They pounded on their desks and floor, then rushed to the door, announcing that they saw a hooded man running outside. The oldest girls had written the notes.[77]

* * * * *

The solution raised as many questions as the mystery:

School officials had also reported seeing the coal move: what of them? Was every student involved in the hoax? And what of George, who saw a lump of coal jump in his hand? Burning bookcases and books in a lonely schoolhouse could easily have led to disaster. Would the children take such a risk?

The science fiction author Nelson Bond crowed in Thayer's magazine:

> Of course, it is a common occurrence for little school-children in Bismarck, N.D., to start throwing lumps of coal at their teacher!
>
> And of course . . . the moment the coal started flying at Schoolmistress Rebel's head, Mrs. Rebel instantly took off her glasses to keep them from being damaged.
>
> Having experimented, I know how easy it is to set fire to a bookcase.[78]

Never mind the inconsistencies—the story was a Fortean delicacy as presented.

Because in dispensing with the idea that the weird events were caused by a poltergeist, by adolescent girls with wild talents, the "witch girls" of Fort's reports and science fiction tales, the authorities had to admit that the lie detector had failed when it declared Mrs. Rebel and those eight kids truthful. Officials implicitly conceded that "a group of backwoods infants and adolescents had made a monkey of the 'lie-detector.'"

"We Forteans win either way in this deal," said Thayer.[79]

Sinister Barrier

Thayer's inconsistency led to someone else picking up the Fortean baton in the years before World War II. Eric Frank Russell ran with it, using Fort, as other science fiction writers would, to expand his imagination beyond the confines of religion and of science. Although materialistic and, in his way, rationalistic, Russell refused disenchantment. The stories he wrote situated the marvelous at the heart of an otherwise cynical view of humans and their so-called achievements. Nor did he restrict himself to obviously fictional tales, also turning out essays that, like Fort's writings, blended fiction with questionable fact. Startling rains fell on Liverpool, or perhaps objects were teleported into the city, just below the notice of most people going about their business in that sprawling, industrial town. Earthquakes and volcanic eruptions seemed impressive, but scientists did not know half of the forces that created them and would have choked if they did. Russell thrust Fort upon friends, acquaintances, and readers, told them that his books would shock them out of complacency.

Russell was three years younger than Thayer, born 6 January 1905 in Berkshire, England. His father was an instructor at the Royal Military Academy, and Russell spent some of his youth in Sudan, then a British colony. He worked as a telephone operator, surveyor, draughtsman. At some point, he discovered the writings of Robert Ingersoll, a nineteenth-century American orator and Freethinker.[80] Ingersoll was associated with the magazine *Truth Seeker*, which was one in a constellation of responses to modernity, alongside disparate organizations—agnostic, atheistic, mystical, Christian—each trying to reinvent religion in light of historical criticisms of Christianity and the rise of science.[81] T. Swann Harding had been part of a reworking of this tradition in the 1920s, then often called humanism, as had Harry Elmer Barnes and Maynard Shipley.[82] Russell seems to have taken from Ingersoll's infidelism a deep respect for science and an aversion to Christianity, particularly Roman Catholicism. He was viciously anticlerical through much of his life, as well as anti-Semitic.

In 1930, Russell married Ellen Broadhurst; four years later, on his twenty-ninth birthday, they had their only child, Erica. The family lived in Liverpool, later moving to Hooton. Russell was a commercial traveler and technical writer for a steel company. He was fascinated by rockets and rocketry and, in the early 1930s, joined the recently formed British Interplanetary Society. The name was not an intentional echo of Fort's mock society, but the similarity was nonetheless apposite. At the time, rocketry was dismissed by scientists as fantasy, relegated to amateurs who debated rocket design, argued over the best fuel, built models, and tried to get them to fly, more often than not causing explosions.[83] Indeed, at least one member of the BIS, Arthur C. Clarke, was a fan of Fort's writings. But there was also a concerted effort to separate rocketry from Fort and others who seemed cranks, fantasists. One of the group's cofounders, P. E. Cleator, was a correspondent of H. L. Mencken, and they discussed with head-shaking disapproval Thayer's society, his attempts to enroll Mencken or at least enshrine him as a Fortean idol.[84] Even Clarke, while impressed by Fort's writing style, disdained his childish cosmology.[85] Over time, the BIS moved further and further away from Fort, connecting with professional engineers; Clarke provided the mathematical foundation for geostationary satellites.[86] The BIS and its members transformed space from a place of imagination to a rationalized, even colonized arena: disenchanted.

Russell's alliance with the rocketeers, however, led him to science fiction: there was a significant overlap between the BIS and Britain's nascent science fiction community. Russell helped stitch together the far-flung fans on a 1934 trip from Liverpool to London. Around the same time, he transformed himself from fan to writer. Russell saw several of his stories published—and plenty rejected—over the next few years. One batch consisted of weird tales, fantasies that dwelled on the uncanny. Another was science fiction proper; many of these stories appeared in the American magazine *Astounding Science Fiction*, Russell having learned the editor's peccadillos and mastered writing to his standards. Britain's science fiction community and publications lagged America's, and Russell's breaking into *Astounding* was considered impressive by his peers.[87] (Clarke, too, would write for the magazine.) By the end of the 1930s, he was a leading light in British science fiction and respected by American readers.

Russell had several opportunities to discover Fort. The Gollancz edition of *Lo!* was reviewed in the seminal science fiction magazine *Amazing Stories*, which he likely read.[88] In 1934, *Astounding* serialized the book, but, as he recalled, Russell "read it with little interest."[89] A few years later, though, he came across a used copy of *Lo!* in a secondhand bookstore and was "inspired with a missionary's zeal."[90] Russell started to collect Fort's works—a difficult task, and

Eric Frank Russell and Bea Mahaffey, a science fiction fan and editor, at the 1953 British National SF Convention. (Photo by Norman Shorrock. Courtesy of Alan Shorrock.)

not just because he was in Britain. American enthusiasts had trouble hunting down Fort's books as well: used copies were going for about four dollars (the equivalent of $90 in 2023). One collector took a year to find a copy of *New Lands* for sale.[91] Russell first discovered a copy of *The Outcast Manufacturers*, followed by *The Book of the*

Damned. He paid an exorbitant $27.50 for *Wild Talents*. However, he could not find *New Lands*.[92] When he heard about Thayer's society, he joined and was listed as a regional correspondent in the very first issue of *The Fortean*.

Thrilled by Fort's style—as Clarke was—Russell was most affected by the books' attacks on science, which he saw as expanding the purview of secularism: Fort, he said, "dared to place himself in the same relationship to dogmatic science as the late Colonel Robert Ingersoll stood to organized religion."[93] No longer confined by scientific conceptions of reality, Russell crafted a personal philosophy that ratified his prejudices. A friend said he'd "reverted to Roman Catholic science, and constructed the universe about the Betelgeuse of [his] ego."[94] Far from disenchantment, modernity liberated Russell to "ignore all logic, be totally destructive, and enjoy the freedom of the undeniably batty."[95] He was not completely without scruple, though. Russell still honored science—if scientists recognized that their theories were temporary, vulnerable to challenge and change.[96] He remained an atheist, although he held that perspective more lightly now.[97] His Fortean skepticism also seems to have done little to soften his racism. Russell respected some portions of Chinese culture—finding folk aphorisms enlightening, the emphasis on introspection a welcome reprieve from Western empiricism—but excoriated Italians, French, Jews.[98] While Russell, like Fort before him, could be coy about what exactly he believed, his Forteanism was not recursive—that is, he believed firmly in his skepticism. His friend concluded that Russell had "grown so used to sneering at the face of authority that [he] no longer dare[d] look in a mirror."[99]

Having remade his worldview to accommodate Fortean skepticism, Russell naturally turned to Fort in a moment of crisis. In 1937, *Astounding* came under the control of a new editor, John W. Campbell. Campbell had very specific ideas about how science fiction should be written, and its didactic purposes; Russell, more focused on entertainment than instruction, had trouble selling to Campbell, finally looking to Fort for inspiration, a prod to his imagination.[100] "I think we're property," he read in *The Book of the Damned*, and from that sentence spun a novel-length story, which he called "Forbidden

Acres." Russell was reading American crime pulps at the time, and the tale bore their imprint, wisecracking heroes confronted with a series of mysterious murders.[101] His culprits, however, proved not to be humans but glowing blue orbs called Vitons, which were earth's true owners, feeding on the energies given off by earthlings. Vitons had evaded detection for centuries because they were invisible—existing beyond a "sinister barrier" in the human visual spectrum. Scientists whose work approached that barrier too closely (threatening to cross into forbidden acres) were killed.

"Forbidden Acres" was thoroughly Fortean, even beyond its premise—"a notably chilling novel," said the science fiction author and reviewer Anthony Boucher, "coordinating Fort's theories into one dire concept."[102] In Russell's narrative, the Vitons' existence is suggested by a series of newspaper clippings, presented in a Fortean scrapbook, and their presence explains several Fortean anomalies, including the disappearance of Amelia Earhart. The Vitons, like aliens in Fort's books, are fish swimming in an atmospheric ocean; humanity lives at the bottom of this celestial aquarium, unknowingly manipulated. Everyone prays for peace, but the planet is forever at war because the Vitons organize aggression, the better to feast upon human emotion. Thus, Russell reduced the diversity of Fortean anomalies to a single cause, and history to a conspiracy. Still, paranoia was not enough to make sense of the universe; it took the Fortean imagination to grasp the true nature of reality. After the hero exposes the Vitons, the story is reported in the *New York Herald Tribune*, citing twenty thousand reports culled from four hundred issues of Thayer's magazine. That was high hopes for a magazine that would not last much more than sixty.

Initially, Russell wrote the story as a tragedy, not knowing that "selling Terra short was anathema to Campbell."[103] The editor could not accept Russell's premise that humans were of cosmic unimportance, saved from the Vitons only by a yet-greater power. He rejected the story. Undaunted, Russell "gulped heavily and tackled the task of deforteanising the ass end."[104] In the revised version, humanity saves itself, casting off the Vitons, regaining control of its history. Changing the story's title from its original "Forbidden Acres" to "Sinister Barrier," Russell submitted the manuscript again, and this time it was accepted. Campbell was starting a new magazine, *Unknown*, a more fantastic companion to *Astounding*, specializing in the kind of uncanny stories Russell also wrote. "Sinister Barrier" kicked off its first issue, appearing on American newsstands the second Friday of March 1939. It was widely applauded. Science fiction historian Sam Moskowitz says, "The reader approbation given the story turned Eric Frank Russell into a major science fiction figure overnight."[105] Later, *Sinister Barrier* was sold as a novel.

On the proceeds from the story's sale, Russell traveled to New York. He met Campbell and other American science fiction luminaries, including Edmond Hamilton. Hamilton had corresponded with Fort and, in 1931, written a story based on the same Fortean sentence as "Sinister Barrier." Apparently, there was no animosity between the two men, no recriminations or suggestion of plagiarism. Hamilton gifted Russell a copy of *New Lands*, completing his collection.[106] Russell also visited Thayer. By this time, Russell was well established with the Fortean Society, but the meeting solidified the relationship. Russell was tall and commanding, a compelling raconteur; he impressed Thayer, turned him into a fan and champion. On his return to England, they started a long-lasting correspondence, Thayer effusive about the raft of clippings Russell sent. Most issues of *The Fortean* had a column devoted to Russell's reports, and comments thereon. Russell became the clearinghouse for Forteans in Britain and the rest of the world, as he was able to accept foreign currencies. He was clearly established as the society's second-in-command, Thayer thinking of him as the spiritual successor of Dreiser.

Second in command he may have been within the society, but Russell was first among Forteans. Nominally, he avoided Fortean themes in his fiction after "Sinister Barrier"; he worried that with so many science fiction writers mining Fort there would be more cases of duplication. But Fortean sentiments nonetheless found their way into his storytelling.[107] He also evangelized in person, as he made the rounds of Britain's science fiction hubs, and in a series of essays published in both amateur and professional magazines.[108] These encouraged readers to stretch their imagination, to consider reality from new perspectives. They were meant to be taken both seriously and literally—perhaps. The first essay appeared only a few months after "Sinister Barrier," also in *Unknown*. "Over the Border" starts from a premise like his novel's: there is a barrier to human thought and human senses beyond which lie strange, literally unimaginable things—which, however, become visible to those willing to vault the barrier. Russell then marshals clippings he had gathered in 1938 to suggest—in a Fortean way—that aquatic visitors are coming from Venus, which explains news accounts of mystery planes seen crashing into the ocean despite there being no evidence of missing planes, no wreckage found. Venusian mining causes earthquakes, volcanic eruptions, and strange "booms," as Thayer later called them. Russell plays with the idea that reports of unidentified objects leaving the ocean are extra-mundane crafts. The language of the article, though, was different than that of the novel, less wisecracking, more Fortean pastiche.[109]

The next month saw a paean to Fort titled "Fort—the Colossus" in the amateur 'zine *Spaceways*. Russell extolled Fort in *Unknown*'s letter columns. Reportedly, he sold a Fortean essay (called "Astral Artillery") to *Astounding*, though it never appeared.[110] In July 1940, *Unknown* published another of his essays. "Spontaneous Frogation" was a short piece that made no attempt to copy Fort's style and focused only on one small event, initially reported by the Liverpool newspaper *Echo*. A colony of frogs unaccountably appeared on a tiny parcel of land in the center of an industrial district. How did they get there? Could they have fallen from the sky? If so, why only on that one spot? Did they escape from a collection or hop through the city at night? If so, how did they negotiate the barriers surrounding the parcel of land? Could frogspawn have been carried to the spot on bird's feet? Or had the amphibians been sent by teleportation?[111]

To the extent that he had a perspective, Russell seemed to prefer a different Fortean notion. Fort had suggested that certain inventions come when the time is right: everyone knew about steam, and yet steam engines were not invented until the very end of the seventeenth century. Similarly, perhaps the suitability of this patch of land—undeveloped, "swampy, with a few small puddles here and there," a "midget haven of peace situated in the very heart of seventy square miles of brick, stone and concrete"—called forth the frogs: was it possible that "life appears of its own accord, in shape to fit existing circumstances, and that this reptilian sanctuary is ground that has undergone 'spontaneous frogation'"? Russell was enough of a Fortean not to take his speculations too seriously. It was the mystery that moved him, the imaginative possibilities. Nobody knew whence the frogs came, but "astonished Liverpudlians know where the frogs are squatting. Singing their songs to the moon by night, surrounded by clattering street cars, yelling telephones, and hammering typewriters by day, these invaders from the unknown go about their own strange, froggish business."

As a new decade dawned, Russell was not only one of Britain's leading science fiction authors but far and away the Commonwealth's—indeed, the world's—most renowned Fortean. The war, however, cut short his mission, as it had Thayer's, though for different reasons. Russell was sucked into the British war machine. He joined the Royal Air Force as a wireless mechanic and radio operator. Based primarily in England, he apparently also spent time in Italy after that country was conquered.[112] His writings, even his visits to science fiction gatherings, did not stop, but they were substantially curtailed. He continued sending clippings for Thayer's magazines, but day-to-day affairs passed to his wife and a Fortean friend, who could not bring the same energy to the job.[113] Fort's two greatest defenders were constrained by the sinister barrier of war. The Fortean Society seemed likely to die a second time, or, at best, to become so marginal it might as well have died, its only monument the omnibus edition.

The Age of Doubt

The end of World War II brought relief and revivification. "Still alive and kicking" became the society's motto, its icon the Manneken Pis: from Olympian heights, the Fortean Society would rain its derision upon the world—and it would reign, too.[114] Renewed, Thayer's pet project grew in size and stature. Society members would continue, as Russell did, to extend humanity's collective imagination, to hold space for wonders and weirdness. With gimlet eye, Forteans would detect and describe the world's hidden skeleton, too often obscured by bureaucrats and blinkered scientists: the marvelous mechanisms that created reality. Forteans posited alternative atomic theories, presented new periodical tables of the elements. Medicine was put on a new theoretical foundation. All of history was misinterpreted, Forteans said, the benighted upholders of convention not recognizing its true drivers, which many thought were spiritual forces, not materialistic ones. Peoples who had been damned, entire cultures and traditions of thought, were said to be essential to understanding the world and its construction. Following Fort, Forteans rescued these damned persons, places, and things to show reality as more complex—more awesome—than officialdom allowed, though what intrigued Forteans, and the uses to which they put the damned, often distinguished them both from Fort and from Thayer.

Thayer married for the third and final time in January 1944. Twelve years his junior, Kathleen Marion McMahon was a singer who'd already been helping Thayer with the society for a few years.[115] In 1946, Russell was demobilized and resumed his work for the society in Britain.[116] That same year, a society member distantly affiliated with the Dreiser estate persuaded Dreiser's widow Helen to bridge the chasm that had separated the writer and the Fortean Society. Helen sent Thayer most of the material Dreiser had collected on Fort so that Thayer could write his own biography.[117]

The next five years would be the society's golden age. "With this issue," Thayer wrote in 1948, "the Fortean Society actively assumes leadership of the entire human race. We have fiddled around long enough."[118] Proving that, indeed, the world had entered a new era, Thayer revised time itself. He introduced a new calendar, calling 1931, the year of the society's founding, Year One, and redividing the year into thirteen months—the extra one called Fort, of course—which allowed each date to fall on the same day of the week every year. In the winter of Year 14, Fortean Style—FS—he renamed his magazine *Doubt*. Thayer and Russell made a concerted effort to expand membership, Thayer advertising on college campuses, encouraging members to evangelize internationally—Germany, France, New Zealand.[119] Russell wrote for a professional magazine called *Tomorrow* that catered to those interested in occultism and the paranormal. The historian William Graebner has called the period between the end of World War II and the beginning of the 1950s the Age of Doubt. Americans in these years "questioned the central assumptions of their culture"—the durability of democracy, progress, science, freedom—creating "a culture that was more pessimistic and cynical than before." "Ambivalence held sway," writes Graebner.[120] Fortean skepticism fit the mood, and the society's attempts to expand were successful. Fortean ideas were labile, could be put to different ends, simultaneously enlarging what seemed possible and challenging the status quo. Though science fiction writers, avant-garde artists, and religious visionaries were the most important adopters of Fort, they were by no means the only ones inspired by him or Thayer's project.

Back in 1872, the British mathematician Augustus de Morgan had written about various scientific heretics—most pseudoscientists but a few (like Galileo) true innovators—celebrating their creativity while disparaging their nonsense theories. With Victorian pomposity, he titled his book *A Budget of Paradoxes*. That description works as well for the Fortean Society, which opened itself to promoters of alternative science and medicine, seekers, cultists, and Theosophists, collectors of the absurd, Native American reformers, conscientious objectors, libertarians, anarchists.

Thayer had started referencing, and publishing, works that promoted alternative scientific theories during the retrenchment that followed "Circus Day Is Over." Their authors insisted that, in important ways, the universe's structure was different from what was generally thought. Among the first was Alfred H. Barley, an astrologer intrigued by the effect of the movement of the earth's axis on the changing position of the north pole.[121] The so-called Drayson

Problem became one of Thayer's favorite heresies, and he advertised Barley's books on the back page of *Doubt* for many years. He also gave space to Albert Crehore and Thomas Graydon, both of whom rederived physical and chemical concepts using different models and mathematics.[122] Crehore distrusted quantum mechanics and instead explained atoms using electrodynamics.[123] Graydon (like Mencken) thought the abstruse formulas of contemporary science untrustworthy and so offered a theory of gravitation and a new periodical table built on basic arithmetic.[124] T. Swann Harding would tire of the Fortean Society and its magazine—"very cryptic and esoteric. Nobody can read it"—but before that introduced Forteanism to his friend Frederick Hammett, a cranky biologist who peppered Thayer with sardonically annotated cuttings from scientific journals and salted his own scientific writing with references to Fort.[125]

Alternative medicine had been intertwined with Fortean thought since Dreiser's dalliance with Albert Abrams's ERA and Fort's curmudgeonly disavowal of therapeutics. Early on, Thayer's society attracted naturopaths and other practitioners marginalized by reforms in the medical profession.[126] As doctoring became wedded to the state and certain practices outlawed, Thayer became increasingly dedicated to spreading the word about alternative cures—even if he disbelieved their advocates' claims.[127] Alternative cancer therapies became a foundational part of his Forteanism, allied with opposition to vaccination, fluoridation of the water supply, food additives, and vivisection. (Russell was, if anything, even more vociferous on the topic of food adulteration and experimentation on animals.) Thayer theorized that tonsillectomies caused polio (at the same time downplaying the severity of polio).[128] The husband-and-wife team of Guy and Nell Rogers published a pamphlet criticizing the germ theory of disease, which proudly quoted Fort and identified themselves as members of the society; Thayer offered the pamphlet for sale.[129] Seeing the society's drift, its curating of all sorts of fringe scientific and medical theories, one magazine concluded, "The Fortean Society bids fair to become the greatest aggregation of Academic Cranks the world has known."[130]

But the society included more than critics of mainstream science and medicine. Religious seekers also flocked to it, as Fort had known they would, drawn by attacks on science that they saw as opening space for their ideas. Cultists and creators of intentional communities found the society a soft spot in a world otherwise hard on them. Claude Dodgin, who renamed himself Maurice Doreal—the Living Buddha—and founded the Brotherhood of the White Temple, had connections to the Fortean Society.[131] So did Wing Anderson, who synthesized a vast array of mystical ideas that had flourished in the nineteenth century into a New Age faith.[132] He established a correspondence school to indoctrinate students into what he called the Essenes of Kosmon. Like Doreal, Anderson provided materials to Thayer's society, books and clippings. Both also formed their own intentional communities. Throughout its history, the Fortean Society attracted practitioners of liminal religions, esoteric forms of Buddhism and Christianity, devotees of New Thought, wielders of magic. All insisted that, hidden to most, were structures and forces that organized reality; the wise would reshape their lives to be congruent with them, not the false reality of bureaucrats and scientists.

The Ur-source for many of these religious traditions was Theosophy—an ecumenical edifice that structured the communities joined by the Barleys, was the basis for Dodgin's and Anderson's cults, and underwrote some of the scientific and medical heresies collected by the society. Theosophy emerged in the nineteenth century, invented by the Russian émigré Helena P. Blavatsky, who collected the paraphernalia of old gods and forged them into a religion that absorbed science. Blavatsky claimed that a fraternity of Ascended Masters living in Tibet—the Great White Brotherhood—preserved the world's wisdom. They passed their knowledge to her, usually through mediumship, though she admitted that the gigantic tomes she wrote were an amalgamation of the adepts' teaching and her own active mind. In *Isis Unveiled: A Master-Key to the Mysteries of Ancient and Modern Science and Theology*, published in 1877, and its sequels, Blavatsky argued for universal, orthogenetic evolution—individual souls improved and humanity advanced, the latter through stages called root races. Earlier root races had been the legendary civilizations of

Mu, Lemuria, Atlantis. Humans had reached a fifth level, she said, and awaited a figure to herald the sixth. After her death, Theosophy fragmented, in part over who was the New Messiah.[133] Theosophical communities, their anthroposophical and metaphysical cognates, blossomed across North America.[134] Some were lodges, some neighborhoods. Some were small gatherings. Some were intentional communities. Often these were demonized as communist, as love cults, because members insisted on the necessity of love in the cold, modern world or frankly discussed sexual matters. They were persecuted. They suffered internal fragmentation. Yet they persisted, sending seeds into the world even after they collapsed. More than a few Theosophists found the Fortean Society an ally—indeed, Theosophy was a pillar of Forteanism as Thayer and Russell recrafted it in the 1940s.

Another support for the reformed society was Don Bloch, an information specialist for the US Fish and Wildlife Service.[135] When he first made acquaintance with the society, Bloch was stationed in Washington, DC; later he was moved, against his will, to Denver, where he would retire and open a bookstore.[136] Even before joining the Fortean Society, Bloch had scrapbooked newspaper reports of the strange; Thayer indexed them and incorporated them into the society's library.[137] As MFS #166, Bloch was a ubiquitous presence in *Doubt*. Thayer said he "displayed a perfect grasp of Fortean aims and an energy in the Society's interests second only to the Founders themselves."[138] Bloch pushed the society to expand and organize. He scoured bookstores, estate sales, Doreal's library for Fortean tomes, suggested new members, took on tasks for Thayer including arranging places for *Doubt* to be sold. Forteans were scattered across the country, their names frequently mentioned in *Doubt*, many of them obviously as dedicated as Bloch; he wanted to correspond with them, but Thayer refused to provide addresses or create a directory, conscientious about members' privacy.[139] Bloch's enthusiasm touched on a contradiction in Thayer: he wanted to evangelize for Fort, to build a vast edifice in his name; he craved legitimacy—but was hampered by his anarchic tendencies and distrust of authority, including his own.

Among the many new topics introduced by Bloch, the one that

consumed the most Fortean time was Iktomi. In 1946, Bloch sent Thayer *America Needs Indians*, a viciously satirical attack on the Bureau of Indian Affairs and the United States' sad history of mistreating Native Americans. Odd, fragmented, subversive, provocative, the book invented new words, played with fonts, and used stream of consciousness in a way that recalls experiments in modernist literature and poetry.[140] The author called himself Iktomi, the name of a Lakota trickster spider. Indians and their ways had been damned by whites, he argued, but only Indians could properly preserve America's natural resources—the foundation upon which the nation was based. Indians understood the real (but obscured) structures of the world. Whites knew not what they did and so destroyed the soil, bringing about through their ignorant actions the Dust Bowl. Thayer was fascinated. Bloch had found scores of copies, all of which Thayer eventually bought for $100, sending some to Russell for placement in British museums. Others he sold, though slowly, the book priced at a whopping $7.50.[141] (One dollar from each sale went to Iktomi.)[142] He tried, unsuccessfully, to get Iktomi's sketches published. He made him an honorary member of the society, calling him an "aboriginal Fortean."[143] Thayer met Iktomi a few times, when he passed through New York, initially finding him a "delight" before souring on the whole matter—a pattern in his life, fervor followed by disdain.[144] Iktomi was a "heartache and a headache," Thayer said in 1947, and five years later complained, "The trouble with him is that he considers himself the Amerind Messiah and all others gentiles."[145] Nonetheless, he advertised *America Needs Indians* on the back page of *Doubt*, alongside Barley's works, for years. Apparently, Thayer never realized that Iktomi was a white man named Ivan Drift, who had, perhaps, been denied a job with the Bureau of Indian Affairs.[146]

Thayer bragged to Helen Dreiser that the Fortean Society advocated equality for "countless minorities," but *Doubt* was quiet on racism and Thayer dismissed anti-Semitism as a petty problem.[147] With Iktomi, he acknowledged Native Americans as fit to manage natural resources but beyond that accorded them no respect.[148] He reserved his esteem for white men, highest among them pacifists. During World War II, tens of thousands of men had declared themselves conscientious objectors. Many were assigned to noncombat roles in the armed services, others given alternate service, such as work in hospitals; some were sent to camps, where they did hard labor; others were jailed.[149] Among those imprisoned was Harry Leon Wilson Jr., son of the author, whose case enraged Thayer; he wrote about him in issue 8, walking a fine line between sedition and expression of Fortean pacifism.[150] As Thayer saw it, "conchies" practiced Forteanism, skepticism of martial politics, and were punished for their dissent.[151] They had been damned by society, quite literally, and should be rescued, their ideas enshrined. Defense of COs became a major plank in Thayer's reformed Forteanism, and conscientious objectors became a new source of support.[152] In their downtime, encamped conchies published newspapers and pamphlets. Thayer swapped subscriptions to *Doubt* for these, exposing COs to Fortean thought and society members to excerpts and references to conchie ideas.[153] He offered free membership to any conscientious objector.[154] Backing of his pacifism even came from some who donned military uniforms.[155] The science fiction writer and MFS James Blish took up the cause of conchies, campaigning against what he called the new slavery.[156] Forteanism encouraged some to worry over the political structures of the world, what they were and how, like any Fortean category, they might be dissolved and reformed.

Exchanging publications was only one link in a web of connections between COs and the Fortean Society. Conscientious objectors usually rooted their pacifism in religion (not Wilson, though, which was one reason he was sent to jail), most commonly Quakerism or one of the peace churches—the society's lawyer and defender of conchies, Julien Cornell, was a Quaker, for example.[157] There were Theosophical COs, too, and they found the society doubly interesting, as both pacifist and skeptical of science.[158] Politically, conscientious objectors and their allies sketched out a left-libertarian politics that wedded pacifism to civil rights and reinvigorated anarchism, which had been moribund since an FBI crackdown after World War

I.[159] These were Thayer's politics, too, leery of the state and worried over social equality—though paternalistic, he was not unaware of how racism worked in America.[160] Some other Forteans also adopted this perspective, Blish contributing to a pro-CO anarchist magazine, for example.[161] On furloughs, released from jail, or working in alternative service, COs gathered in San Francisco, in New York, and in these places bohemian circles formed, pacifists mixing mystical thought, libertarian politics, and Forteanism.

Riding the wave of enthusiasm, Thayer invented a history for the society, tidied its operations, tried to impose order on its members, and clarified its meaning. From those associated with events in 1931 he plucked thirteen names—including his own—whom he deemed Founders. Those deceased or in his bad graces were replaced: Dreiser by Russell, Tarkington by Hammett, Barnes by Harding (despite his bad-mouthing of the society). Woollcott and Stern were replaced too, as was Hammett when he passed. Thayer remained secretary— humbly styled Your Secretary, or YS. New categories of affiliation were created, honoring nonmembers who displayed what Thayer considered Fortean skepticism, past heroes with, at most, a slight acquaintance with Fort or his writings (some having died long before the society was founded). He kept trying to convince Mencken to join, sadly concluding, "He has too much of the old-fashioned Village-atheist's respect for 'urinalysis.'"[162] *Doubt* appeared quarterly throughout the second half of the 1940s, accompanied by a list of books the society sold—Barley's, Crehore's, Iktomi's, Russell's, and many others. Thayer published a perpetual calendar, most of the months decorated by Iktomi sketches.

He also paid on a marker laid down in the introduction to Fort's omnibus edition: the establishment of a "university of disbelief," what he called the Fortean University (provocatively abbreviated FU).[163] The campus was "co-extensive with our student body"—every beer hall and library across the world. But there were formal aspects. He accepted money in support—Sussman sent him twenty-five bucks. The course of study codified Forteanism à la Thayer into thirteen categories, and any contradictions were the point: "No subject is taught as true by F. U.," Thayer explained. "Each is advanced as a question profitable to explore," a means by which structures possibly organizing society—political, economic, scientific—might be identified.[164] The membership responded, some matriculating with essays that he published in *Doubt*. Others arranged themselves into regional chapters—Chapter Two in San Francisco, Three in Chicago, Four in Dallas, Six in London (no Chapter Five was announced).[165] Chapter Two, he crowed to Russell in late 1948, "is going great guns."[166] Bay Area members had formed their chapter on April Fool's Day that year, each attendee of the initial meeting signing a ledger titled, of course, "The Book of the Damned." Thayer advertised the society in a pamphlet called "The Red Cross of the Human Mind."[167] The Fortean Society was mental "antitoxin," neutralizing dogma, bringing health—bringing doubt—to all with ears to hear, eyes to see. Forteanism was "the natural state of man," Thayer said, and doubt the authentic state of the human soul.[168]

Trimming Course, or Lamps

"Don't do so much for the Society that it irks you," Thayer told Russell in the spring of 1948. "If it isn't a pleasure, it isn't worth doing."[169] As the decade turned, however, Thayer ignored his own advice. Driven by some inexplicable urge, he continued past the point of exhaustion and growing disgust. Thayer thought the world corrupt, fallen from a better state, and so built around himself a fortress, part of which was the Fortean Society itself. From this protected place, he theorized power and its properties, how it was wielded, by whom—and against whom. But the members who were the very bricks in his bunker were so very obstinate, required too much tending. They undid him, while the world derided him (or, worse, displayed disinterest). To be sure, all was not irritation; some joy snuck past his walls. Happily, in the piles of mail, he discovered some members who agreed with him, whose slant on the news made him laugh out loud. There were those who shared his politics of aggrievement,

who extended his ideas or built Fortean castles that could stand, as
he did, against the exercises of the powerful. But these did not fully
compensate for the aggravations, and as the 1950s slipped toward its
end, Thayer contemplated giving up, tearing down the society, his
walls, and heading off into the sunset, the world as hellish as it had
ever been, the powerful regulating life, setting flame to liberty, but
he, at last, no longer required to fight, untouched by the sordidness
of it all. He could not escape the strange impulse, though, or allow
the elites to manipulate the levers of power uncontested.

The height of civilization, Thayer thought, was Western Europe in the sixteenth century, when jesters like Rabelais walked the
earth.[170] Of all the books he wrote, his favorite was *Rabelais for Children*, an attempt to create a new generation of jesters. This was not
antimodernism. Thayer knew there was no way back to that time,
no way to fully rescue its spirit. His was, rather, a declensionist view.
He endured culture at its nadir. The world was being pumped full
of nasty chemicals. Thayer and Russell bonded over a shared distaste for modern food, denatured then reconstituted with preser-
vatives and so-called vitamins (another hoax).[171] Atomic weaponry
was terrible, Thayer concluded, though he could not figure out how.
Sometimes he thought it too was a hoax, dynamite promoted by
government officials into an existential threat to cow an ignorant
populace into compliance; other times, he worried that radiation
was a sinister substance infecting the entire globe.[172] The FBI "did-
dle[d]" the constitution—and hounded him, him who stood in Rabelais's shadow.[173] Copies of *Doubt* were turned over to the FBI as
seditious in 1945, 1947, 1951.[174] Certain he was being monitored, his
phone tapped, Thayer ensconced himself in a virtual bunker.[175] No
one could visit unannounced, call him for last-minute plans. Even
his secretary could not reach him by phone.[176] He told Ezra Pound:

> I NEVER see a newspaper or hear the radio or go to the movies, so
> that I never know what has people exercised until it filters through
> the barriers I have built against it. I have no face-to-face social con-
> tacts, so I never have to discuss Greece or Russia. My entire associ-

ation with my fellow worms is by correspondence with members of the Society. The members read the papers for me—and send clippings they think are of greatest Fortean interest.[177]

Not that he trusted the papers. Borrowing the word from Pound, he called them "wypers" because they wiped away the truth.

Rabelais lived when liberty was in full flower, Thayer thought, while he had to find a way to live with freedom trampled. He needed to be an especially clever pantaloon, to show how the wisdom of the fool could catch the powerful at their own game and improve the world in some madcap way. In 1945, he started promoting what he called the Perpetual Peace Program, a joke based on a book he'd written (*The Greek*) that was nonetheless meant seriously. War was a hoax, he maintained, arranged by amoral elites who traded the bodies of young men for filthy lucre. Give the plutocrats their giant programs, he conceded, allow them to profit from government contracts, the interest on enormous loans—but without the death and destruction. Invent boondoggles that supported not war but peace: a cyclotron in every high school, one standing army to translate every book into every language and a second to sell the books on every street corner.[178] The PPP grew over the years, Thayer adding new planks, some suggested by members.[179] A journalist at the *Chicago Tribune* wrote approvingly of the proposal, but in all it gained few adherents; rather, the years after World War II saw the military reaching into domestic life, reorganizing education, the economy.[180]

Yet there were moments of levity, times and people that bucked up his morale. "We've got wrinkles nobody ever thought before," Thayer boasted.[181] He gleefully read of J. D. Stern losing a $50,000 libel case—karmic retribution for Stern's shabby treatment of him.[182] In 1948, adman Thayer moved from J. Walter Thompson to Sullivan, Stauffer, Colwell & Bayles, later earning a six-month sabbatical, which he spent writing his albatross of a novel and vacationing in Nantucket, his favorite place.[183] "Nantucket is where good Forteans go when they die," he said.[184] And there were good Forteans, diamonds amid the coal dust. He delighted in several loyal members, the clever clippings they sent, their wicked commentary. George Bowring tracked down a four-legged fish photographed in Los Angeles, dipped it in formaldehyde, mailed it to Thayer. "Alas, that expedient was not sufficiently preservative . . . [T]he package had attracted a lot of attention in the office of the express company before delivery could be made. In short, the package was opened by the expressmen and the Forteana destroyed."[185]

Among Thayer's favorite Forteans was Art Castillo, a science fiction fan and artist who started contributing clippings on the usual array of Fortean topics in the late 1940s, and then sketches. These were tossed off, Castillo said, drawn on whatever was handy.[186] But Thayer and Russell loved them. The sketches headed columns throughout *Doubt*'s run and decorated many covers. Castillo designed the society's logo, a question mark inside a tree: doubt as the fruit of knowledge. Thayer made these into stickers, sold them to members to affix to mail, a dark shadow of the Easter Seals.[187] "Castillo is a real find," Thayer said in 1951.[188] Russell explored the possibility of Castillo illustrating his books.[189] The three men shared a central concern: the world was organized against the lone white male. Scientists, warmongers, religious prudes, the dictatorial state, and craven businessmen formed a unified system of domination. For *Doubt* 25, Castillo illustrated the conspiracy as a snarling Cerberus, one head each for science, religion, and the state (which included the military and businesses). "A visual manifesto of the Society's contemporary aims," he explained.[190] Forteans stood against this triad, pledging their allegiance to what Thayer called "the Religion of Self Respect."[191]

That notion of Forteanism as a kind of antireligion solidified the society's connection with Freethinkers and humanists. Thayer exchanged ideas, articles, and advertisements with such organizations, prominent among them the American Humanist Association.[192] The AHA was headed by Edwin H. Wilson, who had been instrumental in the resurrection of humanism. Befitting the name Freethinkers, the movement was fractious. Wilson and the AHA emerged as a leader of one branch—the religious, as opposed to secular, humanists.[193] Through the humanist grapevine, he learned of the Fortean Society, was intrigued by its commitment to doubt, skepticism, and perpetual dissent.[194] The AHA's magazine, the *Humanist*, ran Thayer's article "Charles Fort and the Religion of Self Respect."[195] Wilson did have reservations, thinking the society too negative—"Doubt justifies itself best when it leads to new truth in the service of man"—but Thayer continued to number the society among Freethinkers, taking pride in their accomplishments.[196]

The mid-1950s also saw a culmination of Thayer's writing career, bringing (momentary) relief and midlife introspection. He'd conceived the idea to write about the Mona Lisa in the late 1930s or early 1940s, and several times since had been close to publication, until advertising work intervened, or he returned, pencil in hand, to revise, expand, research.[197] The book grew enormously: not a single novel, but a cycle of twenty-one, probably 3.5 million words in all, he predicted.[198] Thayer taught himself Italian, sank around $6,000 into research, and so acutely imagined the setting that Italy looked familiar when he first visited in 1952.[199] He eventually submitted forty-seven thousand manuscript pages—the first three volumes—to Dial, which published them on 8 June 1956. Mona Lisa was not even born by the end of the 1,257 printed pages, but Thayer promised the next installment in nine months.[200] Dial advertised *Mona Lisa* heavily, with bookstores putting up window displays; no dime-store novel this, *Mona Lisa* sold for $12.50.[201]

"Sixteen years ago, there was a popular American superstition to the effect that I knew more about women than any other articulate male since Solomon and before Kinsey," wrote Thayer in the foreword.[202] His fiction repeatedly focused on women—likely it was his rewriting of *The Three Musketeers* from the perspective of a minor female character that gave him the confidence to take on Mona Lisa. But his purported understanding of women never included considering them as full human beings. He attributed impure motives to them, admitted their desire for sex but did not invest them with the complexity of emotion, contradiction, foolishness, and nobility he imagined for his male characters. A fifteen-year lento did nothing to improve his insights. Women remained bundles of biological imperatives, their actions described in the pulp argot he'd perfected in the 1930s.[203] "A drool trickles from the wiseguy, smoking-car prose," said one reviewer.[204] The book was overstuffed with sex, orgies upon orgies. Reviewers admitted that Thayer wrote with brio but were daunted by the book's size and bored by its plot. The newspaper in his hometown, Freeport, Illinois, was embarrassed, if also proud that a local product had made good.[205] Thayer was equally flustered by his natal city, reading clippings from its newspaper with "mixed emotions," saddened that his youthful compatriots should see him in such a negative light, buoyed by the aching pull of the past. He sent a positive review to the *Standard* and asked:

> If you should find space to reprint a line or two of that, the survivors among my teachers and classmates at Third Ward School might blush the less to greet me the next time we meet on Stevenson Street. There were Miss Friday (Freitag) and Miss O'Connor, and the two prettiest girls were Honor Throwe (blonde) and Blanche Miller (brunette). What a memory![206]

Though in his fifties now, Thayer still saw himself as a juvenile nose-thumber. "I am the dirty boy," he admitted. "But I can't help it. I'm incorrigible."[207]

Too often, though, his impudence was bogged down by the dull necessities of life. He regularized some society functions, dealing with correspondence only one day a month, for example, and still the job exhausted him. Some of the aggravations were ongoing. Members sending clippings taped them closed, despite his requests that they fold articles to show the titles, forcing Thayer to open the same item hundreds of times. He scrambled to get *The Books of Charles Fort* into Great Britain, stymied by postwar rationing and import limitations. For a moment, he thought there might be a British edition, but Waveney Girvan, a Fortean who worked in a British publishing house and took the lead on the project, never pulled it off. Several stalwart members left the society, depriving Thayer of an easy source of material. There were loud gripes about choices he made, demands that he expand his operations—"make gold out of water," Thayer quipped.[208] Few recognized that the society was funded not by dues but by Thayer's largesse; he rarely dropped anyone for nonpayment, a "suicidal practice."[209] Members also resisted his schemes. Thayer planned to sell a Fortean holiday card in 1950, an illustration of James Joyce's poem "Joking Jesus." "Very amusing and sacrilegious," he promised.[210] Many took offense. "Man alive," said one, "Christ was not only a great dissenter, his followers have dissented among themselves ever since he died dissenting." Thayer was confused—as he had been several times before, when members suggested Jesus as a posthumous Fortean—unable to reconcile Fortean skepticism and Christian faith, but he relented.[211] The strain between Thayer and the membership brought out his nasty side. In private correspondence, he viciously mocked some of his most committed supporters as muddleheaded kooks, told Russell to ignore their endless (and endlessly long) letters explicating outrageous, esoteric theories.

Nor did Thayer feel supported by the new batch of Founders he had installed. Once more the sticking point was politics. Thayer took special exception to the Civil Defense program, which created civilian auxiliaries to the military—pilots deputized to search the skies for threats, wardens to drill neighborhoods on procedures to follow in case of nuclear attack, publishers to put out pamphlets on surviving atomic war.[212] Thayer wanted to take out advertisements in newspaper dailies opposing the martial regimentation of everyday life.[213] But, mindful of the "Circus Day" imbroglio, he polled the Founders before buying space. Two-thirds opposed his plans.[214] Stung, but maintaining a brave face, he said, "I see that it is un-Fortean even to prefer freedom. Who are we to say slavery is not preferable?"[215]

Through the 1950s, the balance of pleasure and pain that came with running the society tilted toward the irksome. "We should trim our course or our lamps," he told Russell, "I know where to steer, but I'm not sure that I enjoy steering as much as I used to."[216] The pleasures, many though they were, did not compensate for the pain. Castillo's tenure ended in disaster. He either refused induction into

the army after being drafted or he deserted. Without any other support, without money, he telegraphed Thayer at the Fortean Society's
address while on the run. Thayer was dismayed, worried about the
legal consequences. He cut Castillo off, stopped running his sketches
and essays even after Castillo resolved the case by taking alternative
service.[217] Not that the separation stopped the FBI from bedevil-
ing Thayer. He'd baited the bureau with an advertising pamphlet
trumpeting that there were G-men on the society's rolls and telling
the creator of a different perpetual calendar that FBI members out-
numbered communists two to one. The calendar missionary alerted
the bureau to Thayer's claims, and in 1956 the FBI opened another
investigation. Thayer stopped circulating the flyer, though he neither gave the FBI any names nor retracted his statement, as had been
asked.[218] After harassing him a bit, the FBI decided he was "just out
to make trouble" and not worth the effort.[219] As the bureau ended
its investigation, one agent saw a negative review of *Mona Lisa* and
sneered, "Thayer is a pretty typical example of some of the misfits
who delight in taking pot shots at the Bureau."[220]

The society was a literal weight. In 1955, Thayer had to move
his Fortean archives, boxes and boxes of books and correspondence.
"That's all. I'm dead," he told Russell when he was done.[221] He took
a vacation in Nantucket, one of the few breaks he allowed himself in
the mid-1950s, his work, Kathleen's, society business having forced
him to repeatedly cancel planned vacations. In time, Thayer lost interest in the Fortean University and instead arranged Forteana into
seventy-one categories. He asked each society member to pick a topic,
track it in the newspapers, and write a summary essay.[222] Only a couple took up the challenge. Thayer attempted one himself but found
it "an awful lot of work."[223] Everything associated with the society
was an awful lot of work. *Doubt*'s schedule became irregular, and he
stopped using the Fortean calendar. He also stopped selling books
through the society, even his opus, probably because he was trying
to negotiate an exit.[224] "After 26 years of my stewardship it needs
a change of pace," Thayer confided to Russell.[225] He felt old—so
many of his friends were dying—but still too busy: he wanted to

write the biography of Fort, was embarrassed that he could not meet his promised deadline for the next installment of *Mona Lisa*.[226] Overwhelmed, Thayer swallowed his pride, beseeched his enemy for relief: he applied to the Ford Foundation for a grant, money to hire staff, someone else to take over.[227] Unsurprisingly, the foundation rejected his application. Driven by his strange commitment to the Fortean Society, Thayer rededicated himself, resumed using his dating system, returned *Doubt* to its regular schedule. Once more he put on the "harness."[228]

Cherchez la Femme

Red-headed, of course, but slight—unnoticed. That was Wanet McNeil.

She was born in 1935. There were probably good times. There were definitely bad times. A brother was born, World War II started—and her parents divorced. Wanet's mother, Leona, married Lloyd Eagle and moved into a trailer in Bloomington, Illinois. There wasn't enough room for the children. Wanet and Arthur Jr. went with their father—the three McNeils moved in with his sister and her husband, the Willeys, on a farm near Macomb.

The Willeys were older, married late, never had children. Charlie was tall, Lulu short, stolid. They had a good-sized farm, six buildings, including a house, a barn, a chicken coop. They had livestock. Charley'd been working it his whole life. He'd quit school in third grade, Lulu in fifth. Already, Wanet, in sixth grade, had more education than either of them.[229] And she did not want to be there, on that farm. She wanted to be with her mother. Charley and Lulu, she would say, didn't let her buy pretty clothes.

Mid-August, it's hot and humid in Illinois. Sometimes the sky lights up with fireworks. Sometimes the houses do, too. But not like this. One hundred, two hundred fires. They'd start in the wallpaper, spread. Start in cabinets. Curtains ignited. Brown spots, unbidden, smoldered. The five-room cottage, home to the pair of Willeys and trio of McNeils, burned down on a Saturday. The family moved to a garage. The garage burned down, and they moved to a tent, taking supper in the milk house, where they hung paper against the blowing wind. Even as they sat in that small dining area, fires started on the wall.[230]

Sitting outside the tent one day, Charley and Lula saw their second barn burn. It only took twenty-six minutes. Newspapers noticed. The Macomb fire chief told reporters he was completely mystified. The state deputy fire marshal started an investigation.[231]

What could it be? they all wondered. Charley said he didn't have an enemy in the world. Sure, he had insurance, but only $1,800, far less than what the place was worth on any rightful reckoning.[232] Newspapers liked the word *ghostly*, liked the word *phantom*. Others gussied up guesses as science. The fly spray the family used was highly flammable. Maybe it saturated the walls. Could it be radioactivity? That's what Lewis C. Gust, Office of Technical Progress at Wright Field, said. "Suppose," he speculated, "you had material that could be ignited by radio and you wanted to test it for sabotage value. Wouldn't you pick some out-of-the-way place, like the Willey farm, to make the test?"[233]

The Willeys and the McNeils left the home where Charley had lived his long life, moved into a vacant house nearby. Crowds gathered round the old place, hoping in vain to see sparks fly. Scientists studied the wall-paper, found nothing unusual, looked for gas in the ground, found none. The state fire marshal thought it was a firebug, but Charley Willey said it couldn't have been—he'd have seen an arsonist.[234]

Then another Saturday. Lulu, canning tomatoes, found the shelf paper in the kitchen cupboard ablaze. More and more fires at the new place. The family wasn't sure how many.[235]

"I'm finding it quite a trick to form axiomata generalia (so to speak)," Thayer told Russell. "Precedents which are not laws, ties which do not bind, quite a trick."[236]

Charley took the lie detector, so did Arthur and Arthur Jr. But Lulu refused. As did the Willey's niece, Wanet McNeil.[237]

Said W. L. Janney, conscientious objector, in a letter to Thayer, "There really ought to be some all-embracing term for the person (adolescent girl?) who is the center of any disturbance, whether she cracks her big toes or entertains visitors from Pluto, who like to shy stuff at her."[238]

They'd cracked the case, solved the mystery of why fires "shot out of walls like lighted gas out of a blow torch." The fire marshal said, "The fires were set by an arsonist. We know who he is and we're trying to get evidence that will stand up in court." He told reporters that the arsonist had refused a lie detector test. He also said that the arsonist was "not any member of the Willey's [sic] family."[239]

"Never—or hardly ever—have we thought that a Fortean Law could be formulated on any subject," Thayer conceded in *Doubt*, "but that just shows how little we were prepared for Macomb, Illinois, and its fires. After a thorough study of the data, we give you Fort's Law for Explaining Mysterious Fires. It is *Cherchez la Wonet*."[240]

Newspapers weren't sure what to make of the girl. Her name was variously reported: Juanette, Juanett, Wanet, Janet, Wonet, McNeil or McNeal; one gave her the middle name Ethyl. She was twelve, or maybe thirteen. But the color of her hair was consistent. If it was reported, it was always red.

Thayer gleefully explained to the membership, "That is to say that no matter what the real circumstances (which nobody ever knows) and no matter what fictional circumstances the wypers invent (which is all you or I ever have to go by), sooner or later an adolescent female will be found or invented to 'confess' or to 'be accused of' or to 'explain' the otherwise inexplicable. That is Wonet.

"The wypers build the fires, and the wypers put them out. Blessed be the Name of the Wypers."[241]

Wanet said she'd started the fires when no one was looking, touching matches to the wall because she hoped to get back to her mother. A psychiatrist in Chicago examined Wanet, declared her fully normal—just hurt by her parents' divorce.[242] The Willeys didn't press charges. "She's just a child. We don't really know why she did it, but we don't blame her."[243] But the solution proved inadequate, too many details left unexplained: the spreading brown spots, the fires that started when someone else was in the room.

"In the Macomb case," Thayer wrote, "the hysterical rewrite men— probably a hundred miles from the farm—larded their original tale with so much bafflement that not even miracles or magic could have got them out of it. So that even when they spring Wonet nobody believed them."[244]

And after the solution? Wanet slipped away again, unnoticed, mostly. She went to live with her maternal grandmother, nearer her mom. She took a job as a saleswoman, married Russell Beier, a World War II vet, in 1961. They had two sons and two daughters, in-laws, grandchildren. She went by Micki.[245]

Wanet had her troubles. She and Russell divorced. In 1989, she was caught forging checks.[246] She died in 2006.[247]

Pantagraph (Bloomington, IL), 30 March 1979, page 35:

Normal Firemen Thursday night were called to a mattress fire in the basement of the Micki and Russell Beier house at 601 S. Linden St., Normal.

Mrs. Beier said her husband was alone at home when she arrived at the residence to find smoke pouring out of the basement and kitchen.

"The kids and I left to get some school supplies," she said. "We pulled into the driveway and saw smoke coming out of the basement." She said that she could see smoke coming from the side door of the two-story home by way of the basement and kitchen.

Normal fire Capt. John McAtee said the fire was in the mattress. Firemen Friday said they suspected smoking materials had been used carelessly.[248]

"Only Thee and Me Is Not Nuts"

The 1950s saw the crumbling of the relationship that sustained the Fortean Society, the friendship between Tiffany Thayer and Eric Frank Russell. Yet again, the trouble was politics. Thayer and Russell shared a distaste for their country's governments, and agreed in their sexism, their anti-Semitism, their casual racism. "Only thee and me is not nuts," Thayer said to Russell: the two of them saw the world's true, hellish face—and mocked it. But over the years, they pushed out from these mutual commitments in different directions, toward annoyance, anger, finally coolness.

During World War II, Russell associated the society with the British publisher King, Littlewood & King, and its magazines and bookstores (in York and London), all called Tomorrow.[249] The publisher imported for sale *The Books of Charles Fort*—though slowly, tortuously so—which the bookstores sold, alongside copies of *Doubt*. Russell advertised the society in *Tomorrow* and published his essays there. Many of these were straight Forteana: alien invasions, strange objects in the sky, the dangers of lie detectors and fingerprinting, wild talents as the expression of genes passed on to humans by giants, mysterious fires.[250] In time, though, his writing for *Tomorrow* turned political. Russell was critical of the British government, its propaganda and diminishment of the individual, offering praise for Germany and qualified support of Nazism: German fascism had its faults, he admitted, but so did democracy. He had his own Fortean notions about how power moved through the world, and how it should. Russell's politics meshed easily with those of *Tomorrow*, which, between its covers, expressed one nexus of modern thought. The editor, Norman Vincent Dagg, was a fascist sympathizer and monarchist, to which he added an enthusiasm for avant-garde arts and mysticism—the occult, Theosophy, parapsychology.[251] King, Littlewood & King's list disseminated this amalgamation to the British reading public.

Tomorrow's audience was ripe for Fortean preachifying, and until the magazine folded in 1950, a steady stream of readers reached out to Russell, were sent examination copies of *Doubt*, and often converted.

Waveney Girvan, for instance, was a fascist agitator before being drawn into the Fortean flock and attempting to publish a British edition of *The Books*.[252] The new members pushed the Fortean Society further toward the mystical and also toward reactionary politics—sometimes faster than the society could accommodate. Russell endorsed Holocaust denial, for instance, but *Doubt* never acknowledged that heresy.[253] Thayer also mixed with fascists, in both his personal life and society business.[254] (He too was skeptical of the Holocaust.)[255] One of the society's earliest foreign correspondents was Heinz Kloss, a Nazi propagandist who wanted to create a pan-German culture that included Germans in America.[256] Ultimately, however, Thayer was too anarchic, too undisciplined to fully give himself to fascism; his flirtation with the political philosophy, like his anti-Semitism, was dangerous but notional, not violent. Fascism was an apostasy, a damned set of ideas, and therefore worth consideration, but he refused to pledge allegiance to it. Russell tried to interest Thayer in the authoritarian People's League, but Thayer saw only vigilantism.[257] Dagg vexed him. He considered tossing him out of the society, exasperated at his allegiance to hierarchy, but settled for "light[ing] a fire in my heart to Trotsky . . . every time Dagg takes pen in hand."[258]

Some of the differences between Thayer and Russell came out in their interactions with one of the society's most consistent supporters, Judith L. Gee (née Griebler), an astrologer, mystic—she thought she could control the weather with her mind—and reader of *Tomorrow*.[259] Gee was also Jewish. She joined the Fortean Society in 1946, drawn by Russell's proselytizing in *Tomorrow*, establishing herself as a redoubtable member. Her name appeared in *Doubt* sixty-seven times (second most among women). So prolific were her contributions that she occasionally warranted her own column. Thayer was content to swap letters with Russell mocking Gee as a ditzy woman and reflexive Jewish partisan but otherwise did not respond to her voluminous correspondence.[260] Russell, as was his wont, argued with her by letter. They tussled over Dagg's editorship and Russell's skepticism regarding the Holocaust. She challenged his racism as un-Fortean and rejected his contention that Jews caused the Holocaust by making themselves targets, "the Chosen People." The real drivers of history's ills, she insisted, were northern Europeans, spreading slavery, anti-Semitism, misogyny around the globe. Thayer summarized the antagonism to another Fortean: Gee was "a Jewess who wears a hearing aid and hates Russell, who hates Jews."[261]

Though significant, these differences were not enough to drive a wedge between Thayer and Russell; both enjoyed rough-and-tumble wrangling. It was, however, a precondition for animosity. The inciting incident came from Ben Hecht. Since anointing himself the first Fortean, Hecht had gone on to become a celebrated author and an even more celebrated Hollywood scriptwriter.[262] For most of his life, he'd been blithely unconcerned about his Jewish identity, but World War II and the horrors wrought by German anti-Semitism had radicalized him. He became a staunch advocate for a Jewish homeland in the Middle East, funded and cheered Jewish votaries fighting the British there. In the spring of 1947, he paid for a newspaper ad addressed to "the Terrorists of Palestine":

> Every time you blow up a British arsenal, or wreck a British jail, or send a British railroad train sky high, or rob a British bank, or let go with your bombs and guns at the British betrayers and invaders of your homeland, the Jews of America make a little holiday in their hearts.[263]

Russell was apoplectic. He demanded that Thayer kick Hecht out of the society. Thayer refused. "I'm not going to quarrel with you," Thayer said. "The Fortean view seems to me to be that anybody witless enough to put on a uniform . . . is fair game."[264] And Thayer could hardly justify eighty-sixing Hecht when Dagg was a member in good standing. (Gee told Russell he should buck up and endure Hecht as she endured Dagg.)[265] Rather than defenestrate anyone, Thayer opted to chastise both Dagg and Hecht in *Doubt*, fatuously pointing out that since Jews and Arabs were both Semitic peoples, Hecht was himself acting as an anti-Semite.[266] The pedanticism preserved a space for Thayer's own anti-Semitism—equating his distrust of Jews with Jewish aggression toward Palestinians—but it did nothing to assuage Russell, who continued to curse Hecht in his correspondence with Thayer.[267] The rift between the two Forteans was real, deep, and unbridgeable.

Rereading their old letters while moving the archives, Thayer was surprised at how much Russell's early ardor had dimmed.[268] Russell denied losing enthusiasm, but in 1956 he shifted many of his duties to Thayer, asked that his name be removed from the masthead—this at a time Thayer was considering pulling back from the society.[269] It became increasingly difficult to maintain the Fortean Society's foothold in the British Isles. Raising prices on the magazine further reduced sales.[270] Thayer insisted the problems were surmountable: newsstand sales, never brisk, could be compensated by subscriptions; *The Books* cost dear, he admitted, but they could be bought there, the protestations of English Forteans to the contrary.[271] Russell could not quit, though, driven by an impetus as indecipherable as Thayer's, committed to the society despite everything. Maybe that was how both proved they were not nuts.

Amid these difficulties, Russell had friends, acquaintances whispering in his ear about the absurdity of his Forteanism, the society, Thayer. How could he sustain his racism, his old friend Sam Youd asked, if he claimed to suspend judgment on all matters of fact? The firmness of his biases belied his skepticism.[272] It was the same complaint Gee leveled about his anti-Semitism. In long letters, the science fiction editor John W. Campbell argued that "Fort was such a thoroughly unstable, unreasonable, unselective, and unintegrated personality that I couldn't trust his data."[273] The society erected in his name was organized lunacy. The science fiction writer James Blish called it "a clearing house for people who think they are being controlled by radio by little green (Negro-Jewish-Russian) men."[274] And Blish had once been an admirer of both the Fortean Society and Thayer! Dagg said Thayer probably had Jewish ancestors.[275] Look at Thayer's confusion about the bomb, said another science fiction author, L. Sprague De Camp.[276] Atomic weaponry was an existential threat as well as a true marvel, yet Thayer denied its reality. De Camp was expanding his science fiction writing to include critical evaluations of pseudoscientific beliefs, like the claim that there were lost continents, Atlantis and Mu—these were not examples of an expansive modernity but throwbacks to benighted superstition.[277] So was Thayer's Forteanism. The secretary of the Fortean Society was a paranoid perpetuator of illogic, a hater of the world for hate's sake. "In his anti-dogmatism, he is fanatically dogmatic," Russell's friend Fred Shroyer said in 1956.[278]

By the end of the decade, the relationship was fraying, the Fortean Society's continuation tenuous. Russell neglected society duties to review his extensive files and write his own book of Forteana, *Great World Mysteries*, which he published in 1957. It was a series of essays, several of which sought to solve Fortean puzzles. As *Mona Lisa* had Thayer's, *Great World Mysteries* capstoned Russell's writing career. Frustratingly, though, Russell failed to get the support from the science fiction community he expected. Russell and Campbell discussed spinning off chapters as articles for *Astounding*, but Campbell ended up passing on most of them, forcing Russell to sell them to lesser magazines. More surprising was Thayer's cool reception. Russell sent Tiffany and Kathleen an advance copy, which Thayer got to only slowly, then commented sharply, "Since when have Forteans written the explanations? I thought that was the function of Science—and ours was to shoot them full of holes."[279] Thayer had happily flogged Russell's Fortean novels, *Sinister Barrier* and its sequel, of a sort, *Dreadful Sanctuary*, but barely mentioned *Great World Mysteries* in *Doubt*.

Later that year, tensions came to a head. In October, the Russian satellite Sputnik 1 orbited the planet. Russell thought this a signal achievement in human history; he'd started his science fiction career as a member of the British Interplanetary Society, and here was proof that, long ago, he'd been correct in thinking humans could enter space. Of course, Russell's enthusiasm was alloyed with Fortean skepticism—he still enjoyed pricking holes in the swollen egos of astronomers. An essay by him about Sputnik in the science fiction magazine *Amazing* even included kind words for Thayer—and Hecht![280] But Russell thought Fortean mockery was best reserved for those who said Sputnik was impossible, that Russia was naught but "a nation of ignorant peasants."[281] Sputnik was a marvel, just the thing for Forteans. The satellite was an example of how science could be enchanting.

Thayer, though, found in the press reports about Sputnik only more pettifoggery from the powerful. He said so in the pages of *Doubt*.[282] There was no proof Sputnik had launched, let alone orbited the planet. To Russell he wrote:

> I'm not going to argue Sputs with you. I'm sorry I stepped on your science-fiction big toe. The entire moral of the Sput issue is merely that accepting orbits as fact is an act of faith, pure and simple. In order to "believe" you must trust your scientists as you trust your priests, for the quality of their "evidence" is of exactly the same order and they are the sole arbiters of its validity.[283]

Russell, refusing to abandon his sense of wonder, accused Thayer of anti-Russian chauvinism. Thayer replied, "It would not have mattered a tinker's damn to me what nation went through the gyrations chronicled in the Sput issue of *Doubt*. My reaction would have been the same. If you wish to believe without a vestige of evidence that is your privilege. I can't."[284] Russell still did not resign from the Fortean Society, but neither was he prepared for reconciliation. Versifying, Thayer tried to make a joke of what increasingly seemed the society's legacy—to be outlived by the kooks, the cranks from whom Fort had desperately wanted to be disassociated. But it works as an epitaph, too:

> Alas, enthusiasm is not as durable
> as Gibraltar
> and most of the sparklers
> are flashes in the pan
> —or Waveney Girvan![285]

Martin Gardner and the Fads and Fallacies of Science

Martin Gardner was up late, after midnight, changing his son's diaper.[286] Probably he was very tired. Still, he recalled much later, he turned on the radio, perhaps to pass the time as he soothed the baby back to sleep. It was the late 1950s, and Gardner worked for the children's magazine *Humpty Dumpty*, writing poems, teaching himself geometry so he could invent folding activities for readers. Recently, he'd built on these talents for a new venture, a column on mathematical games for *Scientific American*— the gig that would make him famous. He also freelanced. His off hours he spent on hobbies carried over from his Oklahoma childhood—practicing magic with a circle of friends he'd met in New York City, palling around with science fiction fans.

While his wife Charlotte slept, Martin listened to Long John Nebel's radio program on WOR. Nebel spent the hours midnight to five-thirty in the morning interviewing people on strange topics—flying saucers, Theosophy, Fortean anomalies. Science fiction writers were frequent guests. This night, he talked to John W. Campbell, probably about one of several controversial subjects Campbell championed: psionics, Dianetics. During the conversation, Gardner's book *In the Name of Science*— published in August 1957—was mentioned. In it, Gardner assailed all sorts of what he considered pseudoscientific nonsense, including the subjects Nebel discussed, the ideas Campbell promoted. Campbell was dismissive, calling Gardner a liar.

Gardner's ideas about science and its competitors were forged in Chicago. He'd come north to attend the University of Chicago, carrying with him his family's evangelism; he shed this identity (after a brief experiment with Pentecostalism) under the influence of his science classes and a cadre of European philosophers who'd fled the continent, its martial cannibalism, for America's safe harbor.[287] He also discovered Mencken. Out of this timber, Gardner built a particular kind of skepticism. Mencken's ironic pose—and ironic prose—had taught T. Swann Harding, Benjamin DeCasseres, Tiffany Thayer to puncture the pretensions of science. Gardner, rather, picked up on Mencken's distrust of science's weird doppelgängers. Gardner realized that his own fundamentalism was built on pseudoscience, a book of pretend knowledge that disparaged scientific geology in favor of creationism. He saw danger neither in science's extending beyond its capacities nor in religion per se—the latter supposed threat, the conceit of nineteenth-century skeptics and secularists like

Ingersoll, continued by Mencken. (Gardner remained a lifelong believer in God.) Rather, nonsense, fake science—those undermined modernity, threatened to send the world back into an era of stupefying superstition. A few years after his move, Gardner explored these ideas in an essay for the *Antioch Review*. "The Hermit Scientist" argued that working alone had been adaptive during the Middle Ages, a defense against church dogmatism, but now was the mark of the pseudoscientist, cut off from the mainstream of science.[288]

The essay appeared just after Thayer published "Charles Fort and the Religion of Self Respect" in Edwin Wilson's *Humanist*. At the time, the outlines of skepticism were capacious enough to include Forteanism. (Wilson also published an article about science fiction by Miriam Allen De Ford that praised Fort's influence on the genre.)[289] But some were starting to worry that Thayer strayed beyond the borders of rationalism into obscurantism.[290] His attacks on science gave succor to the religious, Fred Shroyer told Russell. "The scientists are under attack by the witch doctors and their ilk, and I think it is time we [Forteans] support [the scientists]."[291] De Camp championed a skepticism defined by its allegiance to science and wariness of Fortean nihilism.[292] So did Bergen Evans, the professor of English at Northwestern University who dismissed rains of fish and dissected modern pseudoscience in his 1946 book *The Natural History of Nonsense*. Gardner's developing notions fit this mode, though he seems to have barely considered Forteanism before 1950, when his piece appeared in the *Antioch Review*.

Over the following decade, Gardner and Thayer became more familiar with each other. In 1952, Gardner published *In the Name of Science*, which expanded upon "The Hermit Scientist," blasting all sorts of prevalent pseudosciences, among them Thayer's project. "If we lived in an age in which the majority of citizens had a clear comprehension of science," he wrote, "there might be some point in preserving an organization to remind scientists of their limitations. The astrology magazines on the stands and the sales of Velikovsky's books are sufficient reminders of how far we are from such an age."[293] Thayer offered nothing but an "ignorant sneer."[294] Fort, though—Fort was more surprising. Gardner appreciated puzzles, loved Lewis Carroll—his annotation of Carroll's Wonderland books was the most successful thing he ever wrote—and saw in Fort the same playfulness, delight:

> [Fort] was far from an ignorant man, and his discussion of such top-
> ics as the "principle of uncertainty" in modern quantum theory indi-
> cates a firm grasp of the subject matter. . . . Even when Fort made a
> scientific boner, as he did occasionally, it is hard to know whether he
> made it deliberately or without knowing better.[295]

Forteans had similarly mixed opinions about Gardner. Norman MacBeth, an Anthroposophist, lawyer, and Fortean, charged Gardner with being callow, inconsistent. His interpretation of Fort and the Fortean Society, said MacBeth, was obtuse: on the one hand, Gardner admitted that Fort's self-defined task of rescuing the damned had value, that checking on the claims of science was helpful. And yet he dismissed the society's utility. For MacBeth, here was proof that Gardner had his thumb on the scale; he would stand with the staid, the acceptable, against the challenging, not because of the evidence but because he was committed to supporting science.[296] On the contrary, Thayer recommended Gardner's book—the only skeptical book written in defense of science he found worthwhile.[297] "The author got a little sore at YS, and took a few snipes at him, but we don't hold that against him. The book is full of meat, and the observations on Fort are in general very fair-minded and admiring. By all means, read it."[298] (Following through on the recommendation, the society sold the book.)

In the Name of Science disappeared into a black hole. Five years later, Dover republished it under a new title. Fads and Fallacies in the Name of Science took off, a success Gardner attributed to Long John Nebel, whose unceasing attacks worked as advertising.[299] Gardner went on to fame, re-directing the skeptical movement, away from infidelism, away from the concerns Harding, DeCasseres, and Thayer expressed over the hubris of science, narrowing the focus to ridiculing pseudoscience. He honed his ironic pen, adopting Mencken's slogan: one horselaugh is worth ten thousand syllogisms. So potent was this version of skepticism that Nebel modeled his own first book on Fads and Fallacies.[300] In the early 1970s, Edwin Wilson of the AHA joined with philosophers, science fiction writers, and others to update the Humanist Manifesto, first published in 1933. Some signatories, united by their objections to newspaper astrology columns, soon met, and from this convention came the Committee for the Scientific Investigation of Claims of the Paranormal, dedicated to disproving pseudoscientific theories. Gardner was a founding member. CSICOP worried over not only astrology but also other paranormal, occult, esoteric practices.[301]

Jostling marked CSICOP's first decade, over the parameters of skepticism, whether to investigate the paranormal with a Fortean open-mindedness or, as Gardner advocated, to debunk it and preserve modernity from the barbarism of superstition.[302] For the most part, Gardner's vision prevailed. One critic of CSICOP said, "Gardner has . . . been the single most powerful antagonist of the paranormal in the second half of the twentieth century."[303] Yet Gardner could no more escape Fort than he could rid American culture of fascination with the paranormal. Over the years, Gardner referred to Fort in his voluminous writings—not often, but that's what makes the scattered name-checks worth noting. Either Gardner read and reread Fort's books, or Fort had become part of his mental furniture.

By his own admission, Gardner was a Mysterian, someone who thought consciousness could never be explained by humans or fully replicated by computers. He also believed in the power of prayer but thought that it should not be investigated by scientific methods. Gardner was drawn to strict scientific rationalism but was reluctant to embrace strident rationalism. He wanted to preserve a space for mystical, paranormal forces to interact. He was a gatekeeper, of a sort, who wanted to make sure that neither science nor mysticism overstepped its bounds. His relationship to Fort and Forteanism was parallel to his relationship with God and fundamentalism. Just as Gardner wanted to preserve his belief in God, prayer, and the afterlife while disparaging fundamentalism and the overreach of Christian politics into the wider intellectual world, so he wanted to preserve his appreciation of Fort while dismissing what might be called fundamentalist Forteans and keeping their views out of the intellectual world. He was a Fortean but of a very specific type, unwilling to associate with other Forteans. Skepticism had changed, arranging itself not against religion, as had once been the case, nor against religion and science, as Russell advocated, but against pseudo-science. Fort and Forteanism could be seen as allies in this fight, if used judiciously. There was yet room for the Bronx philosopher in the revised skeptical movement.

3

The Motor of History

... There is a vast
internal life, a sea or organism
full of sounds & memoried
objects swimming or sunk
in the great fall of it as,
when one further
ring of the 9 bounding
Earth & Heaven runs
into the daughter of God's
particular place, cave, palace—a tail
of Ocean ...
CHARLES OLSON, "Celestial Evening, October 1967," 1975

Robert Spencer Carr demobbed in 1946, having joined America's war effort two years previous, aged thirty-six. After a stint in New York City working with a documentarian, he traveled to New Mexico. Carr had gone through many "incarnations," as a friend called them—struggling short story writer, acclaimed novelist, communist, bum husband and father, Hollywood filmmaker, soldier—and now he felt another coming on.[1] He bought an abandoned tavern "on a dead end road in a dying forest beyond a ghost town behind an abandoned cemetery." In this "American Tibet," he founded a Fortean lamasery dedicated to the Order of the Rainbow; he started writing Fortean science fiction.[2] The lamasery would not last, but his dalliance with Forteanism persisted for a decade. In late 1947, one of his stories appeared in the *Saturday Evening Post*.

In the late 1940s, science fiction was bursting from genre magazines and pulps into the mainstream—slick magazines, books, radio shows, soon cineplexes and television. Carr's was only one example of this inundation. Science fiction became a central part of American popular culture, the nation's self-conception. "The dreams our stuff is made of," as the science fiction author Thomas Disch put it, he and fellow authors having imagined rockets and atomic bombs and automated homes and computers, the material and social texture of midcentury life.[3] Central to these dreams was Forteanism. "It has been said that no truly complete understanding of science fiction is possible without at least a nodding acquaintance with the works of Charles Fort," wrote one genre editor.[4]

Thayer was ecstatic at Carr's success, saw it as proof that Forteanism was poised to take over the world. To Thayer's mind, the novelette, "conceived in Forteanism," was "dedicated to the proposition that thinking can be made painless, even thrilling."[5] He enjoined Don Bloch to send a letter praising the story to the *Post*'s editor, the better to keep the magazine open to further Fortean fiction. The two, science fiction and Forteanism, remained tightly coupled deep into the 1950s until, inevitably, Thayer turned on the genre, once more coming to despise what he had loved, and the science fiction community became equally contemptuous of Thayer and his society.

What brought science fiction and Forteanism together was a shared interest in the anomalous. As Fort had spun imaginative explanations of the strange occurrences he gathered, so too could science fiction authors, unveiling, in the process, the forces shaping human life, the machinery of history, its hidden mechanisms. Even more tantalizing was the possibility that, once revealed, these forces could be controlled, just as humans now controlled the atom, flight, the increasingly complex machines of everyday life. History need not just happen, the future come unbidden; rather, human evolution might be controlled, the future created or at least predicted. From Fort, from the society founded in his name, came the hope that humanity might discover potent new sciences, immense new powers, incredible new worlds. Not an atavism, Fort and his ideas were thoroughly modern, keys to progress.

Scientific Romance

Just after he finished reading "Y" in the fall of 1915, Dreiser told Fort, "My impression is that you are out-Verning Verne. Talk about scientific imagination!" Dreiser was not always a reliable assessor of Fort's intentions and literary alliances, but on this day he was perspicacious. "I wonder you don't put this second book in the form of a romance. If you did it would create a sensation."[6] Fort's turn from realistic fiction toward an exploration of scientific anomalies owed much to Jules Verne, H. G. Wells, Ambrose Bierce. Science might disenchant the world, yet it also created new ways to experience awe, expanding "the human imagination with previously inconceivable ideas of movement and magnitude."[7] Writers willing to engage science could extend imagination's frontier further still. Scientific romance—*roman scientifique* in French—was a typical designation for such fiction, and Fort stood foursquare in the tradition, building on Bierce, obsessed with scientific romance's motifs, borrowing from contemporary practitioners, though twisting the genre toward his own purposes: elucidating the hidden structure of reality, a practice adopted by later writers working in the same vein.

Scientific romance paralleled literary realism. Both were responses to the irruptions of the nineteenth century: the creation of science, the invention of a historical consciousness, secularism, industrial capitalism, a plethora of economic and technical disruptions of quotidian life. But they explored different repercussions, one the realistic worlds wrought by these changes, the other impossible provinces and modes of life that did not exist, perhaps could not. Central to romance was the anomalous. Anomalies exposed the skeleton of social life.[8] Monstrous anomalies revealed the world to be hell, or one of its anterooms; Bierce, for instance, wrote of mysterious disappearances, strange rains, all elements of a "cosmic diabolism."[9] Anomalies might also suggest that history had taken a wrong turn, or that history's motor could be controlled, directed (or not) toward better ends. Or anomalies might be the point of a story, an excuse to play with the exceptional but impose no order.[10] In the early twentieth century, the French writer Maurice Renard tried to set an agenda for *roman scientifique*, an approach that was influential (though not without competing visions for the genre) in Francophone and Anglophone countries. Science had marked the modern world, a machine for creating truth, but its explanations by necessity created realms of the unknown, things not yet explained, "strange ambiguous domains."[11] Writers of romance, Renard said, were to scope out these realms, apply the logic of science in explaining them—and invent marvels to populate them. Or, as Fort would do, discover the things and speculate on the mundane forces that brought them into being, a shadow of science, a parallel of modernity that mixed rationalism with wonder.

Renard implemented his ideas in 1911's *Le Péril Bleu*. The story opens with an anomaly: a series of mysterious disappearances. At first, characters think the events a joke but soon blood and body parts rain from the sky. Earth's hellish scaffolding is disclosed: invisible beings inhabit the upper atmosphere, gossamer to the point of transparency. These creatures capture and experiment upon humans. *Le Péril Bleu* transformed the upper atmosphere into the domain of the strange, a common trope, but romancers also explored other arenas—underground, in the sea, in the future. A grimoire of fairies and demons and devils, monsters and imps, frolicked through the modern world.[12] Renard's story was adapted into English by John Raphael, joining other romances in the profusion of magazines sold on both sides of the Atlantic. The American writer George Allan England (at the time "second only to [Edgar Rice] Burroughs as the most popular writer of scientific romances in the pulps") wrote his own version of Renard's tale.[13] "Empire of the Air" was published as a four-part serial in late 1914.

Fort almost certainly read at least one of these stories, but even if he did not, it was clear that he ploughed the same fields, as would his heirs: *Le Péril Bleu* was a clear antecedent to Edmond Hamilton's "Earth Owners" and Eric Frank Russell's *Sinister Barrier*.[14] Fort, too, was preoccupied with strange objects in the sky, fretted

that the world belonged to some nefarious species. He was fascinated by Bierce, who himself disappeared mysteriously in 1913. Perhaps Bierce possessed some wild talent and had teleported himself
into another dimension, Fort speculated in *Lo!*[15] Or, perhaps he had
been captured by an Ambrose collector who also snatched Ambrose
Small, a Toronto millionaire who disappeared in 1919.[16] Not content
to expand science's borders, Fort used anomalies to explode science
and from the wreckage create something else: the world as a strange,
ambiguous domain populated by things that science had damned.
Other romancers tugged at the robes of realism, larding their stories with paratextual apparatus, maps, diaries, letters that hinted at
the narrative's veracity, but in the end their tales were clearly fiction,
the addenda meant to be approached ironically.[17] Fort went further in his intermingling of the true and the imaginary, inventing
what he called a "non-fictional fiction." His ideas were meant to be
taken seriously.[18] He had uncovered the hidden, bizarre but rational
structure of the universe, revealed the stuff of scientific romance as
real and subject to laws that themselves had been hidden until he
explained—or expressed—them. No wonder, then, that Dreiser sent
Lo! to H. G. Wells and assumed he would be amazed.

Scientific romance fragmented in the 1920s, shattering into high
fantasy, science fiction, crime fiction, and horror. (As Russell showed,
though, even two decades on, the boundaries between genres were
porous.) Perhaps American horror's highest expression during the
decade was the magazine *Weird Tales*, around which formed a circle
of writers, among them H. P. Lovecraft, Robert E. Howard, Clark
Ashton Smith—and Robert Spencer Carr, who sold his first story to
the "Unique Magazine."[19] These writers alloyed scientific romance's
focus on the anomalous with the decadent movement in literature,
which had originated in Poe, then passed through Bierce and a bevy
of French authors. Lovecraft most clearly articulated the group's
philosophy. Science had "stripped the world of glamour, wonder,
and all those illusions of heroism."[20] Humanity was alone in the universe, confined by natural laws, its existence without ultimate meaning. He wrote of humans discovering alien gods, curses, obscene scientific practices that startled exactly because they were impossible.

The influence of *Weird Tales* stretched beyond the magazine into American literary culture more broadly. Several of Thayer's novels belonged to this lineage, and Claude Kendall hired a *Weird Tales* author as an associate editor in 1936.[21] Three years later, the author August Derleth cofounded the publisher Arkham House (named after a fictional Lovecraftian town) to put out anthologies of stories from the magazine's authors. *Weird Tales* itself declined in the 1930s, losing its key authors and founding editor, but Arkham House canonized the tradition.[22]

Fort's ideas ran as a red thread through *Weird Tales*. Hamilton was among its most popular authors, and his story "The Earth Owners" ran in its pages.[23] Lovecraft discovered Fort in 1927, but found him too credulous.[24] To be of any worth, Lovecraft thought, Fort's books needed reinterpretation by a skilled artisan who could transform his yammering into art.[25] In October 1930, Clark Ashton Smith read *The Book of the Damned*, having checked it out by mail from the state library. Annoyed by the epigrammatic style, Smith nonetheless felt his "innate romanticism" released. Fort encouraged his hope that materialist science had not fully characterized the universe's workings, that reality contained more objects than science recognized. Perhaps, he mused, he really had seen the apparition of a "bowed and muffled woman, weeping or at least sorrow-stricken, which appeared one night in a corner of my bedroom." Perhaps he really had experienced something momentous "when a dark, lightless and silent object passed over us against the stars with projectile-like speed." Poverty deepened by the Depression had recently forced him to try his hand at writing fiction for the magazine market, and Fort proved a bigger inspiration than "any of the orthodox crew."[26] Fort spurred him to imagine interstellar travel among prehistoric earthlings and other such wonders for *Weird Tales* and similar magazines.

While horror developed from stories that used anomalies to plumb the world's hellish depths, science fiction came out of tales in which the anomalous exposed the chugging engine of history, its gears and governors made available for tinkering.[27] Science fiction was a form of rational wonder, awe of the marvelous fused to an insistence that the world was explicable, its mechanisms knowable, even controllable. The genre harnessed the same mixture of emotions as modern magic, astonishment at legerdemain bound to an understanding that magicians employ sleight-of-hand, not occult powers.[28] That does not mean science fiction was always a handmaiden to science. Stories could be critical of the world's organization, recognizing it as out of synch with reason: Harry Leon Wilson spent his last years rereading Bierce and struggling to compose a story for the *Saturday Evening Post* called "The Stranger," about a Martian whose outsider perspective makes clear humanity's stupidity; he died in 1939, though, unable to concentrate long enough to finish the tale.[29] Science fiction authors could also build stories around the magical, the paranormal. The impetus of history might be intelligible, but its mechanism might be based on paranormal principles or controllable only by magic. Theosophy, with its swirling together of science and religion, influenced a lot of mid-twentieth-century science fiction, inspiring tales in which the universe's invisible motor moves humanity toward a more enlightened, spiritual plane of existence. Reason and the paranormal were tethered together in modernity.

Amazing Stories is typically described as the first science fiction magazine. Started by the inventor and engineering enthusiast Hugo Gernsback in 1926, its first issue reprinted stories by authors such as Poe, Verne, Wells, and George Allan England, whose "Thing from— 'Outside'" had first appeared in Gernsback's earlier magazine *Science and Invention*. Eventually, Gernsback assembled a cadre of regular writers—and competitors, too, his success breeding imitators. He used the letter column to promote interactions among readers and, over time, a community took shape, what came to be known as fandom.[30] Fans were attentive to their place in history and the role science fiction played in human evolution. The stories they read were certainly entertaining, engaging—but they were more than escapism. The genre trained them to look for life's engine and imagine ways to manipulate it. They were not at the mercy of the unaccountable forces of destiny but could create tomorrow. American society

The THING from -"OUTSIDE"
By George Allen England

"... Out of the door crept something like a man. A queer, broken, bent-over thing: a thing crippled, shrunken and flabby, that whined. This thing—yes, it was still Marr—crunched down at one side, quivering, whimpering. It moved its hands as a crushed ant moves its antennae; jerkily, without significance. . . ."

THEY sat about their camp-fire, that little party of Americans retreating southward from Hudson Bay before the on-coming menace of the great cold. Sat there, stolid under the awe of the North, under the uneasiness that the day's trek had laid upon their souls. The three men smoked. The two women huddled close to each other. Fireglow picked their faces from the gloom of night among the dwarf firs. A splashing murmur told of the Albany River's haste to escape from the wilderness, and reach the Bay.

"I don't see what there was in a mere circular print on a rock-ledge to make our guides desert," said Professor Thorburn. His voice was as dry as his whole personality. "Most extraordinary."

"*They* knew what it was, all right," answered Jandron, geologist of the party. "So do I." He rubbed his cropped mustache grayly. "I've seen prints like that before. That was on the Labrador. And I've seen things happen,

---

HERE is an extraordinary story by the well-known magazine writer, George Allan England. This story should be read quite carefully, and it is necessary to use one's imagination in reading it.

The theme of Mr. England's story is unusual and extraordinary. If we can take insects and put them upon the dissecting table in order to study their anatomy, is there a good reason why some super-intelligence cannot do the same thing with us humans?

It may be taken as a certainty that Intelligence, as we understand it, is not only of our earth. It is also not necessary to presume that Intelligence may have its setting only in a body of flesh and blood.

There is no reason for disbelieving that a Super-Intelligence might not reside in gases or invisible structures, something which we of today cannot even imagine.

---

where they were."

"Something surely happened to our guides, before they'd got a mile into the bush," put in the Professor's wife; while Vivian, her sister, gazed into the fire that revealed her as a beauty, not to be spoiled even by a tam and a rough-knit sweater. "Men don't shoot wildly, and scream like that, unless—"

"They're all three dead now, anyhow," put in Jandron. "So they're out of harm's way. While we—well, we're two hundred and fifty wicked miles from the C. P. R. rails."

"Forget it, Jandy!" said Marr, the journalist. "We're just suffering from an attack of nerves, that's all. Give me a fill of 'bacey. Thanks. We'll all be better in the morning. Ho-hum! Now, speaking of spooks and such—"

He launched into an account of how he had once exposed a fraudulent spiritualist, thus proving—to his own satisfaction—that nothing existed beyond the scope of mankind's everyday life. But nobody gave him much heed. And silence

67

The first issue of *Amazing Stories* (April 1926) included, among reprints of Wells, Verne, and Poe, "The Thing from—'Outside,'" a story by George Allan England (the byline misspells his middle name) inspired by Fort. An editorial introduction uses Fortean language, intentionally or not, evoking superintelligences that dissect humans. The characters in the story confront this superintelligence in one of earth's ambiguous realms, a place not yet fully explored.

itself was waking to the future, an increasing number of magazines using that word—future—in their article titles, replacing antiquarian concerns with dreams of what could be.[31] Science fiction fans saw themselves as at the vanguard of American culture, argonauts of the future.

As he'd been for the *Weird Tales* circle, but even more significantly, Fort was a source of plots for science fiction authors. It is more than symbolic that a story using his name, England's "The Thing from—'Outside,'" appeared in the first issue of the first science fiction magazine. A rumor circulated that F. Orlin Tremaine, who took over a struggling *Astounding* in 1933, bought the rights to *Lo!* so as to farm out ideas to his regular writers.[32] (The book was serialized in the magazine from April to November 1934.) Certainly, John W. Campbell, who succeeded Tremaine, thought Fort's omnibus was a sourcebook for science fiction. He reviewed *The Books of Charles Fort* in both *Astounding* and *Unknown*. "It probably averages one science-fiction or fantasy plot idea to the page," he said.[33] (A. Bertram Chandler, a seaman, picked up *The Books* when his ship docked in New York, wrote a story based on a fragment as he steamed toward New Zealand, and sold the tale to Campbell when next in New York.)[34] Gernsback's original conception of the genre emphasized gadgets, technological breakthroughs, a form of storytelling that persisted over the decades. Fort fed a different stream.[35] Fortean anomalies keyed characters in science fiction to the operation of history's engine and ways to control it—rather than, as in horror tales, life's hellishness. Through the window opened by the strange, they saw reality's hidden structures, one of three ways (along with expanding the imagination and exposing the actions of power) that Fort was useful to science fiction. Science fictioneers hoped to control those forces, to drive the world toward a future rather than allow themselves to be driven. Methods of control varied, from the mechanistic and rational to the magical and occult. What mattered to Campbell and (to a lesser extent) other leaders in scientifiction, as the genre was sometimes named, was the discovery of history's blueprint, what Campbell called "universal operating principles."[36]

Thayer and Russell tied tight the strings that bound Fortean-ism to science fiction, another source of support after the "Circus Day" brouhaha. Scientifiction, Thayer said in his magazine, was the most vigorous Fortean art yet developed.[37] Plans to advertise *The Books* in *Unknown* fell through, but that did not stop Thayer and Russell from canvassing among science fiction fans.[38] Thayer hit a jackpot at the University of Minnesota, where he found a number of Fortean scientifiction fans, chief among them Jesse Douglass, who'd been excited by "Circus Day."[39] A. L. Joquel in southern California was similarly moved by issue 6 of *The Fortean* (which he reprinted in his own 'zine) and preached Fortean ideals to the Los Angeles Science Fiction Society.[40] Thayer's cultivation of fans was matched by their interest in the Fortean Society, the reasons for overlap multiple: political, Theosophical, scientific, philosophical. The serialization of *Lo!* in *Astounding* was the sixth most favored piece that year, generating about eighty letters to the editor, among the most of any feature, commentary running into 1936.[41] Science fiction fans put out their own amateur Fortean 'zines as well. From Milwaukee emerged *Frontier*, a fanzine by Donn Brazier and Paul Klingbiel featuring news reports of anomalies, callouts to the soci-ety (as well as discussions of Lovecraft, science fiction). A question-naire circulated among science fiction fans that tried to measure the contours of fandom included among its queries "Are you a follower of the late Charles Fort?"[42] The Fortean subculture within fandom was robust enough that a fan encyclopedia included an entry on Fort.

Natural History in the Water Pipes

"To those disciples of Bacchus I will offer additional justification to swear off water, by the compiled accounts of the creatures swallowed accidentally in loathsome, unsterilized water which lacks the germ-killing, purifying qualities of 100% proof hootch. In fact, some of the following are fish stories, but not the kind that phrase usually implies; the whale could swallow Jonah, so I guess the reverse is just as possible."[43]

In his midthirties, George Wetzel interspersed manual labor—mechanic, greaser—with hours in the library, the Enoch Pratt Free Library's Maryland room and the Peabody Institute Library, reading old newspapers, looking for Fortean events.[44] He compiled a selection into a short pamphlet.[45] Thayer reported in *Doubt* 48:

> MFS George Wetzel has written and produced an 18 karat Fortean item of which only 70 copies exist. It is Natural History in the Water Pipes, and concerns itself with the eels and fish and other critters which have frequently been drunk by humans in tap water. What's more, the material is handled in a manner reminiscent of Fort himself. YS got three solid laughs out of its 8 mimeographed pages, fully documented.
>
> It's a home-made job, put together with wire staples, rather too full of typographical errors and so on, but one feels sure you will treasure it, all the same. Suppose you send 80 cents. Money back on request.[46]

Minor though they were, these tales pointed toward a major Fortean conclusion: the structure of reality was mostly hidden, its true contours suggested by the anomalous.

> "In 1898, a man turned on a spigot and filled a bucket with seventeen [eels]. This is not the most remarkable account, though it does edge close to that of the 'winged eel.'"[47]

Wetzel's interest in the fringe had begun in his midteens, when he discovered science fiction, *Weird Tales*. He joined a science fiction club, planned a 'zine, passed Forteana to Donn Brazier's *Frontier*. A hiatus: 6 June 1942, a few days after he turned twenty-one, Wetzel, with three years of high school to his name, joined the Army Air Force; discharged 26 November 1945.

1874: "A 38 inch long eel just taken out of a water pipe could only have
gotten in when no larger than a darning needle, as the strainer across
one water main intake from the reservoir had holes in it only about .5
inch in diameter. The eel therefore had inhabited the water main for
approximately 18 to 21 years! As other later accounts of monstrous
size eels are compiled in the bibliography, one can see this was no
exception but rather a common occurrence."[48]

Returned to the Baltimore area, Wetzel reconnected with Brazier—
his postwar 'zine *Ember* also focused on Fortean matters—sending him
material on the Salem witch trials, a brief musing on "the emotional basis
of the melodic line."[49] He read Blavatsky's *Isis Unveiled*: "a gold mine of
outré facts and assumptions."[50] Especially intriguing was her claim that
tunnels beneath the Pacific Ocean connected now-disappeared islands,
accounting for the spread of ideas from the Lemurian and Atlantean root
races.

A ten-year-old boy suffered baffling spasms until June 1859, when he
vomited up a live frog, two inches long. In 1883, a laborer drinking
from a spigot swallowed something big, spent the night with cramps.
A doctor treated him for a week until administering an emetic. The
man vomited blood, then a three-inch lizard, "alive and kicking."[51]

Theosophy never inspired Wetzel the way Lovecraft did, though. He
tried his hand at Lovecraftian fiction, never more than amateurish, placing his stories in 'zines, producing a mimeographed collection in the mid-
1950s. He read through Lovecraft's influences—Nathaniel Hawthorne,
Edgar Allan Poe—discovering a possible origin for Poe's "A Descent into
the Maelström," completing one of Poe's unfinished stories.[52] He traveled
north to Philadelphia, the Franklin Institute, which housed the Library
of Amateur Journalism, looking for anonymous, pseudonymous works
by Lovecraft, compiling the most detailed bibliography to date of Lovecraft's writings.[53] He wrote, for August Derleth's Arkham House, analyses
of Lovecraft's work, arguing that many of the stories comprised an inte-

grated narrative arc, chapters in what was essentially a novel of earth's alternative history.[54]

His exploration of Lovecraft and Lovecraftian fiction continued until the end of the 1950s, then ceased for the better part of a decade. He married around this time, which may account for some redirection of energy. More pointedly, he fell out with Derleth and other fans. As Wetzel had it, Derleth tapped him to update his bibliography for publication, only to reject it, instead publishing one by another writer, which was mostly just a copy of Wetzel's. (Wetzel took his revenge by untangling the skein of confusion around Lovecraft's copyrights, which Derleth claimed to own but which Wetzel showed belonged to no one.) Wetzel was further stung by complaints from other connoisseurs of weird tales about his bigotry, his anti-Semitism; they set postal authorities to investigating him for sending illegal things through the mails.[55]

In the 1870s, Baltimore's water engineer blamed the fishy taste of the city's water on a sudden change in the weather; a year later, he said the problem was the defective plumbing installed in houses. Later still, he noted that when the weather changed cold, few complaints came to him—though Wetzel noticed no seasonal lag in newspaper reports of marine life coughed up by the pipes.[56]

Forteana, Wetzel contended, underwrote Lovecraft's philosophy of horror in two ways: first, Fort offered a materialistic explanation for the supernatural, which Lovecraft adopted. The ancient deities that populated his stories were not gods in the traditional sense but manifestations of the stuff that made up the universe.[57] Second, Fortean reports—folklore, word-of-mouth—could be refined into tales of the uncanny, just as Hawthorne and Lovecraft had transformed New England gossip into literature. Wetzel wanted to do the same for Baltimore, hence those hours at the library. He collected stories of witchcraft—in Salem, locally—uncovering a lost manuscript by one of Lovecraft's peers, putting together a pamphlet on a Maryland healer, Michael Zittle Sr., who in the 1700s wrote a conjure book. Wetzel spent a decade tracking down various editions, commentaries.[58] These formed a basis for his own story, "The Saga of Mr. Cushwa":[59]

> One other thing I recall is my mother's trying to break one of my brothers (when very small) from drinking with his mouth to the spigot, telling him he might swallow a snake that might be in the pipe, adding such had happened to someone years and years ago. Such word-of-mouth stories about stuff in the water pipes used to interest me slightly. Now that I know the truth about them and have documented them to my own satisfaction, I am terrifically astonished that more word-of-mouth stories are not current.[60]

In December 1954 came another pamphlet, "Baltimore Subterranean," twenty-seven mimeographed pages on the Clement Street Caverns, the labyrinth under Federal Hill, tunnels under Battery Avenue, beer vaults, mines, and "other curious earth voids," complete with maps. Wetzel supplemented his library research with field reports:[61]

> This research has solved, indirectly, one mystery that has long puzzled me. A neighbor I knew had a cat, which had a habit of sniffing at street corner fire plugs, attempting to crawl into sewers, and gazing fixedly at the plumbing. If your cat acts this way, consult a plumber.[62]

The mid-1950s saw, as well, publication of his manuscript "The Sea Serpent—Its Existence Proved" in the 'zine *Umbra*. As with his survey of Baltimore's cryptic natural history, his study of sea serpents relied on newspaper reports, the *Baltimore American*, the *Sun*, from 1875 to 1954—almost a century's worth of data proving "the undoubted existence of sea serpents."[63]

> An "Aztec nightmare" was ripped from the water pipe at the Baltimore City Health Department in 1902 by a plumber fixing a trickling spigot: "springing all over the room on a glittering tail and flapping wings that seemed untried for years." The plumber bludgeoned it to

death. The city chemist tasked with inspecting the creature of "Fortean character" foreswore drinking from the public water supply.[64]

New Sciences, New Powers, New Worlds

Bruited about the letter columns of *Astounding* in 1934 was the idea that Fort's writings might be the prologue of future sciences. Anomalies did not contradict science but existed on the frontiers of knowledge, to be subsumed in due time. Maynard Shipley, three years before the serialization of *Lo!*, had made that exact point in the *New York Times*. It was an important interpretation, put forward by many other science fiction aficionados. Fort expanded the imagination along two dimensions. His collection of weird things could be the source of plots. More importantly, they could prod the creation of new, legitimate sciences, might even open new worlds to exploration. "If only we could find the pattern hidden there among the vast jumble of facts," Campbell said, the omnibus edition "probably contains the root truths of about four new sciences." The Fortean Society, too, might spark the creation of new sciences, might be, as Shipley said of Fort, "the enzyme orthodox science most needs" to catalyze something novel, to efface the boundary between the world and its engine, granting humanity access to hidden mechanisms.[65] One had only to approach Fortean anomalies with open curiosity and limber imagination. Taking control of history's motor was challenging, though, even frightening, Campbell would say. New ideas could be so startling as to cause the imagination to seize. Best, then, to present them as fiction: superficially entertaining, deeply nourishing.[66] Already, the science fiction community took credit for prophesying the importance of rockets, sticking to the idea when mainstream science rejected it; one story in *Astounding* had so convincingly described the development of atomic weapons that the FBI investigated the author, worried about the possibility of espionage.[67] Fort, who rifled through old periodicals, who ridiculed astronomy, was a doorway to the future.

James Blish articulated this vision of Forteanism as future science in the early 1940s. He'd discovered science fiction just before Fort's death, when he was about nine, paid entry into the fan community a few years later when, aged fourteen, he started his own 'zine.[68] Apparently, he corresponded with Thayer around that time, too, who encouraged his writerly ambitions. (The letters were lost to a flood in 1955.)[69] Blish attended Rutgers University, hung around with the seminal science fiction group the Futurians, was inducted into the army. He joined the Fortean Society sometime before the US entry into World War II, soon becoming involved in a tussle over the meaning of the society.

In 1942, the Futurian Donald Wollheim published a short story that gently mocked Fortean cosmology: an astronaut rips a hole in the gelatinous curtain that surrounds the universe.[70] The tale, written under a pseudonym, appeared in *Science Fiction Quarterly*, which was edited by another Futurian and Wollheim's sometime agent. In the following issue, the editor (also using a pseudonym) highlighted the story's Fortean elements—praising Fort for his keenness of mind—but cleaved them from the society:

> When you try to regiment cynicism or criticism, or even organize it to the extent that a formal society would do, you merely start the groove which eventually becomes a rut—you end up with a clique far more dogmatic, idiotic, and mystical than those aspects of the behaviour of certain "scientists" which Fort continually attacked.[71]

Blish, from his station at Ft. Dix, sent a reply, explaining that the Fortean Society was not out to debunk (that was an occasional side-effect of its real work) but to investigate. Its files bulged with nearly a hundred thousand reports of inexplicable events and these, as much as Fort's books, were the seeds of new sciences. "That's our job and it is a herculean one," Blish wrote. "To take all these inexplicable facts . . . and make a pattern or patterns out of them." Certainly, some eccentrics Thayer promoted were wrong, he conceded—as Fort was sometimes wrong—but that only proved the need for more investigation. Other Fortean mavericks had caught the tiger's tail: Thomas Graydon's chemical theories "work out most spectacularly," he insisted. Members only "foamed at the mouth" when they finally lost patience over science's refusal to give their ideas a fair hearing; otherwise, they were quite reasonable.[72] The editor, at least in print, was persuaded by Blish's summary of the Fortean mentality. Perhaps the magazine could run more Fortean pieces, including nonfiction ones, he mused, though the magazine stopped publishing before the next issue appeared.

Others saw in Fortean anomalies a more radical potential than Blish allowed: that to take them seriously not only conjured into existence new sciences, powers, and worlds, but proved all conventional science wrongheaded. Anomalies exposed the world's hidden mechanisms, but there was no reason to assume—as Blish did, and Campbell seemed to—that the tools for tinkering were those of science. Magic might do as well. Fort, after all, had seen the two, science and magic, as the same.[73] Follow that thread and the world itself was turned upside down. Frederick Hehr, an immigrant engineer, told readers of *Astounding* that Fortean theories inverted "theories of matter," uncovering the esoteric laws that organized gravity and revealing the hidden secrets of ether.[74] Robert Spencer Carr made the point in another story, "The Coming of the Little People," which began with a series of anomalies that were eventually discovered to be the work of fairies, desperate to save the earth from nuclear annihilation. Such anticipations explain the appeal to the science fiction community of Theosophy, what with its emphasis on the role of the spiritual and the occult—as well as science—in history's unfolding. Myriad spiritual forces drove humanity, orthogenetically, through its root stages and toward a time when the noumenal and phenomenal realms would merge.

The idea not only that Fortean anomalies point toward history's hidden motor but that the engine could be controlled by paranormal means inspired perhaps the most famous, most important science fiction writer of the twentieth century, Robert Heinlein.

Born in 1907, Heinlein served in the navy until 1934 when tuberculosis forced him out on a pension, only twenty-seven years old. After some false starts, another flareup of the disease, he turned to writing.[75] Heinlein was a man of many facets: hardheaded engineer; freethinking apostate from the faith of his Kansas family; seeker of philosopher-kings to rule society, fearful he might not be one himself; crusader against the corruptions of religion, politics, business; alternately moved by desires to save society, reform it, destroy it. Taught to doubt by his loss of faith, not trusting the dominant institutions of social life, he thought what counted as reality was often wrong, wondered if the world was but a projection of his own mind. Science fiction seemed the genre for exploring these issues.[76]

During his apprenticeship, Heinlein met frequently with a group of science fiction authors and fans in southern California known as the Mañana Literary Society to discuss story ideas. Jack Williamson, who'd come west to see a psychotherapist, was there, along with L. Ron Hubbard (one of Campbell's favorite authors, not yet a prophet) and Edmond Hamilton.[77] Southern California was not as intellectually freewheeling as San Francisco, but since the Great War a wave of immigrants from the Midwest, shaken by the rapidity of social changes, had opened the region to metaphysical religions; the Mañana Society, and American science fiction more broadly, overlapped with this burgeoning metaphysical community.[78] Also among the Mañana members was Jack Parsons, a rocket engineer who helped found the Jet Propulsion Laboratory and who was an occultist.[79] Parsons practiced sex magic to bring about the birth of a savior. In time, Heinlein graduated to *Astounding*, became its leading author—about 20 percent of the words that appeared in the magazine during 1941 emerged from his typewriter, though many were published pseudonymously.[80]

True to his engineering background, Heinlein could be hardnosed about scientific accuracy, but many of his stories wore proudly their Fortean allegiances. His favorite among them was "Lost Legacy," about parapsychologists who discover a secret society on California's Mt. Shasta—including Ambrose Bierce, who'd not disappeared in Mexico but come to live there—waging psychic war against the forces of human depravity.[81] Flabby though it was as a narrative and so stuffed with the pseudoscientific that even his fans rolled their eyes, "Lost Legacy" represented uncut Fortean science fiction: earth's history was organized around a battle between groups using magical—material, not spiritual—weapons. Also in 1941, he wrote "Creation Took Eight Days." A Fortean tale in the mode of *Le Péril Bleu*, the story told of explorers whose investigation of anomalies leads to the discovery that aliens control human life, the owners as far beyond human intelligence as humans are beyond fish. Like Russell, Heinlein sometimes had trouble with Campbell's capriciousness; the editor rejected "Lost Legacy," forcing Heinlein to sell it to a lesser magazine. Campbell also returned "Creation" sans check, but by that time Heinlein was feeling his oats. He'd met Campbell in New York, established his dominance in the pairing to the point that, as one historian glossed, "he would write what he wanted, when he wanted, and the way he wanted."[82] Campbell didn't like the ending of "Creation," as depressing as that of Eric Frank Russell's "Forbidden Acres," but conceded and published it as written when Heinlein seemed to retire in protest.[83] Came then from California a steady stream of stories, many of them synthesizing Forteanism, the paranormal, and Campbell's insistence that stories unveil universal operating principles.

Fredric Brown sent to *Astounding* and other science fiction magazines copious stories along the same lines: Fortean anomalies that revealed magical ways of controlling history's motor. Born in 1906 in Cincinnati, Brown went to college in Indiana, then settled in Milwaukee, where he worked as a printer and proofreader. A slight man with a tidy mustache, he connected with the city's very active science fiction community and, while suffering through the Depression, started experimenting with short stories. He hit his stride in the 1940s, becoming one of magazine fiction's most popular authors. Brown turned out science fiction, fantasy, mysteries, and mixtures of those genres. He was known for his wordplay and trick endings; other authors admired his craftsmanship. Innovating through contrarianism, he eschewed space opera to focus on the small-bore, highlighted the limits of science, and imagined the failures of rocketry. Like Heinlein, Brown was uncertain of reality, its contours and creator, but he resisted solipsism, thought instead that reality might be the invention of some cosmic author.[84] A fundamentalist, in the sense offered by the sociologist Simon Locke—reality authored by a singular being outside of time and space.

Fort was central to this philosophy, and it is no coincidence that Brown's work became more salable, more consistent after he encountered *The Books*. His novel *What Mad Universe* had "a philosophy behind it," his biographer wrote, that was "almost Fortean: everything is true somewhere, so no anomaly can be dismissed."[85] Other stories were equally Fortean. *Compliments of a Fiend*, a 1950 novel, had one of Brown's recurrent characters, Am Hunter, go missing. At first his nephew and usual partner in solving mysteries, Ed Hunter, thinks Am (short for Ambrose) is the victim of a Fortean antagonist—that an Ambrose collector is on the loose. The real mystery is rooted in everyday ill-doings, but the solution is reached thanks to a name inscribed in a copy of Fort's *Book of the Damned*.

Perhaps the story best representing Brown's Fortean philosophy is "The Angelic Angleworm," written in September 1941—about five months after the appearance of the Fort omnibus—and published in February 1943's *Unknown* (by then retitled *Unknown Worlds*). Printer Charlie Wills—a play on Fort's name—puzzles over a succession of anomalies (winged worms, ducks from nowhere) until he deduces that the problem is with the press that is printing the universe's story, a regular glitch that he uses to reach heaven, meet the universal author, and correct the problem. Humanity lives in a Fortean fantasy, Earth's owner essentially a newspaper office. And out of the work and imagination of its authors and editors and publisher emerges reality, in all its strangeness. Science had often failed to fulfill its promises; rockets remained grounded. But there were rules, which could be glimpsed when one paid attention to the anomalous, the damned. Once identified, these universal principles were amenable to control, if the right (magical, paranormal) tools were used.

Not everyone thought these ideas needed to be coated in fiction to ease their swallowing. More than a few were willing to insist on their factuality, to underscore the "non" in Fort's nonfictional fiction. Sam Merwin, editor of several science fiction magazines, admirable rival to Campbell, followed Fort in believing that no science was permanently true—every foundational theory would be superseded, as Einsteinian physics had replaced Newtonian.[86] Merwin had come to believe that Fort's speculations were on the right track, and said so in his editorials. "I am not much given to prophecy," Fort had written, "but I'll take this chance—that if England loses India, we may expect hard winters in England." As it happened, the winter after passage of the Indian Independence Act, New (not old) England suffered the Great Blizzard of 1947.[87] Merwin pointed out, as well, that astrology had become a practical science, in line with Fort's claim that what had once been superstition would become accepted knowledge, if only temporarily. "Radio stations," Merwin said in 1951, "had recently determined that static was caused by the positions of the planets." At least in this fairly trivial way, life on earth was modulated by the position of heavenly bodies.[88] Fort was clearly on his mind at the time, and the next year *Astounding* published a story of his that fictionalized Fortean cosmology.[89] Others in the science fiction community insisted that Fortean notions of space were true, the sublunar realm resisting rationalization by scientists, remaining an ambiguous domain of the exceptional and strange.

The aptness of applying Fortean and science fiction ideas to reality in a nonfictional way was especially attractive to those of a Theosophical bent. Fortean anomalies could be used as proof that the universe was organized according to Theosophical principles, history driven through a series of stages by ethereal forces understood by an elect that compiled and coordinated the esoteric knowledge of various religious traditions. This was a New Age interpretation of reality, anomalies pointing to a vast system that correlated manifold religious traditions. A. L. Joquel was an avatar of this nexus. A science fiction fan, promoter of rocketry, founder of the the "Un-Intellectual Brotherhood of Anti-Science," anarchist, CO (a "traitorous little bastard," in Heinlein's opinion), dedicated Fortean, and member of southern California's spiritual community, he worked at Manly P. Hall's metaphysical library, where he unearthed new authors for Thayer's delectation, even discovered an unknown short story by Fort.[90] His research formed the basis of a series of Theosophical essays for the magazine *Theosophia* later compiled into *The Challenge of Space*. Joquel argued that science's promised future was too narrow, excluding the sacred: modernity would not disenchant but deepen humanity's understanding of the numinous. Exploration of space was ushering in the next stage of Theosophical evolution, when spiritual values would catch up to technical capacity and the universe would come to understand itself in its full glory. Fortean anomalies were keys to this evolution. "By putting researchers on the track of new concepts, the vast mountain of unexplained phenomena may mark the grave of dogmatism in science, and be a guidepost to the new vistas of nature about which we are just beginning to gain a comprehension and an understanding."[91] Science was a stage, something that had to be overcome as the Theosophical plan unfolded. Fort had brought the next phase into view by showing science's contradictions. Joquel hoped to lead humanity further into that new space.

I Think We're Property

Science fiction writers found in Fort the tools for excavating reality's hidden form, the motor of history. Forteanism was, as well, a goad to the science fiction imagination, an inspiration for creating the future. Fort was a prophet, not only of the unexplained, as Fort's biographer, the science fiction writer Damon Knight, had it, but of modernity. In addition, Fortean science fiction was defined by its focus on how power moved around and through the universe's secret structures. Contemplation of power could lead toward fundamentalism, the belief in a single spiritual being controlling all of history, or New Agery, which also saw reality as animated by the spiritual, though not by a singular figure. Power could be imagined as operating in Fortean ways, similar to New Agery in its insistence that the universe was quickened by multifarious forces, but different from both New Age theorizing and fundamentalism in its assertion that those forces were mundane. For the most part, though, science fictioneers—like avant-garde artists, like UFOlogists—tended, in their analysis of power, toward the conspiratorial: a singular cabal, using material means, controlled everything. "I think we're property," Fort wrote, and science fiction authors and fans imagined he might be right. They wondered who those owners were and what powers they used to devise reality.

Heinlein's "Creation Took Eight Days," published in March 1942's *Astounding* as "Goldfish Bowl," described a Fortean conspiracy. In 1947, H. Beam Piper published "He Walked around Horses," about a mysterious disappearance, which he elaborated in subsequent stories into a vast tale of multiple, parallel timelines and a police force that prevented those on other timelines from learning to jump tracks—a conspiracy of chronologers, as it were. A year after Piper's first paratime story appeared in *Astounding*, Russell published in the same magazine another Fortean conspiracy. As with "Sinister Barrier," "Dreadful Sanctuary" begins with a mystery: not who was killing scientists, but why rockets kept exploding before reaching Venus. Through a haze of Fortean references—mysterious teleportations, elites manipulating public opinion for their financial benefit, *cherchez la femme*—the solution seems to take shape. Two competing cults battle for supremacy, rocketry a pawn in their secret war. Both agree that earth is the solar system's insane asylum, cut off from the rest of space by a "dreadful barrier." One group wants to keep humanity earthbound to protect extraterrestrial planets from contamination; they become saboteurs. Vying against them is a cult that believes humanity is the offspring of an alien species; they push rocketry forward so that humans might visit their natal planet, force their cosmic parents to recognize them. As it turns out, both groups are wrong. The story ends with two rockets leaving earth's atmosphere, one of which reaches the moon, the other headed for Venus. The story was published as a novel by Fantasy Press in 1951, and again in 1963, the latter edition offering a tragic conclusion: none of the rockets succeed in the end.

As with Fortean science fiction generally, not everyone was content to treat the conspiracies delineated in the pages of pulpish magazines as purely imaginary, nor to accept them merely ironically: some thought them a genuine description of modernity, enchantment not the opposite of rationalism but its fulfillment. Some thought Lovecraft's stories of ancient gods described a real pantheon whose actions intersected with human life.[92] Jack Parsons, the rocket engineer and occult leader, took literally a story Jack Williamson wrote about witchcraft.[93] Maurice borrowed liberally from *Weird Tales* in his reconstruction of Theosophy. His writings are impossible to date, but a coherent mythology can be deduced from them. He said he'd visited Theosophical hotspots, Mt. Shasta (supposedly home to remnant Atlanteans or Lemurians and the scene of Heinlein's "Lost Legacy") as well as Tibet. There, Doreal claimed to have learned earth's secret history. Once upon a time, there'd been an epic battle between humanity and the shadow people, half human, half serpent: beings clearly taken from the stories of Robert E. Howard, part of the *Weird Tales* circle. The reptilians had been defeated, Doreal sometimes said, or, disguised as humans, had insinuated themselves into positions of power. They were real, indubitably so. Doreal worried that earth teetered on the brink of another fateful battle. He established a commune in part to survive the coming war.[94]

Norman (a.k.a. David) Markham also worried about the end of the world and the Fortean conspiracy that would bring it about.[95] Markham lived a hardscrabble, peripatetic life, science fiction and its allies a constant presence.[96] He was especially drawn to Russell and his Forteanism. Russell had floated the idea that Venusians regularly visited earth to harvest resources, in the process causing earthquakes and other "natural" disasters.[97] Markham had developed a similar theory after "anatomiz[ing] Fort's encyclopedic lore with an eye toward attempting to solve the riddle of their origin."[98] Fortean anomalies occurred most often, he concluded, when Mars and Venus closely approached the earth, leading him to hypothesize that Martians and Venusians took advantage of these times to visit earth, abduct ships and planes, and cause other turmoil. Markham continued his study of correlations over the next two decades, through all his life's tribulations, inundating Thayer with material, applying to the Fortean University with an essay based on his theories, beseeching Russell and Bloch for help publishing a book explicating the interplanetary conspiracy, sending letters outlining his ideas to newspapers, magazines, the US government. It was imperative that humanity understand its place in the solar system so that it might contravene the march toward annihilation and spur, instead, rocket trips to Venus or Mars, the meeting of advanced civilizations, a chance for earthlings to learn—as Theosophists did at the feet of Tibet's elite—the secrets of existence.[99]

In the late 1940s, Ray Palmer arranged Forteanism, Theosophy, and conspiratorial theorizing into what was then science fiction's most influential cause célèbre, the so-called Shaver Mystery.[100] Palmer was, like Fredric Brown, a veteran of Milwaukee fandom. Born in 1910, a few months after Halley's comet was most visible, he swore to the end of his days he remembered witnessing the comet from the arms of his grandmother, staring through the nursery window into the night sky.[101] When he was seven, Palmer was hit by a truck; a series of painful surgeries left him hunchbacked and short, never growing beyond about four feet tall. Palmer found refuge in science fiction, his imagination taking flight while his body remained bedbound. In the late 1930s, he assumed control of *Amazing Stories*, the cradle of American science fiction, then struggling, and revived it by turning to space opera. While *Astounding* then set the standard, Palmer's magazine had its fans. During the war, his world was shaken when he discovered *Oahspe*, his wife Marjorie and he staying up all night reading it together. Supposedly channeled from spirits by a New York dentist named John Ballou Newbrough, the would-be new Bible synthesized nineteenth-century metaphysical thought. Over nine hundred pages—complex, confusing—*Oahspe* purported to tell the secret history of the world, the stages of mankind, the bureaucratic organization of heaven's many gods, the prophecy of a glorious age destined to arrive at the middle of the twentieth century.[102]

Not long afterward, Palmer received a letter from Richard Shaver, a Pennsylvania laborer who said he'd uncovered the motor that drove history.[103] Over the next several years, Palmer and Shaver massaged Shaver's ideas into a grand theory, explicated in a cycle of stories, essays, and editorial columns: deep caverns, they wrote, riddled the earth, housing Deros, *de*trimental *ro*bots, which had taken refuge after losing an interstellar battle with the beneficent Teros (in*te*grative *ro*bots). The Deros and Teros had a secret language, which Shaver had deciphered. Via mysterious rays, Deros foisted hurt and heartache on humanity. The tales told by Shaver and Palmer incorporated Theosophical elements, space opera, and *Oahspe*. Shaver insisted the stories that ran in *Amazing Stories* were true; Palmer was coy.[104] Sometimes he seemed a genuine proponent, imagining a vast circulatory system in which *Oahspe* offered evidence for the Shaver Mystery, which in turn verified *Oahspe*. At other times he seemed to wink. He spent a day at the annual science fiction convention in 1949 and left some attendees with the impression that the Shaver Mystery was just a play for readership.[105] That was Martin Gardner's conclusion, too. *Amazing Stories* was based in Chicago and the two men, Gardner and Palmer, had met at science fiction gatherings, Gardner coming away with the sense that Palmer had "the personality of a professional con artist."[106] Whatever Palmer's ultimate feelings, the Shaver Mystery worked in his favor. *Amazing Stories* became hugely popular.

Central to the Shaver Mystery was Fort, and thus many Shaverians were also Forteans. Some Forteans struggled to reconcile the mystery with Fort.[107] Others were certain there was a concordance and brought Shaverian insights to Thayer's society. "Charles Fort was one of those who came closest to guessing, or knowing the mysteries contained in the artificial cave world beneath the Earth's surface," wrote one of the myth's expositors.[108] Palmer dissolved Fort's philosophy and the reports of anomalies he had collected into the paranormal matrix of his own thoughts, which increasingly approached *Oahspe* in their abstruseness. One of his stable of authors explained fafrotskies—frogs, toads, fishes, snakes, lizards, stones, warm water falling from clear skies—by recourse to intricately ordered theories about space: outer space, far from a vacuum, abounds with life in many forms, and in its far reaches, great civilizations. These elements are arranged musically, by octaves; by vibrating at the correct frequency one may commute between octaves via a "frequency elevator." Thought itself creates vibrations, which in turn bring into existence matter; Fortean rains occur, then, when a vibration sent into space carries objects down the shaft of the frequency elevator.[109]

Late in the 1940s, *Amazing Stories* moved its editorial offices to New York, leaving Palmer behind in the Chicago area. He ran a bookstore in Evanston, started his own publishing house, cofounded a new magazine, *Fate*, which turned *Amazing Stories* inside out: whereas the Shaver Mystery appeared as fictional tales communicating real truths, *Fate* printed ostensibly nonfictional stories that told of fantastic, impossible things. "True Stories of the Strange and the Unknown" was its tagline.[110] To get the magazine off the ground, Palmer bought a mailing list from the man who'd published *Oahspe*. He edited *Fate* under the pseudonym Robert N. Webster but eventually was elbowed out by his partners and drifted toward the fringe of the science fiction and paranormal communities. His publishing house put out books, most of which he wrote, and magazines on flying saucers, contactees, spiritualists. Despite these changes, his decreasing clout, Palmer did not quit on Shaver or *Oahspe* or Fort; rather, he continued developing their ideas until his dying days.

During the heyday of Shaverism, a cadre of science fiction writers elaborated on its themes in stories and essays published both in genre magazines and metaphysical ones. Perhaps most influential was Vincent Gaddis. Having grown up in a family touched by the Pentecostal Holiness movement, charismatic and evangelical, Gad-

Back cover of *Amazing Stories*, December 1947, promising readers "details of this mystery . . . on page 175"—though no such essay appears in the issue. Whatever argument the missing article may have made, the image was clearly suggestive of Fort. Fortean ideas were deeply intertwined with science fiction, particularly as the genre was developed under *Amazing*'s editor Ray Palmer.

dis worried the world had grown "dark without Christ," given itself over to materialism, needed celestial intervention.[111] But Gaddis
was also a fan of the pulps—*Weird Tales, Amazing Stories*—and
thought the pivot of the current crisis was not Christ but creatures
that lived "beyond the etheric veil," maybe Teros, maybe something
stranger still.[112] Into his interpretation of the Shaver Mystery, Gaddis
wove strands of Theosophy and Fortean speculation. Like Fort, he
compiled newspaper clippings, eight thousand by the end of World
War II, some of which supported the Shaver Mystery, others that
proved Fortean hypotheses.[113] He cited the work of H. T. Wilkins, a
Fortean who'd supposedly ransacked the Theosophical archives and
uncovered a vast number of references to inhabited underground
caverns.[114] Gaddis transformed Tibet from the home of the Theosophical Great White Brotherhood into the center of Deros activity.[115] That the Deros—or whatever one called them—were harmful
was beyond doubt. Cases of spontaneous combustion proved that
people were being attacked.[116] War was here, civilization unraveling.

In February 1947, *Amazing Stories* published his "Strange Secrets
of the Sea." Cataloging mysterious disappearances of ships, Gaddis
eschewed conclusions, alluding rather to Eric Frank Russell's Fortean speculation in *Unknown*: Venusians mined the oceans, caused
earthquakes, volcanoes, mysterious lights in the sky.[117] Through the
efforts of Gaddis, the Shaver Mystery was projected from earth's
bowels into space; neighboring planets became home to the Deros
and Teros, their alien equivalents: the earth's true owners.[118] That
was the point of Gaddis's "Visitors from the Void," published by
Amazing Stories in June 1947, which collected accounts of weird
things in the atmosphere, suggesting something—or someone—
unknown was behind mysterious crashes, phantom planes, mysterious rays that stopped plane engines, unaccountable lights,
slow-moving meteors.[119] Malevolent and benign beings from near
space—New Lands, as Fort would say—were fighting for the soul
of humanity. A conspiracy of dimensions unimaginable was afoot,
and humanity was only beginning to grasp that its fate lay in someone else's hands.

His Objectives Fade in the West

The relationship between the Fortean Society and fandom was not without tension. Russell came to dislike fans, though he'd once been one himself. Fans shared magazines and books, robbing money from publishers and authors, then sent hectoring letters telling writers how to do their job.[120] He cautioned Thayer that fans could be bane as much as boon to the society, and Thayer came to agree. "They do take a lot of time without much thought-profit," Thayer complained.[121] By the mid-1950s, Thayer worried that scientifiction obscured the full range of Fortean possibilities, kept Fort's ideas tethered to a single genre of writing. "Marvel is only one facet of this growing crystal, and by no means the most important one."[122] For their part, some fans concluded that Thayer was too dogmatic. Jesse Douglass, who had found the Fortean Society while a student at the University of Minnesota, got fed up with Thayer's paranoid ways. He'd joined the US Public Health Service and wanted to visit the society's archives when the ship to which he was assigned docked in New York. Thayer refused; only he was allowed in the archives.[123]

A substantial number of science fiction connoisseurs disdained Fort and his namesake society from the very beginning. *Amazing Stories* panned *Lo!*[124] Plenty voiced complaints about the book's serialization in *Astounding*. "It's awful."[125] "Surely only the hyper credulous could read with enjoyment such a confused muddle!"[126] "Utterly senseless."[127] Isaac Asimov said, "It irritated the devil out of me, since to me it seemed an incoherent mass of quotations from newspapers out of which ridiculous conclusions were drawn."[128] Another science fiction fan wrote a parody of Fort in *Weird Tales* (owning later that the satire might not have been sharp enough).[129] *Lo!* bombed in England, and while the weight of British science fictioneers' opinion would, in time, tilt toward approbation, there remained dissenters.[130] Views of the society could be harsher still; Fort, at least, had the virtue of suggesting plots—Asimov would write Fortean tales for *Astounding*—but the society seemed without reason. Robert Spencer Carr cajoled a friend from his *Weird Tales* days to join, but after contributing to *Doubt*, the friend decided Thayer's heretical dogmatism was boring.[131] The author Richard Matheson agreed.[132] Introduced to Fort by George Haas, an important member of San Francisco's Chapter Two, Matheson wrote a cycle of stories based on Fort, but he too was disgusted by Thayer's philippics and jeremiads. Over the years, these objections increased, a rift growing between the society and the science fiction community. "The rank and file of Forteans do not go along with Thayer and do not have his sneering attitude," Haas said. "We believe one can be a good Fortean, and probably a better Fortean, without belonging to the society."[133]

Blish, Campbell, and others came to realize, over the course of the 1940s and into the next decade, that no new science would emerge from Forteanism. Charles Fort had been too lazy to do the hard work of proving his hypotheses, Campbell lectured Russell, and had so offended scientists that none bothered to investigate his ideas.[134] Forteanism was a dead end—better sources of new science were at hand. Blish thought Thayer equally slothful, too easily tempted by crackpottery away from the labor of converting the inexplicable into science—if indeed, Thayer would have even seen profit in building something scientific. Blish had taken a job as a writer and editor of food and drug trade magazines, which necessitated reading the scientific literature, and he was impressed that the best scientists possessed a Fortean cast of mind, combining respect for facts with skepticism. "One needn't feel that the statements of authority are final, but certainly one can accept any answer that seems to be reasonable, even if authority makes it," he said to Russell. "No?"[135] In 1949, at the same moment Blish was complaining to Russell, his wife Virginia Kidd published in her 'zine *Frappe* a blast undermining the very basis of Forteanism, written by the science fiction fan Norman L. Knight. One could not correlate strange events with the periodicity of planetary orbits, Knight argued, if one also claimed astronomers and their theories were ridiculous. Besides, science itself, with its infinite space and parallel dimensions, was far more wonderful than anything Forteans proposed.[136]

Both Blish and Campbell appealed to Russell, asked him to push the Fortean Society in more fruitful directions. You're reasonable, Blish implied, and could get the society back on its track. Intransigent as ever, Russell sent Blish an essay for his postwar 'zine *Tumbrils* titled "How High Is the Sky," which argued, first, that the moon was a barren rock, unworthy of the millions of dollars it would take, on conventional science's own accounting, to send a rocket there, and, second, that a pair of heterodox Italian astronomers had proved, by abstruse reasoning, that the moon was quite close and likely inhabited. Blish ran the piece but annotated it with peevish footnotes that insisted upon the reliability of scientific reasoning and pointed out the holes in Russell's argument.[137] (The essay was reprinted in *Tomorrow* without annotations.) Blish considered expressing his reformed views in answer to Thayer's contest to write an essay for the *Humanist* on Forteanism and the Religion of Self Respect but ultimately punted. He'd given up on Fort and Forteanism, though Thayer's breezy prose in *Doubt* still tickled him. He mocked the society in his 1952 novel *Jack of Eagles*. Thayer was transformed into Cartier Taylor, "a slickly handsome man past middle age in the process of going to seed," his Fortean group cursed by "a special bias toward the idiotic." Science, not Fort, promised a way forward.[138]

Campbell thought he'd discovered just that path to the future, a replacement for Forteanism. In the late 1940s, L. Ron Hubbard, one of his favorite authors, brought what Campbell saw as a new science to him, a mélange of nineteenth-century metaphysical religions and Freudianism. Hubbard called his creation Dianetics. The mind, Hubbard reasoned, comprised two halves, analytical and reactive. Audited by someone steeped in Hubbard's system, one could be liberated from the reactive mind, the warping effect of emotional wounds, and sculpted into a rational being. Repeatedly rebuffed by professional scientists, Campbell and Hubbard opted to introduce Dianetics via science fiction, *Astounding* running an explication of Hubbard's theories in its pages.[139] "I firmly believe this technique can cure cancer," Campbell told Robert Heinlein. "This is, I am certain, the greatest story in the world—far bigger than the atomic bomb."[140] Hubbard's science became immensely popular, elbowing aside the moribund Shaver Mystery, spreading beyond the pulpy confines of science fiction to America, the world at large. *Dianetics: The Science of the Mind* sold tens of thousands of copies in a few weeks, was translated into other languages. Gardner called it "a nationwide cult of incredible proportions."[141] Forteans also took to the practice—Hubbard was repeatedly nominated to be a Fortean Fellow. "Many," Thayer said, "are enthusiastic, hailing the subject as way beyond psychiatry, etc., etc."[142] Thayer sold Hubbard's book even as he thought Hubbard "a monster," no different than Palmer and Shaver, his prose a soporific.[143]

Astounding was recentered on Dianetics, Campbell farming out ideas to favored writers who would develop the new science as well as the study of the mind's powers: "I deeply want to attract the attention of physical scientists, whom I respect for good and sufficient reason (they've produced results in making things actually work)."[144] He asked Russell to write Dianetics fables: if a technology existed offering immortality in exchange for removing all irrational prejudices, would anyone use it? What about a response to Heinlein's "Goldfish Bowl"—humans not as property but as children of intergalactic beings, scarred by their parents, overcoming their absurdities by self-auditing?[145] Russell refused: Dianetics was hogwash, nor did he care to be sane, prizing his crotchets. He offered, instead, articles on his Fortean research into levitation, teleportation, telepathy.[146] These Campbell rejected.

Within a few years, Dianetics collapsed under its own weight: Hubbard was accused of embezzlement; professional scrutiny by the American Medical Association and other mainstream groups boxed his claims as airy promises; his demonstrations went badly awry. Campbell was frustrated. Hubbard's personal life imploded publicly, with accusations of adultery, bigamy, and abuse played out prominently in the press—hardly the actions of a thoroughly rational man.[147] Hubbard regrouped, injected his Dianetics with space opera, inventing a sweeping history of the universe marred by intergalactic conflict and Theosophical striving toward perfection, what would become Scientology.[148] Campbell dropped Dianetics, though not its ultimate aim, which, as he saw it, was to mature the human race. The goal of his magazine now was to promote all manner of mental sciences under the umbrella of what Campbell called "Psi." Fort was no source of novel sciences but proof of Campbell's theories—and a warning. Campbell had "a hunch that Fort was scared blue with pink polka-dots" from a childhood experience in which he'd unleashed the powers of psi.[149] Fort pulled back from the confrontation, though, refusing to investigate his discovery, opting instead for brick-throwing. Fort and Forteans became stand-ins for Campbell's increasing frustration with Hubbard, his disinterest in attracting scientific notice, his contentment with lambasting scientists. Likely, this interpretation of Fort was why Campbell refused to publish Russell's Fortean speculations: in them, wild talents remained wild, universal laws uncontrolled. *Astounding* was out to change the world, formalize new sciences, and in that endeavor, Fort was useless. He was the past, to be forgotten or subsumed, not emulated.

Thayer's vexation with science fiction reached a crescendo in the late 1940s, and then again another a decade later. In a moment of pique, he read the chapters out of the society in the late 1940s: "The Chapters have no 'official' standing whatsoever with the Fortean Society," he wrote in *Doubt*. "They are social or study groups, subject only to such rules as they themselves agree to observe." It wasn't just the chapters that set him on edge, though: he refused the very idea of organizing. "It is expressly because nobody connected with the Society—and Your Secretary least of all—ever has attempted to 'organize' them that these admirably non-joining, flaming individuals do not mind being entered upon our rolls, and welcome the opportunity to make this one exception among societies as such."[150] Part of the problem, he told Russell, was the Shaverites. "Otherwise sensible people have made a fetish of that nonsense, and they, together with table-tippers, can dominate a gathering, both groups hooting PREJUDICE, CLOSED MINDS, at us if we scoff."[151] Far more consequential seems to have been the activities of Chapter Two. The Bay Area Forteans were busily mixing interests in science fiction, weird tales, and Forteanism in a manner that Thayer did not like.

Members of Chapter Two had tried to forge a relationship with Thayer, who they venerated as Chapter One. Kirk Drussai, the group's "bugler," sent Thayer the minutes of eleven meetings between the chapter's founding in April 1948 and the summer of 1949.[152] "Meetings that last all night and so on," Thayer said.[153] Kirk's wife Garen Drussai, a budding science fiction author, published a pacifist parable in *Doubt*, which Thayer thought pointed toward new forms of Fortean art, beyond the confines of science fiction.[154] Chapter Two investigated damned facts, dutifully collected local reports, and even gathered ice that fell on Oakland in the middle of summer.[155] These reports they passed on to Thayer. They probed a weird footnote in the history of Stanford University. Thomas Stanford, brother of the university's founder, had been a devoted spiritualist, his activity in the field increasing after the death of his wife. He worked with the Australian medium Charles Bailey, who, during séances, conjured objects from different lands, so-called apports that traveled through the astral plane to the table where Thomas and company sat.[156] The apports were willed to Stanford University, and at least one of the Chapter Two members, Robert Barbour Johnson, had seen them on display at its museum. *Fate* did a story on the apports, which prompted a letter from the school denying that such things existed. The usual damning of anomalies, Chapter Two's members thought, and some of them wrote a letter to *Fate* affirming the objects' existence. (And, indeed, they are in the university's archives.)[157]

As Johnson recalled, the tiff in the pages of *Fate* provoked Thayer's ire. He not only dismissed the chapters generally but supposedly sent a letter to the Bay Area Forteans belittling their activities, insisting that the society was focused on political rebellions, not spiritualist apports. Johnson's memory is not always to be trusted, but there is confirmatory evidence from Haas, who recalled that Thayer kicked them out of the society. For a time, there had been interest in starting their own magazine, *No Doubt!* Nothing came of that plan, but neither did the group give up on Forteanism. Some erstwhile members read *Fate* until they saw Palmer's magazine "degenerat[ing] into a regular dream book."[158]

Reportedly, Johnson had once given a "near-perfect" lecture on Forteanism at a Chapter Two meeting.[159] In light of Thayer's axing the chapter, he lectured on Forteanism once more, this time before Berkeley's Elves', Gnomes' and Little Men's Science-Fiction, Chowder and Marching Society. A writer of weird tales and science fiction, Johnson noted that Fort was central to both genres, his skepticism and imagination an inspiration. But Thayer perverted Fort's cause, turning toward pacifism, to anarchy.[160] Membership in his society was not worth the two-dollar dues, Haas said. Johnson claimed never to have purchased another copy of *Doubt*:

> So the prophet is without honor in his own society. The "gargantuan laughter" is stilled, and Fort's name declines in the West.[161]

Johnson's lecture was published in the Elves' 'zine *The Rhodomagnetic Digest*, then republished, in 1952, by the professional magazine *Worlds of If* under the title "Charles Fort: His Objectives Fade in the West." Fort remained an icon, the "spiritual father" of modern fantastic literature, in Johnson's estimation, inspiration for at least a dozen novels and hundreds of short stories.[162] But Thayer warranted no such respect. The science fiction and weird writers Johnson knew were turning away from the Fortean Society, and there was, as well, an unease with Thayer's project throughout the science fiction community (although science fictionists remained fundamental to the society).

Not unlike Sussman after the publication of "Circus Day Is Over," Johnson wanted to drive a wedge between the memory of Fort and Thayer's usurpation of him. The problem was, by the 1950s there was no true Fortean perspective, rather a proliferation of Forts. Thayer could lay claim to some of Fort's legacy, as could members of Chapter Two. But their fascination with "the odd, the unusual," their possession of a "Fortean quirk of mind that permitted us to believe that there just might be something to all those stories of flying saucers, that there might actually be sea serpents, that perhaps black magic wasn't a lot of hooey after all, that teleportation might take place," also cherry-picked from Fort's books, ignoring, among much else, the monistic philosophy that undergirded Fort's philosophy.[163] Fort could be marshaled to explain mundane mysteries in multifarious ways, but also in support of modern New Age beliefs, fundamentalisms, and conspiracies. In Fort's writings were both the promise of new sciences and the betrayal of scientific modernity. Thayer's name was most closely associated with Fort's, and his helming the Fortean Society granted him a special status, but his view was only one among many, as were Johnson's and Russell's and Blish's and Campbell's and Heinlein's and Palmer's. Nor was the multiplication of Forteanisms restricted to science fiction. Fort frolicked, as well, in the fields of avant-gardism and UFOlogy.

The Legend of Grandma Fellows

"You'd better read" Fort, said Joseph Henry Jackson in the *San Francisco Chronicle* at the beginning of September 1943. "The time to read him is when a story like that of Grandma Fellows and her rock-shower is fresh in your mind."[164] Her Fortean tale was an X-ray, revealing the skeletal structure of reality underneath the coagulated opinions of authorities.

Anthony Boucher, nearby in Berkeley, snooped around, struck up a correspondence with Miriam Allen De Ford. They were excited by Jackson's story, the attention it might bring to Fort—both had recently written in to the *Saturday Review of Literature* urging readers to pick up Fort's books.[165] Their correspondence exposed other mutual interests: true crime, mysteries, left-leaning sensibilities. De Ford was pushing out from nonfiction and mystery writing into science fiction and fantasy, leaning on Boucher's deeper knowledge of these markets.[166]

The sky was clear when the stones fell, big rocks, pebbles, color washed out, like construction gravel, pelting a little white stucco house on 89th Avenue in Oakland, day and night. The house belonged to Irene Fellows. It was late in August 1943.[167]

Jackson edited the *Chronicle*'s book review page. He'd reviewed *The Books of Charles Fort* in 1941: "Fort's favorite occupation was saying 'No!' or, at the very least, 'Oh, Yeah?'"[168]

Boucher was born William Parker White. A liberal Catholic, thirty-two at the time the stones fell, he had attended USC and the University of California, Berkeley, between bouts of asthma. Diverted from academia, he wrote for the pulps, breaking into *Weird Tales*; he spent time with the Mañana Literary Society on a recuperative trip. Anthony was his confirmation name, Boucher his grandmother's maiden name.

Among De Ford's nonfiction works was a biography of Fort for *Twentieth Century Authors*, a biographical encyclopedia. "Fort was a strange genius, often wrongheaded but with nothing of the charlatan about him."[169]

"Police have kept a vigil under Sgt. Austin Page," reported the Oakland Tribune. *"Yesterday he was standing in the yard, talking with Mrs. Fellows, when a fist-sized rock crashed against a wall, chipping off a piece of stucco from the house. Neither Page nor Mrs. Fellows heard or saw anything until the impact. And afterward nothing could be found to indicate the stone had been hurled by a human hand."*[170]

Fort expanded the mind, Jackson said—toward, not softness (he mocked Heinlein's "Lost Legacy" as mush), but independent thought.[171] "What Fort was after was to remove the halo from the head of science to make people think; to destroy, if possible, the faith of scientists in their own works, thus compelling a general return to the truly scientific principle of 'temporary acceptance.'"[172] Fort identified a problem in fringe writing, noted Jackson: as much as orthodox science, paranormal accounts had evolved into stereotyped explanations. "It was Fort, in fact, who in the pursuit of such happenings grew so sick and tired of the familiar pattern—stones, mystery, police, investigators, frustration and the final invariable report: 'It was all done by naughty small boys!' His comment, one of the most wonderful ever made about such things was, 'If only, just once, somebody would blame it on a naughty old lady!'"

Among Boucher's great enthusiasms was Arthur Conan Doyle's Sherlock Holmes. For those worried about Max Weber's iron cage of rationality—the defeat of authentic life by science, reason, bureaucracy—Holmes offered resistance: how to accept rationality and reason but imbue it with imagination. Historian Michael Saler writes, "Holmes solved cases by relating seemingly discrete facts to a more encompassing and meaningful configuration, whose integuments were derived from a combination of rigorous observation, precise logic, and lively imagination."[173] Forteans, receptive to life's clues, blessed with an imagination, could sum their observations, discover wonders ignored by science.

De Ford appreciated Fort—but not the Fortean Society. Too dogmatic, she thought, too political. Not that she differed with Thayer—she admitted she "might even agree with some of his extra scientific views"—rather she thought the society should be free of such extraneous topics.[174]

America was at war on two fronts, and Oakland a shipbuilding center. But the military could not be blamed for the stones buffeting Grandma Fellows's house. Across town, small rocks flew from a highway bridge toward passersby below—construction workers saw the boys responsible but could not catch them.[175] *The* Tribune *insisted no construction was under way near 1629 89th Avenue.*

Jackson kept his mind open to the possibility of ghosts, poltergeists, teleported stones. Under his editorship, the *Chronicle's* book page included contemplation of parapsychology. After all, the great philosopher and psychologist William James had once expressed belief.[176] When one of his book reviewers was sucked into the military maw, he tapped Boucher to substitute, Boucher whose asthma made him undraftable.

Holmes aficionados enjoyed a game. As the Baker Street Irregulars, they pretended that Holmes was a real person, celebrated his birthday, sought to explain discontinuities in Doyle's books. (Boucher and Jackson were Irregulars.) Of course, they knew Sherlock Holmes was fictional, but they chose, Saler argues, to deploy another response to the grim prospects of modernity: the "ironic imagination."[177] By suspending disbelief, playing with the idea that a fictional character actually exists, they enchanted the world—made it seem magical and wonderful—while not giving up on rationality. The same could be true with Fortean anomalies. Maybe they were real, maybe not: either way, they were wonderful.

After apparently making it out to 89th Avenue once, Boucher never followed up: "A combination of factors (travel, then a long illness, then pure damned inertia)," he conceded to De Ford. "Sorry."[178] Nor did De Ford cross the Oakland–Bay Bridge. Jackson: "I've always been sorry that the shortage of gasoline ration tickets kept me from going out to have a look for myself, and so are others I know hereabouts. We mightn't have caught either a naughty boy or a mischievous old lady, but the firsthand experience would have been worth having."[179]

Fort solved some literary problems, De Ford suggested. This was the golden age of the mystery story—bloodless (mostly), intellectual (unlike noir, which succeeded but did not replace it after the war)—including locked-room mysteries that asked how a seemingly impossible murder could be explained. "Eliminate the impossible. Then if nothing remains, some part of 'impossible' must be possible," Boucher had a character paraphrase Holmes in his book *Rocket to the Morgue*, a roman à clef of the Mañana Literary Society. That insight leads to the exposure of the murderer, whose range of motion is extended by his double-jointedness.

A second gold rush: thousands migrated to Oakland, took jobs in the shipyards. One result: a housing crunch.[180]

Jackson caught wind of Boucher complaining about the *Chronicle*'s meager pay rate and having his opinions clipped by editors. He fired him.[181]

Examining his shelf of "crackpottery," Boucher felt irked at Fort's "cryptic documentation." He thought the Fortean writer R. DeWitt Miller a better documenter.[182] Boucher evinced no interest in the Fortean Society. Like De Ford, he thought Thayer too dogmatic, unwilling to play with ideas.[183] Destruction for the sake of destruction was no virtue, and no anarchist was he.

Fort insinuated himself into De Ford's stories. "Henry Martindale, Great Dane," starts with a woman being hit on the nose by a button—mysterious Fortean rainfall—then considers her husband, transformed into a talking dog, reminiscent of a famous passage from *Wild Talents* in which Fort contemplated a talking dog that disappears into a green vapor.[184]

Irene Fellows, fifty-six, seemed to be estranged from her second husband. Her two daughters from a first marriage seemed to have similar trouble. Vivian married Earl Waid, but their marriage ended in 1945; her sister Lucille divorced her first husband and married Earl's brother. War work and marital changes had people moving in and out of Grandma Fellows's house.

Jackson continued to think about Fort into the 1950s as he reviewed Hereward Carrington, Nandor Fodor, books on poltergeists.[185] But he'd lost track of Forteana, 1943 his only reference to contemporary events in the Bay Area, and that fading.

De Ford, edited by Boucher, produced another biography of Fort, this one for Boucher's *Magazine of Fantasy and Science Fiction*. She recycled the title of her husband's review of *Lo!* but mostly moved beyond fascination with Fort's facts.

Two girls, granddaughters, seem often to have been around Fellows's home: Donna, five, daughter of Lucille and Earl, and Audrey, nine, daughter of Vivian and her first husband. He was a gunner in the Pacific, hospitalized on 10 August 1943.[186] That was the exact moment the stones started to fall. Was Audrey, under intense stress, lashing out? Throwing rocks in anger, in grief? Should one cherchez la femme? Impossible to know. Two years later, she was killed when a navy man hit her with his car as she rode her bike.[187]

Boucher was at loose ends after he was fired. He concentrated on fiction, peppering Campbell's *Astounding* with stories in the vein of *Weird Tales*; he later cofounded the *Magazine of Fantasy and Science Fiction*. Fort moved through his own stories like a ghost: "The Tenderizers" posits some mysterious "they" who control weird writers, making them tell of horrors so "they" can savor the fine bouquet of human fear. After the writers are used up, they're harvested—which accounts for Ambrose Bierce's disappearance.[188]

Irene Fellows's legend persisted in the Fortean community, increasingly untethered from facts. Manly Hall wrote it up in one of his magazines.[189] Years later, Fortean investigator Brad Steiger said a psychic had determined that seagulls were at fault; he also reported an inconclusive investigation by people attached to the University of California.[190] There was no evidence for either statement. The anomaly was slipped into compilations, all the messiness of life stripped away, a free-floating Fortean factoid.

Who Killed Science Fiction?

Late in the 1950s, science fiction fan Earl Kemp, having spent years hearing about the collapse of the science fiction magazine market and confused by a lack of critical insight into the situation, canvassed fans, genre authors, editors, and publishers about the state of the field: Was magazine fiction dead? What caused the collapse? Could it be corrected? In 1960, he sent out a circular collating the responses titled "Who Killed Science Fiction?" Boucher thought the genre needed a new leader: first had come Gernsback, then Campbell, then his own magazine. Someone needed to invent the next big thing. Blish blamed Boucher and his ilk, paying writers too little too slowly, favoring female authors. Russell was phlegmatic: "Tough problems exist—but tough problems always have existed." Campbell insisted all was fine: "Dead?! We're going better than ever before!" Kemp's, though, was the wrong question. The magazine market might never recover, but science fiction was not dead. It had transformed, cheese into beetles, into mice.[191]

Also changing was science fiction's relationship to Forteanism, once strong, now attenuated. The genre's newest iteration made less room for Fort, his ideas and his anomalies, though it did not shut them out entirely. Then, too, developments in science and engineering had made Forteanism seem out of synch with the times and so unworthy of consideration by forward-looking science fictioneers. The major cause of the estrangement, however, was Thayer, who insistently severed ties in a mood at once grim and gleeful. Again, he grew tired of what he had once supported, pushing Forteans to move past science fiction toward some other, richer artistic form. He became paranoid about science fiction and wondered about the motives of its writers. Always given to conspiratorial thinking, Thayer imagined that science fiction was part of a nefarious plot to enthrall humanity, to force minds to kneel, as he said once about education and the press, before the great gods of science, of technology, of the military, of the state. Science fiction enchanted but dangerously so, in the manner of dictators supported by propaganda.[192] Convinced that he was clear-eyed and dedicated to the pursuit of truth, whatever the cost, he now refused any connection with the genre that had long sustained the society.

Science fiction magazines proliferated in the 1930s and 1940s, riding out a series of booms and busts, but the dip in the 1950s seemed especially deep. Initially, paperback books extended the magazines, cheaply repackaging the stories, or revised versions of them. They sold in the same drugstores, on the same newsstands.[193] "Now rather pleased at having sold all my published fiction of any length for the second time," Arthur C. Clarke bragged to Russell in 1948.[194] Kemp wondered if paperbacks could be science fiction's salvation, but there's some evidence they competed with magazines, tempted away their readers, and thus contributed to their disappearance.[195] Magazines also suffered from a drastic rearrangement of the distribution system. The American News Company, which monopolized sales of magazines and paperbacks to newsstands, drugstores, and train stations, went out of business in 1957. Magazines with large audiences—*Astounding, Amazing, Fantasy and Science Fiction*— found new distributors, but these smaller companies, competing against one another, could not afford to carry magazines with only niche audiences the way ANC could, and so refused to distribute many other science fiction magazines.[196]

The science fiction impetus moved beyond paperbacks, too, into glossy magazines, onto the screen, both silver and small. Obscure forces altered the attitudes of people reading genre magazines: fiction seemed indecorous, unfit for a world that could be blown to smithereens. The times demanded truth. And so came into being a host of so-called men's adventure magazines filled with revanchist fantasies and conspiracies about the modern world—still fiction but dressed in the garb of fact. Most of these simply absorbed elements of the old pulps and genre magazines, the same artists, the same writers, editors, publishers, distributors, repackaged as slicks for barbershops, dens.[197] Garen Drussai, of Chapter Two, wrote for the new magazines, as did Vincent Gaddis and Fredric Brown. In southern California, science fiction writers centered on Richard

Matheson and Ray Bradbury moved to screenwriting. Cineplexes were stuffed with science fiction tales—*Thing from Another World*, *The Day the Earth Stood Still*, hundreds of others. Science fiction appeared on television, too. Bradbury and Matheson wrote scripts; their magazine stories—and stories of their progenitors, such as Ambrose Bierce—were adapted for the tube. *The Twilight Zone* debuted in 1959.[198]

Science fiction traveled, as well, internationally and interdimensionally. France was home to Jules Verne and Maurice Renard, grantor of the name *roman scientifique*, but for most of the twentieth century's first half, French literature had turned away from the fantastic. Then, just as the American magazine market cratered, an influx of translations brought science fiction back to France. Translated piecemeal over several years, American science fiction brought a complete system of tropes—a collision of worldviews—different from the native *roman scientifique* (which was undergoing a renaissance of public interest as well). Novelists writing for the Anticipation imprint borrowed from American science fiction as they explored ways to reconstitute French identity after the decimation of war, science fiction stories variously encouraging a return to older modes of being, the embrace of humanistic values and progress, and the questioning of what it means to be human.[199]

Strands of science fiction uncoupled from science, became fully occult, esoteric, wrapped into New Age spiritualities.[200] Science fiction had benefited from colonization and imperial conquest, the discovery of Eastern bodies of knowledge.[201] Synthesized into Theosophy, these other ways of knowing underwrote forms of science fiction focused on magical and esoteric means of controlling earth's evolution. But imperialism also threatened the viability of the East as a site of wonder, as distant places adopted Western technologies and mores. The myths of Tibet, of India, of Shangri-La—like the myth of outer space—"could no longer be trusted . . . to the geographical place; instead it had to be transferred on to what was truly timeless and formless," as the scholar of Tibet Peter Bishop wrote.[202] And so the myths were projected, deeper into space, the past, the future, unreachable dimensions. Science fictional notions with which they had been bound went with them. Thus, space opera underwrote Scientology, which told of an intergalactic war that started long, long ago and far, far away, its ripples structuring earthly lives.

Several times, Thayer had been asked for the rights to make *The Books of Charles Fort* into movies; several times, he refused.[203] Elements of Forteanism worked their way into science fiction movies and television programs, but the full Fortean vision was not interpreted for the screen, not in the 1950s or 1960s. Similarly, Forteanism was carved up and parceled out to the men's adventure magazines. Fort's name appeared occasionally, and the conspiracies that were the magazines' bread-and-butter often rooted themselves in Fortean notions, particularly the paranoia that humanity was owned, or that something nefarious, yet never quite specified, accounted for inexplicable events.[204] But Fort remained a sidelight in these publications, crowded as they were with stories of hunting and war and advertisements for trusses. Nor did Fort's absorption into esoteric theories serve to continue the Fortean tradition; his notions were prominent, but his role was downplayed—a ghostly presence. It was in France, probably, that Fort was featured most prominently during these years. Still, the dissemination of science fiction along its many axes weakened Forteanism, even as it made the philosophy more widespread. Fortean ideas, in the late 1940s, early 1950s, achieved a density in science fiction magazines that would not be repeated in any other medium.

As science fiction weakened its link to Forteanism, Thayer hacked away at the connecting tendrils as well, trying to free Forteanism from the genre. He was tired of the society, trying to get out or, failing that, to rebuild it on a new foundation. Science fiction was too conventional, hackneyed, and boring. He wanted a new Fortean art. Thayer had come of age awed by the provocations of decadent and modernist writers. Hecht had taught him to write.[205] Ezra Pound's poetry had shocked the world in just a few lines.[206] Thayer collected first issues of small-circulation modernist magazines, nostalgic for their experimentation.[207] He'd tried to be equally disruptive with his fiction—look what he did with *Mona Lisa*, writing the creation of the world's most famous portrait as a burlesque. He wanted Fortean painting. Fortean dance. Fortean music. He wanted art expressing the "Fortean viewpoint . . . without flying to Mars or turning Time forwards, backwards or cross-wise." But it was impossible to coax the membership away from science fiction, weird tales. He complained in *Doubt* about the dearth of experimental Fortean art. "We should like to print much more material, but you don't send it in."[208]

As in his relationship with Russell, Sputnik brought the matter to a boil. Science fiction fans indisposed toward Fort or Forteanism or both could point to the satellite as proof that Fort's cosmology was bunk: space was as astronomers had described; engineers were rationalizing, if not quite disenchanting, it.[209] Arthur C. Clarke said:

> If Fort had lived to see men walk on the moon (well, he would have been only 95 . . .) he would have had to eat a good many of his sarcastic words about astronomers. Scepticism is one thing; stupidity is another.[210]

Thayer absorbed the attack, doubled the energy, and returned the opprobrium: not only had Sputnik failed to disprove Forteanism, but the claptrap revealed science fiction writers to be part of a vast conspiracy to hoodwink humanity, to make it more pliable for authoritarians. There was nothing to Sputnik beyond pronunciamentos of the powerful, no proof. The world was willing to believe only because people had been conditioned to do so by science fiction writers. He suspected Robert Spencer Carr was part of the conspiracy, his story in the *Saturday Evening Post* a piece of propaganda, coordinated with Washington, DC, to prepare this moment—this reveal. Sputnik was a long con; science fiction writers were setup men.[211] Anthony Boucher had given up the game, Thayer said, when he wrote in the *New York Times*:

> Science fiction is, first of all, imaginative entertainment, but if it has a more serious function, it is less that of precisely pinpointed

prophecy than that of creating in its readers a climate of acceptance of new wonders and a willingness to think at least one step
ahead. Nineteen fifty-seven is, in all probability, a more significant
date in the history of civilization than 1492. We have stepped into
a new age—and it is the age in which the science fiction reader has
been living all along.[212]

Thayer, like Fort, had long believed the world was hell. In his lifetime,
he'd twice seen it nearly obliterated by war, seen the range of human
expression contracted. Now humanity bent its knee to a fresh devil
called into being by traitorous Forteans; not Castillo's Cerberus, but
something worse. "The human race is now in the maw of the new
Moloch, with credulity at an all-time high."[213] Moloch, the child-
eater—Moloch the very figure of war, devouring the young for the
comfort of the old, the rich. "Asinine gullibility—intentionally—
has reached a new apogee, and we are being asked to believe the most
preposterous lie since the invention of virgin motherhood," Thayer
said. A setback for Forteanism, no doubt, but also a call to arms—
Forteanism against science fiction and the fantasies it had created,
the gauzy dreams it draped between the eyes of the world and reality.
"The world's need for our dissent has never been so great . . . lest our
posterity be born into bondage forever."[214]

Ivan Sanderson and the Investigation of the Unexplained

"I'm now half fish—can swim 200 feet without coming up for air," Arthur
C. Clarke told Eric Frank Russell in July 1956. He'd just spent five months
in the United States, most of that in Florida. Clarke had a bit of Forteana
to report, as well, beyond his transformation into a merman. "Seems that
occasionally the tracks of an enormous bird are observed on one of the
beaches there and the naturalists come running down from the north
with field-glasses and cameras. The whole baffling business has doubt-
less been written up somewhere in the Fortean magazine."[215]

He was right about that. Eight years earlier, in November 1948, Ivan

Sanderson had come down to investigate. Born in Scotland, Sanderson had studied geology and biology at Cambridge, traveled around the world once on his own, then set sail on collecting expeditions for British natural history institutions. These he turned into books that made him famous. During World War II, he served as a press analyst at the British Ministry of Information's New York office, soon moving to America, becoming a citizen, and settling down as a writer and media personality.[216] Contemporary newspaper reports have him saying contradictory things about those Florida tracks: they were a hoax; maybe they were made by some animal; probably they were authentic spoor.[217] Thayer discussed the tracks in *Doubt*, and Sanderson's investigation: "He is variously quoted by interviewers. The tracks are a hoax—the tracks could not be a hoax. In fact he out-Forts Fort in suspending his judgment."[218]

At the time, Sanderson was just dipping his toe into Fortean waters. Eventually, though, he innovated, became a competitor to Thayer's society as much as a product of it. His writing became the basis for a new science, science fiction in the flesh, not just imagining earth's hidden parts but discovering them. He aimed to prove correct Blish's and Campbell's intuition that Fort might provide the root of new sciences. Sanderson probably encountered Fort's writings at Cambridge in the early 1930s (there was a circle of fans at the school) and through the Gollancz edition of *Lo!*, retaining an interest in the man and his ideas for the rest of his life—though, with a few notable exceptions, his enthusiasm was not apparent in his early natural history books. Once in America, Sanderson joined the Fortean Society. He openly contemplated the existence of sea serpents in the *Saturday Evening Post*.[219] Then, for the same magazine, he wrote about the possibility that dinosaurs still lived, galumphing about central Africa.[220] One society member recommended that a department of the Fortean University be named after Sanderson.[221] (Thayer, however, thought the hunt for monsters belonged to Iktomi.)[222] His Forteanism gathered steam through the 1950s, as he wrote for science fiction magazines, *Fate*, the men's adventure magazines—hundreds of articles and a handful of books. He became interested in flying saucers. Sanderson eventually mixed themes developed by Russell, Markham, and Palmer into the theory that UFOs originated in undersea bases and were perhaps responsible for the mysterious disappearances of ships and planes.[223] In 1968, he revisited his Florida investigation for *Fate*. Now he was decisive: the tracks were probably left by a fifteen-foot migratory penguin. He'd seen a large creature frolicking in the surf, he now said.[224]

By this point, Sanderson was the world's most prominent Fortean and no longer content merely to chronicle the anomalous. He sought to found Fortean sciences. Mysterious animals were common in Fortean discourse. Fort mentioned them. Russell too. Thayer wrote about them in *Doubt*. The Loch Ness monster was a standard topic among Forteans. Sanderson raised a scientific umbrella over all of them, called his discipline cryptozoology, the study of hidden animals. More than a little romance suffused cryptozoology, but the study of hidden animals also responded to modern science.[225] Explorers of the past four centuries had identified most of the world's megafauna; over the previous seventy-five years, intensive surveys filled many gaps in zoological knowledge.[226] Left to be explored, then, were the exceptional ones. As the next stage in natural history, cryptozoology embedded itself in the networks of midcentury science. A congeries of scientists—anthropologists, limnologists, engineers—joined Sanderson, employing the technology of the Cold War (maps, aerial photography, sonar) and its system of patronage (governments fretting over international competitors) to hunt wildmen, dinosaurs, the Loch Ness monster.

Cryptozoology remained the core of Sanderson's Forteanism, but he meant to circumscribe the entire field, studying wildmen, talking ants, strange writings, historical mysteries, fafrotskies, all sorts of "things," as he called them, the gravitas of his scientific background making plausible Fortean anomalies—not a single science, then, but all of science fiction transformed into fact, just as the men's adventure magazines inverted the fanciful into the truthful. He created his own organization, the Society for the Investigation of the Unexplained, and put out the magazine *Pursuit* ("Science is the pursuit of the unknown").[227] Sanderson rewrote history, making himself the heir of Forteanism. He appropriated Thayer's Poundian term for newspapers, "wypers," attributed it to Fort, and said that, like Fort, he was writing against the wiping away of the inexplicable by the mainstream.[228] He invented a story about seeing Fort lecture to a huge audience in New York, scandalizing the tony listeners with his outré theories—a role he now saw himself playing.[229] There was some truth to this contention, however sloppy Sanderson could be with facts. His ideas were pure Forteanism, unleavened with New Agery, fundamentalism, or conspiracy. From his perspective, the universe was structured by a myriad of strange, otherwise unknown forces. But these were not spiritual, and he firmly rejected Martin Gardner's suggestion that he dabbled in New Agery.[230] The forces were physical or biological, amenable to scientific investigation and, as science fictioneers hoped, control. Sanderson's goal was to loose the imagination so that the true structure of the universe could be understood and humans could seize control of its complex levers and pulleys.

Sanderson's influence on Forteanism lasted throughout the twentieth century and into the next. But it was built on a shaky foundation. Those mysterious tracks in Florida were fakes. Clarke told Russell, "I was taken into a back room and shown the footprints, neatly built round a pair of boots. Whenever the character who owns them feels like a bit of fun, he puts them on and walks backwards down into the sea."[231] Thirty years after Sanderson's *Fate* publication, the hoaxers admitted their prank publicly.[232] Many of Sanderson's other hypotheses were based on strained interpretations, faulty evidence, strategic credulity, purposeful lies, and exaggerations—he was crafting a writing career, not discovering truths. Sanderson died in 1973, without a single mysterious animal having been discovered by cryptozoologists, nor any of his other theories gaining scientific traction. Other Forteans and like-minded people would take up his ideas, promulgate them; a professional cryptozoological society came into being in the 1980s. But it could not rely on government largesse or the patronage of eccentric individuals and thus was chronically underfunded.[233] Still, Sanderson's ideas circulated through a reconstituted Fortean community, among esotericists and other challengers of the scientific status quo.

Without institutional and material support, cryptozoology, along with the rest of Sanderson's obsessions, were stripped to their core: these were species of science fiction. They mixed imagination and science in search of the secret motive forces of the universe. The difference was, they presented their claims as facts. Predecessors—Poe, Verne, Doyle, Russell—did too, but always with a wink.[234] There was no irony in Sanderson's cryptozoology, nor in his UFOlogical hypotheses. However much a creation of fakery or imagination, they were meant to be taken seriously. Sanderson and his followers were not alone in making such claims. Small but vocal factions, envious of rockets, atomic weapons, and their transformation from fantasy to reality, defended space opera, Lovecraftian theology, tales of Earth's secret owners as legitimate. Novel worlds and new histories of the universe were conjured, not as supplements to the mundane but as replacements. Eldritch spells from *Weird Tales* were said to work.[235] Though even some inventors of this form of science fiction stated that they were responding to the disenchantment of the world, from another perspective these developments furthered modernity. Sanderson did not reject science; he wanted to perfect it: not a retreat, then, rather a culmination. So too with other science fictional improvisations. They sought not refuge in the past, but to absorb modernity into a grander philosophy.

4

The Mermaids Have Come to the Desert

The mermaids have come to the desert
they are setting up a boudoir next to the camel
who lies at their feet of roses

A wall of alabaster is drawn over our heads
by four rainbow men
whose naked figures give off a light
that slowly wriggles upon the sands

I am touched by the marvelous

PHILIP LAMANTIA, "The Touch of the Marvelous," 1944

Antony Borrow was surprised by science fiction. A biochemist by trade, he had an affection for modern literature (though he confessed he struggled with the more recondite poets), especially that which touched on the occult, the mythical.[1] He wrote verse plays and published a small literary magazine, *The Glass*, which he printed on a century-old hand press with the help of his girlfriend, Eva Steinicke.[2] Borrow's interest in the fantastic had not been a gateway to science fiction. He looked down on the genre until convinced to investigate it by the literary critic and writer John Atkins. Much of the genre was "trash," he concluded, but in the best examples there was an admirable "vigour." Fandom, too, astounded him, so like the ferment of early modernism, with its cliques and small magazines—unfortunately now, despite his own efforts, "emasculated."[3] Likely, it was also Atkins who introduced Borrow to Fort; the two men joined the Fortean Society in the late 1940s.

Borrow wasn't a zealous Fortean. He didn't read Fort's books before joining the society, didn't even know how to spell his name (he added an "e" to the end), and had the sweetly naive view that one had to demonstrate bona fides to be admitted.[4] He did contribute some material and maintained his membership through the 1950s, finding something of interest in the magazine, if only Thayer's occasional splenetic assaults on Borrow's bugaboo, psychiatry.[5] Like Borrow, Atkins was often tardy in paying dues, but not so shy about contributing material. Thayer thought him "a dear."[6] Atkins repeatedly asked Thayer and Russell to coauthor an introduction to Fort for his own magazine, *Albion*, but, while Russell finished his part, Thayer never did his.[7] Apparently fed up, Atkins did it himself, publishing his essay in Borrow's *The Glass*. Atkins saw Fort as breaking up science's sclerotic hubris but not challenging the institution itself: Fort only wanted to hold alternative explanations in suspension until one was definitively disproved, at which time he would accede to the conclusion of scientists. What intrigued Atkins about Fort was that moment of suspension, his ability to hold in mind so many possible explanations, some breathtakingly wondrous. Fort, Atkins thought, was a pluralist, possessor of an imagination wide enough to contain multitudes. He placed Fort in a line of writers that culminated in modernism: from Rabelais through Whitman, with a tangent to John Cowper Powys, then to Henry Miller, who "has given a Fortean shove to literary expression."[8]

Atkins's understanding of Fort was idiosyncratic, but his insertion of Fort into the modernist tradition was astute. Fort was in the mainstream of the modernist avant-garde. His prose was consciously modernist, after the same effects as Ezra Pound, James Joyce.[9] He influenced Miller and the novelists Malcolm Lowry and William Gaddis; surrealists and artists associated with the San Francisco Renaissance; the painter Martha Visser't Hooft, the photographer Clarence Laughlin; and French proponents of a movement called fantastic realism. The Fortean Society was also part of the modernist milieux. Thayer supported modern art and involved the society in debates over the canonization of modernism in the late 1940s.

The line dividing modernism from science fiction is often seen as rigid—high art versus low, aesthetic refinement versus debasement, exploration of interiority versus extraterrestrial geographies. But they were close in many ways. Drugstores sold cheap paperback editions of modernist classics and science fiction fixups side by side, with similarly lurid covers.[10] And each would sometimes draw ideas, tropes, and metaphors from the other. Fort, too, united science fiction and modernism; his writing was used as a tool for similar jobs and in service to similar metaphysics. Henry Miller, Lowry, Gaddis, and the surrealists took from Fort methods of expanding the imagination, just as some science fictioneers had. Those involved in the San Francisco Renaissance used Fort's writings to outline society's structures, not, as with science fiction, in hopes of imagining ways to control history's motor, but rather to resist modern institutions, to create autonomous spaces. Advocates of fantastic realism mixed Fort, science fiction, and surrealism to expose the operation of power in society. Cross-cutting these uses of Fort and Forteanism were commitments to differing metaphysical propositions. For Gaddis, Fort supported a species of fundamentalism, the universe controlled by a singular—if ineffable—divine force. Miller, Lowry, and the San Francisco Renaissance artists incorporated Fort into a New Age system in which multifarious, mystical forces organized reality. Surrealists came closest to a pure Forteanism, their focus also on multifarious forces but on mundane ones, materialistic rather than spiritual. The fantastic realists found in Fort evidence for their conspiratorial views, imagining the world shaped by elite cabals of geniuses and mutants.

Revolution of the Word

After he finished his only published novel, *The Outcast Manufacturers*, Fort drifted from realism to other literary commitments, which he would hybridize with scientific romance in his later, more famous

works. Information on what Fort was reading at this time and its
effect on him is thin, but his habits can, at least partially, be recon-
structed from stray comments. This evidence suggests that he was
exploring decadent and modernist writers—as were Benjamin De-
Casseres and Ben Hecht and Tiffany Thayer, H. G. Wells and Ezra
Pound. No wonder, then, that DeCasseres and Hecht were tickled
by his books, that Thayer would venerate him alongside Friedrich
Nietzsche and H. L. Mencken. No wonder, either, that later inher-
itors of the decadent tradition and, after Fort's death, modernists
would find value in what he wrote. Odd though his books were, they
were not without precedent or parallels.

The decadent movement developed in Europe at the end of the
nineteenth century, arising in France and spreading, a response to
the prevailing fear that Western civilization was spent. Science pro-
gressed and new technologies came into being, but decay was obvious, a shadow that would consume the world.[11] The second law
of thermodynamics—whatever its technical definition—seemed to
promise that the universe was inexorably headed toward death.[12] Scientists consistently made new discoveries, but, as Max Weber said,
these came at the expense of prior science, new knowledge erasing
the labor of previous generations.[13] As Fort would conclude, the
world was ruled by the strange orthogenetic god Decomposition.
Rather than despair, however, decadents took delight in decay,
found beauty amid the rubble, made gay the grotesque. DeCasseres
wrote:

> We who strip the petals from the Rose of the World and build
> mosaics and arabesques out of the débris of the ancient theorems
> are forever procreating imps and changelings. Thought breeds
> thought; mood breeds mood; feeling breeds feeling. And so long
> as this continues to be a psychological law the decadent will have
> the last word.[14]

The decadent movement stood against the natural order, skeptical
of science, religion, authority of any kind.[15] There was no ultimate

truth, only an ironic relation to the world. One played with various systems of truth—religion, science, socialism, capitalism—the wise always understanding that, at base, these were hypocritical stances, lying to the world if not also oneself.[16] DeCasseres insisted that the artist's duty was not to represent the world, as realists like Dreiser urged, but to create an aesthetic experience: art for art's sake. Decadent works privileged the effects of language—artificial, fragmented, affected. Even as decadence expired in Europe, it found a home in the novels of Ben Hecht, the short stories of the *Weird Tales* circle.[17] Fort stood foursquare in the tradition, while also pushing toward another frontier.[18]

That new frontier was modernism. By certain standards, decadence is a variety of modernism, but there is also a sense in which a self-conscious modernism, then taking shape in Paris, London, New York City, and Chicago, emerged out of decadence. Fort was not the only writer to transition from decadence to modernism. While he struggled to find his literary voice, the poet Ezra Pound was inventing a new style of prosody. Pound had come up through the lyrical tradition of Victorian and decadent poetry—his native language, in some estimations—but as the world fell into a first global war, he worked toward a "breakthrough in modern poetry."[19] The decadents caught something real: European culture verged on collapse. But their language, the language of most English poetry, was abstruse, cut off from the soil of life. Traditional meters and forms used by the lyrical poets needed to be broken—as some decadents had already been doing. Poets needed to discover a new relationship between language and reality—a charge taken up by modernists generally, those influenced by Pound and those not. Realism had an uncomplicated relationship with language, using words to represent reality: the realists' quarrel was with literature that refused to acknowledge life's harshness, the existence of the poor, but they trusted that language was up to the task of doing so.

Modernists, less certain of the relationship between the word and the world, worried over the problem of human consciousness, how it translated experience into meaning.[20] Modern art focused on human interiority, the actions of the mind. Pound, living in London, the center of an international network of artists, found support for his project in two little magazines, of the kind Antony Borrow would later memorialize: Margaret Anderson's *Little Review* and Harriet Monroe's *Poetry*.[21] Both were based in Chicago, then undergoing a literary renaissance. Hecht was there, a contributor to both publications, and Burton Rascoe and Tiffany Thayer. Pound published some of his most influential writing in *Poetry*, having discovered the language that would mark his work for the rest of his long life.[22] He aimed for precise and concise descriptions of objects, real and imagined, that did not hew to conventional poetic forms. The rhythm itself was to provoke in the reader particular emotions. Poetry, in its focus on the working of the mind, was a kind of science, aiming for an accurate portrayal of how thought arose and behaved. But it was not a dreary science, rather one that evoked intense feelings.

As he revised his alphabetic series into *The Book of the Damned*, Fort's writing evolved in response to these developments. "Some of the best fiction has not been fiction so much as interpretation of actual events," he said, reiterating his commitment to blending fact and fiction while also emphasizing the importance of attending to the mind's working.[23] In his case, he focused on how the brain constructed belief from the stuff of life. Fort was conscientious about the style he adopted, his books exhibiting a "theory of impressionistic writing" that he'd invented, which, he thought, heralded something novel: "Maybe I am a pioneer in a new writing," he told Miriam Allen De Ford and Maynard Shipley.[24] This "jazz style," as one reviewer described it, was affected but, unlike the prose of the decadents, grounded, rooted by concerns with everyday things: reports in the daily papers about animals and the weather and disappearances.[25] A decade before the publication of the modernist manifesto "Revolution of the Word," Fort exemplified its rules: "The imagination in search of a fabulous world is autonomous and unconfined"; "The writer expresses. He does not communicate"; "The plain reader be damned."[26] DeCasseres recognized in Fort similarities to James

Joyce, author of the banned *Ulysses* and *Finnegans Wake*, who was associated with "Revolution of the Word." Joyce's books, obscure and dense, exemplified the Fortean obsessions with doubt and what Fort called "equalization": everything blending into everything else, the universe perfectly balanced.[27]

The historian Peter Gay argues that the core of modernism was heresy: the desire to challenge authority along many axes—social, political, moral, aesthetic, and, more controversially, scientific.[28] Modernism opposed convention. The heretical impulse explains Boni & Liveright's decision to publish Pound, T. S. Eliot, Faulkner—and Fort's *Book of the Damned* and *New Lands*. Iconoclasm also explains why Claude Kendall, excited about tweaking bluenoses, added to his list of decadent novels Thayer's *Thirteen Men* and Fort's *Lo!* and *Wild Talents*. And adherence to the apostatic explains how Sussman could go from Liveright's house to working with Fort to writing a celebrated ad for Joyce's *Ulysses* when it was legalized in 1934.[29] But modernism is large and complex; it cannot be reduced to a single, monolithic expression (even "Revolution of the Word" was only one articulation of modernist intentions) or a single writer. And the links between Forteanism and modernism were more specific than a shared desire to *épater le bourgeois*. Forteanism vined about a branch of modernism that celebrated radical individualism; that made room for the occult, the Theosophical; that sought to remake society via economic reform; and that preferred a reactionary politics.

The modernist turn toward investigating individual subjectivity was a turn away from the liberal social order fragmented by the Great War. "Liberalism decomposed into egoism," notes scholar Michael Levenson, at least for some modernists.[30] DeCasseres, Hecht, Pound, Thayer—all, in their own ways, could be numbered in this line of thought. It's no surprise that some of Pound's early poems were published in a magazine called the *Egoist*, which took its name from the writings of the German Max Stirner, who, in the mid-nineteenth century, outlined a philosophy of egoism.[31] Another German, Friedrich Nietzsche, was often interpreted as supporting egoism as well.[32] He celebrated the Superman who, unlike the great masses, saw the truth of the world and acted in accord with his own will, unconstrained by the social order. While Fort read Nietzsche, it is not known if he wrote consciously as an egoist, but others saw him as a radical individualist. DeCasseres called Fort's "a strange mind, an unattached mind, a trans-sensory mind."[33] The Fortean Society attracted several proponents of Stirner's philosophy, and discussion of him (and of Nietzsche) was not uncommon in the pages of *Doubt*. Thayer even sold Stirner's book.[34] Fortean George Christian Bump said, "To me Forteanism represents the revolt of the realists (the 'individualists') against the church of modern science and the 'men in white.'"[35]

In critiquing the liberal social order, certain strands of modernism integrated the sacred or the occult into the arts.[36] (The modern world was never disenchanted.) As these artists saw the situation, science, bureaucracy, realism, even religion, which could be conventional and rule-bound, were too clinical, too empirical to encompass the full wonder of the universe. Mainstream institutions papered over the untamed, bound with red tape the wild vigor of spiritual forces. The British magazine *New Age*, for example, mixed modernist prose with Theosophical and occult speculation (as well as discussions of politics and economics). *New Age* was one of Pound's favorite outlets, and he was attracted to some discussion of the spiritual realm. He thought the "modern world full of enchantments."[37] Physicists had discovered an invisible realm beyond the senses, a space filled with atoms and protons, neutrons and electrons, which shaped reality. Perhaps this was the spiritual plane. But Pound still hewed, ultimately, to respect for science, his poetry scientific in its activity. He wanted to harvest the wonder of science, incorporate it into his art.[38] Others of his acquaintance seemed to want something more, surrender to Theosophical forces, magical laws, strange gods. A similar modernist configuration was assembled by N. V. Dagg, first in his magazine the *Modern Mystic*, then in *Tomorrow*, which hosted Eric Frank Russell's Fortean essays. Antony Borrow's *The Glass*, while granting no space to economic or political writing, also combined the mystical and the modern. For that matter, *Doubt* encompassed the same constellation of elements, though organized differently. While in *New Age* and *Tomorrow* modernism absorbed the enchanted, including Forteanism, in *Doubt* Forteanism was pre-eminent, Fort's skepticism broad enough to include heretical ideas, heretical art, and the possibility of heretical forces, though none of these were endorsed. At least, most of them were not—they were played with, not taken too seriously.

New Age also introduced the world to C. H. Douglas's social credit theory of economics, and modernists and Forteans to a redoubtably heretical idea. Social credit finessed the conflict between socialism and capitalism. Modern technology produced enough to provide everyone with the basics of life, Douglas argued, but regular citizens lacked the resources to enjoy these products because the financial industry too heavily dunned them; this inequality led to all the ills of the modern world, including war. Nationalize banks, Douglas said, and subsidize consumers so that they could enjoy the fruits of modern industrial capitalism: peace would follow. Pound took up social credit to such an extent that in the 1920s and 1930s economics and politics replaced poetry as his major concern; or, perhaps better put, the two merged such that his promotion of economics was a kind of poetry.[39] The circulation of money was poetry in action, thought Pound and some of his associates, another way in which verse could be scientific, investigating not only psychology but economics; the poetic imagination, by showing how to pleasingly arrange the movement of money, reformed the world.[40] Forteans, too, were intrigued by Douglas's theory. Rumors circulated that British Forteans were at each other's throat over it—Liverpudlians in favor, Londoners critical.[41] I. O. Evans, who investigated the Loch Ness monster, had been one of the leading proponents of social credit in Britain before joining the Fortean Society.[42] One of the few subjects Thayer took seriously in *Doubt* was economic reform. He wanted a useful way to reorganize society such that the individual was freed from the burden of feeding his belly, the better to attend to feeding his mind.[43] Economic ideas associated with Douglas's theory were a central feature of *Doubt*, and Thayer recruited into the society members who advanced social credit, prominent among them Pound's allies Alfredo and Clara Studer and Britain's Duke of Bedford. Social credit, along with other heterodox economic ideas, was one of the departments in the Fortean University and one of the seventy-one categories into which Thayer carved Forteanism in the mid-1950s.

The critique of the liberal order's economic base slid easily into the promotion of illiberal ideas. There was no necessary connection between social credit and authoritarianism, just as there was no necessary connection between modernism and authoritarianism. Many modernist attacks on the liberal order were in the service of leftist politics, socialism, communism, anarcho-syndicalism, left-libertarianism. But dreams of overthrowing the social and political order that created the meat grinder of the Great War led some to rightist politics, to the celebration of strong leaders and a religio-mythical organization of the world.[44] Several of Pound's compatriots were beguiled by fascism. It took no great imagination to use Douglas's ideas as an excuse for anti-Semitism; indeed, Douglas himself showed the way: the economic system that immiserated the laboring classes, he and others said, was controlled by Jewish financiers.[45] In this light, Nazism spoke to the biases of some, and Mussolini seemed to offer hope for the world's reformation. Pound, who moved to Italy, was infatuated with Mussolini and tried to convince him to implement social credit.[46]

Forteanism was not immune to the appeal of authoritarianism; quite the opposite. Forteans were often unprotected by their vaunted skepticism from accepting the word of strong men as truth. In the early 1940s, Blish called himself a "book fascist"—agreeing in principle, not in practice.[47] Thayer ran with a pro-Nazi circle in New York during the same period.[48] His pacifism could look like appeasement, a silent vote for fascism, especially in his alliance with Hastings Russell, the twelfth Duke of Bedford, a social credit advocate who tried to broker a secret peace pact between Britain and Germany and who founded one of Britain's fascist political parties.[49] Eric Frank Russell found much to admire in fascist-adjacent ideas, including Dagg's version of Douglas's ideas.[50] *Tomorrow* was a pipeline through which authoritarians moved into the Fortean Society, counterbalancing Thayer's own penchant for anarchism as well as the socialism, communism, and left-libertarianism of other Forteans.

It is too much to say that Thayer conceived Forteanism purely as a type of modernism. Fort's ideas did other kinds of work for him, too—personal, social, political, scientific, philosophical. But there is no doubt he saw Fort as an avant-gardist, and Fortean aesthetics best represented by modernist experimentation. Nor can it be denied that his own thought represented—admittedly not perfectly—the modernist configuration of appreciation for the avant-garde, radical individualism, maverick economic theorizing, illiberalism, anti-Semitism, and adulation of strong men, even if such admiration inevitably waned. (How remarkable, then, that Judith L. Gee—woman, Jew, progressive—elbowed her way into this place so hostile to her.) Thayer wanted Pound as a member from the society's founding and mourned Mencken's resistance to Fort and Forteanism.[51] He hoped to honor Joyce in the society's Christmas card. When he imagined Fortean arts more authentic than science fiction, he turned to experimental artists, seeing Forteanism exemplified in the architecture of Frank Lloyd Wright and R. Buckminster Fuller, in the writing of Kenneth Lawrence Beaudoin, Carl Zahn, and Felix Riesenberg.[52] He loved to see artists outrage the masses and thought Forteanism (sprouted in the same soil from which grew those veteran agitators Hecht, Mencken, and Pound) served its true function when it inspired similar art.[53] "There's a lot more to Forteanism than just watching the papers for snail blizzards," he lectured.[54]

Character against Chaos

Born Mary Maude Wright in Texas, touched by the muse of poetry, she married, moved to California in the 1920s, emerging as Gertrude Wright, priestess, proprietor of a Theosophical school in Oakland—authorities called it a cult—promising the birth of a new savior, a coming utopia. Arrested for contributing to the delinquency of a minor (accused of arranging a sexual relationship to bring about the redeemer's birth), she was jailed but, ever the magnanimous sage, looked on her captors with pity:

> As I gaze out through prison bars
> Life is not half so gray,
> As souls of men whose bars of hate
> Shut God's pure love away.
> Better the stars through prison bars,
> With only a gleam to guide,
> Than the stinging goad of an open road,
> With conscience at our side.[55]

However stinging, the road proved too tempting. Wright skipped bail for Mexico, later reincarnating as Lilith Lorraine. In the early 1930s, she authored several Theosophically inflected science fiction stories. In the early 1940s—having returned to poetry, now working in a number of genres under several pseudonyms—she founded the Avalon Poetry Shrine in Texas, then, after the war, moved it and herself to Rogers, Arkansas, transforming the institution into the grander Avalon World Arts Academy.[56] Lorraine was a Fortean and a stalwart against modernist poetry, her stance leading Forteans into a larger debate over the place of modernism in American culture. That fight would come much later, though. First would be her own poetic evolution, a hectoring shadow of modernist prosody.

Lorraine's poetry 'zine *Different* circulated among readers of fantastic literature. Something in the publication appealed to Eric Frank Russell. He maintained a file on her, sent a copy of *Different* to John Atkins, another to Antony Borrow. Atkins was less than excited by what he read, admitting he didn't rigorously study his copy, but he was sufficiently intrigued to exchange a subscription to his *Albion* for one to Lorraine's 'zine.[57] Borrow was likewise ambivalent, unsure if he should be grateful or offended that Russell sent him a copy; he opted for grudgingly grateful:

> I was interested by the existence of such a magazine, though not by the actual contents. I think you must have misunderstood my "Glass" completely if you saw any affinity with "Different." And yet I am glad

it exists. It is harmless and honest, though its literary attainment is by no means high.[58]

Borrow's opinion unknowingly echoed Thayer's feeling, expressed in *Doubt*: "Almost anybody can understand the poetry; whether it is all worth printing or not is another question, and one which nobody asked us to answer."[59] Whatever hesitations there might be about the quality of her verse, Lorraine was ensconced within the Fortean community and *Different* was discussed widely. Miriam Allen De Ford belonged to Avalon; Lorraine published the drawings of Ralph Rayburn Phillips, who was associated with Chapter Two; and Clark Ashton Smith—whom she also published—said, "Lilith Lorraine is among the true and rare initiates of what Benjamin de Casseres calls 'The Unearthly Imagination.'"[60] Almost certainly one reason Forteans read Lorraine was because Fortean imagery saturated her poems:

The stars are trapped in a space-web
Swing from sun to sun
Where Time, like an Evil Spider,
Eats them one by one.[61]

She titled one poem "Time Is a Circle," another "Since We Are Property." *Different* had a department devoted to Forteana called "Strange Experiences." She joined the Fortean Society after the war.

Lorraine's postwar life was organized around poetry. Each morning, she awoke, attended to domestic chores (Cleveland, her husband, recently retired, helped around the house), then settled down to write, often finishing poems the same day she started them. She edited and advised young poets, put out her 'zines, and published seven books of poetry and poetry instruction between 1942 and 1947, while also traveling the country to give talks to community groups.[62] She wanted a generation of poets "dedicated to the restoration of wonder, surprise and spiritual adventure to life and literature and to the opening of new frontiers of mind that will give to peace its 'victories more sublime than war.'" Avalon and the publications that emanated from it encouraged all those "who while not discarding the enduring values, courageously interpret them in the language of a changing world."[63]

Lorraine's ambitions were consonant with her cosmic vision, as had been the case when she ran the religious school in California. She saw herself as a Christlike figure saving the world. Her poetic voice was oracular: "Truth is what the gods make it— / And there are many gods; / Beauty is above all gods, / Immutable, immaculate."[64] While a loving force organized the universe, humanity itself did not always fulfill the divine promise, instead setting aside its duty, refusing to evolve, consequently finding itself adrift in the cosmic vastness, alone or destroyed. The atomic bomb incarnated the apocalyptic possibilities of existential error, of humans choosing science and "so-called practical values" over love and the "Golden Rule . . . the only successful law of life."[65] Humans needed, under the tutelage of poets, to transcend their limitations, merge with the universal good in the realm of the eternal feminine, the plane where Christ resided. In poetry hung the fate of the world.

For all that Lorraine had come up through feminism and socialism, her aesthetic was conservative, centered on the eternal verities of truth and beauty. She wanted poems that inspired greatness. Addressing herself to a "vast audience," she sought to create the kind of poems "the soldier carries into battle . . . that the child learns from his text-book, that the pilgrim hurls as a shining weapon at the fearful shapes that close around him in the Valley of the Shadows."[66] Science fiction could be deliquesced into this aspirational humor. Her "poetry of the atomic age" imagined a time "when the moon lies prone beneath our feet, when Mars is a reborn miracle for the joy and the glory of man, when Pluto is only a whistle stop to the outer galaxies of earth's expanding empire."[67] This aesthetic commitment set Lorraine against "the chopped prose, the buzz-saw rhythms, and the intellectual vaporings" of modernist poetry.[68] She appreciated modernists for banishing the slick prosody of Victorian verse, and even their open discussion of sex, although she worried that it could be too frank for an immature public. But she detested its radical implications, the communism that had structured so much poetry in the 1930s, the experiments that destroyed poetic structure, killing lyricism. Her book *Character against Chaos* urged new poets to reject modernist obscurity, which discredited all poetry, rather to embrace clarity and eternal values as a bulwark for democracy.

The Fortean Fantasy

Benjamin DeCasseres, in 1931, noted contemptuously that too much modern thought was literal. "Facts, like a lot of flies, blow into our minds, get stuck on the fly-paper of our recording apparatus, and, after a few hopelessly feeble efforts, die." What the world needed—but found in short supply—was the artistic mind,

> which, instead of being covered with fly-paper to catch and kill facts, is composed of millions of cocoon-cells. The facts that fly into them are incubated and nurtured, and out of them come giant butterflies, dragons, and even little spiders that catch and eat little scientific flies.

Fort's mind belonged to that rarer class (as, DeCasseres thought, did his own). These cocoon-brained sorts were the "yarn spinners" who put facts to work expanding humanity's imaginative reach. DeCasseres was unconcerned with the truth of these different yarns— the world was too vast and unknowable to ever be enclosed by a mere story. "As there is no absolute of space or time, let us imagine anything," he said. The beauty, the thrill of the yarn was what counted. The Fortean fantasy was "the mightiest" of these yarns, "the strangest, the most overwhelming" that he'd studied. Fort's mind swallowed facts, arranged them, according to his genius, into "the strangest things that ever a human being has observed since John of Patmos and his only begotten son, William Blake."[69]

DeCasseres was not the only writer astounded by Fort's imagination. Science fictioneers, certainly, respected his fantasies. So did some modernists. The Fortean fantasy expanded humanity's imaginative reach in several different directions. As Atkins noted, Henry Miller's "psychopathic embrace" of reality, its contradictions, paralleled Fort's attempt to encompass the universal. British novelist Malcolm Lowry also found Fort valuable. Both Miller and Lowry saw in Fort support for their New Age vision, their sense that multifarious spiritual forces orchestrated the world. William Gaddis, similar to Lowry in his writerly style, interpreted Fort differently, as support for his fundamentalism—the idea, expressed in his fiction, that a singular divine force orders life, though its appearance is difficult for humans to recognize. Surrealists, too, were fascinated by Fort and his fantasy. Twenty or so years after it was first put to paper, the Fortean fantasy remained at the bleeding edge of the avant-garde.

Living in Paris during the 1930s, having launched himself from Brooklyn, Henry Miller invented a style of writing to "reveal the inner pattern of events."[70] The poet Robert Duncan said, "He has cut clear thru to the inner world where everything takes place."[71] Inside this architecture, Miller housed the damned, his project the "recording of all that which is omitted in books." Profanity and sexual explicitness were simultaneously a mark of his caddishness and contempt for women, and a means to communicate the universe's deeper meanings. "When obscenity crops up in art, in literature more particularly, it usually functions as a technical device. . . . Its purpose is to awaken, to usher in a sense of reality. In a sense, its use by the artist may be compared to the use of the miraculous by the Masters."[72] By confronting stark reality, one could approach the spiritual realm, "the ecstatic love of the divine."[73]

Miller had long been inspired by scientific romance, which may have led him to Fort, whom he discovered in late 1940.[74] *The Book of the Damned* was "a very queer book," he said, containing "startling data and still more startling beliefs."[75] By this point, he'd been ejected from Europe by war and returned to his home country, which he despised as an air-conditioned nightmare, eventually settling in Big Sur, California, where he lived in a shack, neighbor to other bohemians. Miller's books were banned in America but smuggled in by GIs returning from France, and sold under the table at Ben Abramson's bookstore, alongside Lovecraft (whom Abramson published, as he did Miller) and Thayer and Fort, "People who wrote 'strangely' strange," in the words of bookmaker Jonathan Williams—an interlocking compilation of authors offering alternatives to modernity's supposed disenchantment.[76] Miller was a very different writer than Fort, in style and intention; nonetheless, he fancied himself "another Fort."[77] "People are constantly supplying me with startling facts, amazing events, incredible experiences," confident that he, like Fort, would respect the damned and fit them into his spiraling narratives.[78] The two men each, in their own way, enlarged the scope of the imagination and found in this expansive field new connections between the self and the world.[79] For Miller, these associations were, ultimately, religious; he was constantly staggered by what he saw as the occult and mystical forces at work around him.[80]

The Fortean influence burned even brighter in the celebrated work of Lowry. Lowry learned his Fort at Cambridge in the early 1930s (about the time Ivan Sanderson was there), probably from a borrowed Gollancz edition of *Lo!*[81] "A red letter day in my life," he said. "I know of no writer who has made the inexplicable seem more dramatic than Charles Fort."[82] After university, he followed his wife to New York and likely read the rest of the Fortean corpus at the New York Public library (where, only a few years before, Fort had beetled), inspired by all but *New Lands*. Fort, mixed with his studies of esotericism and mysticism, remained a touchstone as Lowry moved to Hollywood, to Mexico, to Vancouver, trying to outrun the consequences of his drinking—which finally killed him in 1957—and fashion himself into a writer of fiction in the image of D. H. Lawrence, T. S. Eliot, Wallace Stevens. Lowry had several manuscripts under construction at the same time, some of which he planned to fit into a Dantesque trilogy.[83] *In Ballast to the White Sea*, "written while the thrill of discovering Fort's work was still fresh," was to be his *Paradiso*. The book was deeply informed by Fort, not just his compilation of anomalies, but his larger philosophical project to invent an era of the hyphen, to illustrate the interconnectedness of the cosmos.[84]

Fire, like alcohol, dogged Lowry through his adult life. In 1944, the Vancouver shack he shared with his second wife, Margerie, burned, destroying much of his writing. Lowry worried that, as with the printers in Fredric Brown's "The Angelic Angleworm," he had conjured the conflagration by writing the word *flame*.[85] He fictionalized his fear in 1953, shortly after Margerie gave him *The Books* as an anniversary present. Lowry, working on the novel that would become *October Ferry to Gabriola*, composed a vignette in a rush of writing to prove to his anxious publisher that he was progressing.[86] The main character, Ethan Llewelyn, similarly victimized by a series of fires, reads Fort ("obviously a genius if ever there was one") and comes to realize he's haunted by a malevolent force, one of the myriad spiritual powers that influence human life.[87]

Lowry had not lost everything in the Vancouver fire; from the burning shack he rescued his manuscript of *Under the Volcano*, the trilogy's *Inferno*. He and Margerie worked together on the novel, she paring down his excesses, clarifying his themes. The central character was transformed from a simple drunkard into a node in a network of esoteric symbols. *Under the Volcano* was published to much acclaim in 1947. References to Fort were pruned during the writing, leaving a faint Forteanism: a reference to teleportation, to wild talents, a vague suggestion of vampires in Trinidad. But mere editing was not enough to remove Fort from Lowry's modernism, nor was death. *October Ferry to Gabriola* came out in 1970; a manuscript of *In Ballast to the White Sea* was found and later published. The vignette about Llewelyn's flammable life appeared as a short story under the title "The Element Follows You Around, Sir," in 1964.[88]

A few years before Lowry died came another novel, often compared to his—complex, allusive, "a veritable Katchenjunga [*sic*] . . . of a book," in Lowry's own phrase—that also made use of Fort.[89] William Gaddis started writing *The Recognitions* shortly after World War II as he traveled through New York, Europe, and Latin America. He tried to read *Under the Volcano* but put it down, having "found it coming both too close to home and too far from what I thought I was trying to do."[90] At some point in his ceaseless foraging through Western literature, Gaddis came across Fort; exactly when is not known.

Massively long, intricately ordered, *The Recognitions*, among many other efforts, registered the exhaustion of modernism as a literary genre.[91] "Make it new," Pound advised modernist writers, and the avant-garde had done just that for a half century by the time Gaddis's tome appeared in 1955; novelty was increasingly hard to come by.[92] *The Recognitions* was also a profoundly religious work, about the difficulty of finding the sacred in a secular world, and it was in support of this theme that Gaddis turned to Fort.[93] He quoted, then repeated, Fort's conclusion that maybe "we're fished for" as a kind of fundamentalist prayer.[94] The initial reference to Fort comes when the book's main character, Wyatt Gwyon, is metaphorically drowning. Humanity, the phrase suggests, has been cut off from God, submerged. But the phrase is also hopeful. Gwyon realizes that one portal to the ineffable modern God is art. When he views Picasso's *Night Fishing in Antibes*, he accesses a "reality that we never see."[95] Fishing, then, connects one to the divine, an act of reconciliation: God wants to catch and number every individual.[96] Not Fort's extraterrestrials, but a transcendental God was the earth's owner. Gaddis's book, though, sunk like a stone, hardly recognized itself, except by cultish fans, for another two decades.

Around the same time as Miller and Lowry, the surrealists (who influenced both those novelists) also discovered Fort and absorbed him into their project. Surrealism, like the various pulp genres, like Fort's books, was rooted in awe at the anomalous. For surrealists, the marvelous, the fantastic, the wonderful, are probes into the individual's psyche, the deepest layers of the subconscious, which operates according to strange logics. Surrealist art (poetry, painting, novels, films) rigorously investigates the unconscious, dream states. Very different in execution, surrealism was like Pound's poetry in aiming to capture the authentic working of thought. As with Fort, the surrealists emphasized the worldly mechanism of multitudinous forces. Rooted in materialism as well as the immaterial world of dreams, surrealists allied with communists, socialists, left-libertarians in promoting revolution and anticolonialism.[97] In the 1930s, early 1940s, surrealism spread, its adherents forced from war-torn Europe, many taking refuge in New York.[98]

In October 1942, the Whitelaw Reid mansion on Madison Avenue, at 50th Street, hosted an exhibition, *First Papers of Surrealism*, organized by André Breton, leader of the movement and author of its manifesto.[99] The catalog included an essay by the journalist Robert Allerton Parker, "Explorers of the Pluriverse," which extolled American pulps as indigenous examples of surrealism. Pulps, like formal surrealist art, played with the anomalous; their scrutiny of history's motor or the world's hellish scaffolding, Parker thought, were not descriptions of external reality, but dreams: the stories exposed the working of the mind and offered respite from the drudgery of everyday life. "These homegrown eccentrics of ours," he said, were "doughty defenders of the subjective from the regimented invasion and standardized error of the external world."[100] Surrealism and pulps both explored the point where the material and immaterial worlds met.

Parker was particularly inspired by Fort, "the Socrates of the Bronx." As he saw it, Fort was a collagist who assembled "unconsidered, yet disconcerting, trifles" into an idiosyncratic "picture of a paradoxical and highly unpredictable universe."[101] Fort was a mythologist, and at that very instant, Breton was in search of a new mythology, a grand story that would circumscribe the irruptions of the moment and reify its subconscious and unconscious angst. Breton endeavored to create such a myth himself, one that unknowingly echoed Fort's. A few months before the exhibit, in June 1942, he'd overseen publication of the first issue of a new magazine, *VVV*, that included his piece on what were called in English "the Great Invisibles." These were beings, invisible to the human eye, that lived among humanity, having domesticated people in the way people have domesticated dogs—like Renard's Blue Peril or Russell's Vitons—but benign.[102] While Breton derided the pulp stories Parker celebrated, and never seems to have read Fort, his surrealistic fantasy buttressed a conviction of other surrealists: that Fort had invented a myth of the modern age, his writing a montage broad enough to encompass the era's contradictions. Parker made the point again in the next issue of *VVV*, extending his investigations to include Lovecraft and his vision of a universe inhabited by incomprehensible Elder Gods.[103]

Eagerly reading those issues of *VVV* was Philip Lamantia, who would become one of his generation's leading poets. Born in 1927 to Sicilian immigrants in San Francisco's Excelsior district, he was touched by surrealism while viewing exhibitions of Dali and Miró at the San Francisco Museum of Art. Astounded, he read everything he could find on surrealism in the museum's small library (where he came across *VVV*) and the larger public one. His childhood in one of the city's Italian enclaves had overflowed with wonders. He stayed up late at night listening to horror shows on the radio.[104] He made scrapbooks of Ripley's *Believe It or Not* comic strips. He read Lovecraft, Clark Ashton Smith. (A half century on, the latter prompted him to compare the world of imagination to a separate sphere of existence, the "Mundus Imaginalis," he called it, borrowing an old Arabic idea, a place to be explored as diligently as any physical frontier.)[105] Lamantia melded these influences into an erotically charged, surrealistic vocabulary. Shortly after attending the exhibits, he sent a clutch of poetry to the surrealist magazine *View*, then quit high school and moved to New York City, where he took a job as an assistant at the magazine. Lamantia was sixteen.

Lamantia's Forteanism was subtle.[106] Parker called Fort "a connoisseur of the incredible . . . the alogical, the illogical, the analogical, the neological."[107] The same could be said of Lamantia, his poems held together by such ligaments. While in New York City, he learned alchemical symbolism, which he used to organize some of his poems, images linked by means other than conventional logic.[108] Fort was another of these means, a way to uncover the world's many wonders. The poet Kenneth Rexroth said:

> What is sought in Alchemy or the Hermetic Books, or irrational fads like flying saucers, is the basic pattern of the human mind in symbolic garb, as it presents itself to the individual believer, and behind that, in the enduring structures of the human organism itself.[109]

Lamantia continued with this task through a tumultuous decade. He fell out with Breton and the surrealists, returned to San Francisco and made his way from materialism toward belief in a Supreme Being.[110] But even adrift from surrealism, his poetry's "main objective" was to raise "awareness of the most hidden levels of sensibility, vision & being."[111] He was intent on expanding the imagination enough to include the many objects that had been damned to the unconscious. His poems from this era, not published until 2008, had a Fortean quality, the lines quick, slashing, dashed, less introspective and more assertive than his early ones:

> The marigold space empties
> Into solar hieroglyphs.

Or, from an untitled poem:

> Flame gates open to water gongs:
> Roll to the wayward depth
> Before & After the caves of Sol[112]

Islands of Freedom

Lamantia returned to San Francisco in 1946. The city was controlled by a moneyed elite, but amid their conservative provincialism were what he called "islands of freedom," where artists gathered. These were inheritances of San Francisco's bohemian past, the storied Barbary Coast of the last century. Through the 1940s, COs, disaffected soldiers, and starry-eyed youths came to San Francisco from around the country, toured its islands. There was the Monkey Block at 658 Montgomery, with its warren of rooms that had housed artists and financiers, hosted Ambrose Bierce and George Sterling and Jack London and Mark Twain. North Beach had its bars. Across the relatively new Bay Bridge was Berkeley, its university, and down the coast Miller's shack and areas artists had bivouacked after the 1906 earthquake. Chapter Two's Robert Barbour Johnson joined the city's artistic kings and queens and jacks at these places and was himself a stop on the itinerary of visitors. A veteran of circus life, he built miniature replicas of circuses, exacting in their detail, and taught his cats to jump through hoops, like trained lions.[113]

Out of these spaces coalesced what would be called the San Francisco Renaissance, an artistic ferment that marked the city in the late 1940s and early 1950s. Influenced by surrealism, but growing beyond it, Bay Area *litterateurs* and visual artists added a commitment to various mysticisms, agitated for anarchism, left-libertarianism, and pacifism. Visual artists at the California School of Fine Arts took up abstract expressionism, which abjured the social realism of the Depression and realistic depictions of America associated with dangerous patriotism, finding the instinctual expression of individual emotion a better way to capture relief at the end of the war. Said one San Francisco abstract expressionist:

> We had all been fed up with regimentation, with being put in a uniform and told what to do. We were looking for a way out of that discipline—a way to be individual, a way to be human.[114]

San Francisco artists attended to the hidden structures of the mind, as did the surrealists, but also to hidden and not-so-hidden social structures. Unlike science fictioneers, those most central to the Renaissance (themselves not unfamiliar with science fiction and its tropes), looked not for ways to control the forces of history—the growing power of the state, the military, corporations—but to resist them. They carved from the harshness of the world spaces of liberty, New Age places where one could experience the spiritual forces denied by increasingly dominant bureaucratism. Forteanism was part of this artistic movement, a tool for countering modern determinisms, for building sites of autonomy and sustaining the islands of freedom.

Godfather to the San Francisco Renaissance was Kenneth Rexroth. He'd come to San Francisco in 1927, aged twenty-two, already having lived in Chicago during its literary Renaissance and in New York's bohemian Greenwich Village, visited Mexico, Paris, camped and tramped across the American West. Here was a place empty of artistic pretensions, he decided, untouched by developments in Europe and New York: a place that he and his wife Andrée, a painter, could shape in their image.[115] Rexroth's name circulated through avant-garde and leftist circles, but his ambition to make California's artistic society an equal of those in New York and Paris went unfulfilled, Rexroth thrown into a funk after Andrée died in 1940.[116]

Then, during World War II, prospects brightened. Out of Berkeley came the poets Robert Duncan and Jack Spicer, both taught by a new professor there, Josephine Miles. Lamantia made his name. Rexroth was a CO, doing alternative service at a mental hospital, and, along with Miller, attracted other COs with artistic aspirations.[117] The physicist and artist Bern Porter, horrified by the atomic bomb, quit the Manhattan Project (he'd been working at Berkeley) to devote his time to publishing avant-garde works and experimenting with photography. Madeline Gleason, a poet who'd come to San Francisco in the 1930s as part of the WPA, returned after the war and organized an important poetry reading at a gallery on Gough Street in April 1947.[118]

Also emerging from the university was George Leite, a dynamo who kept the San Francisco Renaissance moving. Passionate about avant-garde art, Leite put out literary magazines while still a student, explored Theosophy and mysticism.[119] He was a follower of the maverick psychiatrist Wilhelm Reich.[120] Leite knew Porter and Lamantia and, through him, Rexroth. He ran a workshop on experimental writing at Berkeley (from which he dropped out) and organized film festivals, intrigued by the radical potential of the visual arts as well as the written word. Porter introduced Leite to Miller, and Leite moved to Big Sur, where he lived in a shack with his wife, children, and, in time, his mistress too.[121] He idolized Miller: "You have become whole and real in the way Christ must have been to the apostles," he told him. (Miller shrugged: "I know they are making quite a cult of me—I can't help that—it's the writing that causes it.")[122] Leite lectured on modernist art, including surrealism, at the communist Labor School—surrealism was essential to him, pushing art away from the social realism of the 1930s and toward the abstract.[123]

He opened a bookstore, daliel's, named for his son, always styled with a lowercase *d*, which became an important hub of the artistic community. It sold copies of Reich's work alongside Theosophical and mystical books. In 1944, he and Porter started *Circle*, the signal magazine of the San Francisco Renaissance. Over the next several years, they published Rexroth and Miles and Duncan and Spicer and Miller and Pound, as well as the experimental photography of George Barrows and Porter.

Upon returning to the city, Lamantia started his own magazine, *Ark*, though it lasted but a single issue. More enduringly, he and Duncan convinced Rexroth to start a salon, held "on the top floor of a house in the Fillmore District owned by a branch of the Old Jewish Workmen's Circle," where fifty or so people met regularly—COs, poets, radicals—to smoke marijuana and mull the state of the world; Lamantia and Duncan had attended similar gatherings in New York City.[124] They called themselves the Libertarian Circle, and out of their discussions emerged an organizing theory for the San Francisco Renaissance that both summarized the intention of the artists working across the Bay Area and acted as a manifesto for the future of the avant-garde. They mostly sloughed off commitments to communism without losing a dedication to radical politics—anarchism, pacifism, gay rights, left-libertarianism. Driven by a religious impulse, they refused to fall into mainstream faith; rather they maintained fidelity to mysticism and alternative religious traditions. In his own conscientious objector statement, Lamantia argued that humanity's original sin was turning away from the Supreme Being and developing an intellect: "The XIX century dream of man's final perfection through scientific progress has become, at least theoretically, realized." The dream, though, proved a nightmare, perfection appearing "in the form of the most monstrous weapon of all time: The Atomic Bomb." Lamantia wanted to be "free from . . . the State," which he deemed a "social evil."[125] Politics and art were united in a duty to create, from the imagination, a sense of wonder that stood against the increasing power of the military, the government, corporations, and science. Historian Richard Cándida Smith writes:

The imagination, manifested in its highest form in the aesthetic act, became the most stable source of personal freedom in a world otherwise deterministic and frightening. An absolutized privacy, no matter what the costs, turned out to be the most radical defense against the claims of public order. Bohemian enclaves developed a repertoire of self-images that proved to have appeal to the collective imagination far beyond the limited boundaries of the art and poetry worlds. As feelings of powerlessness spread, the aesthetic avant-garde provided an antidote.[126]

The San Francisco Renaissance suffered from external plaints, internal strife, and its own success. From without, conservative voices in and around the city castigated the new bohemians as a "cult of sex and anarchy." Miller, the *San Francisco Examiner* declared, headed a "hate cult," the inverse of Lilith Lorraine's earlier transgression and potentially even more threatening to public order.[127] Meanwhile, within: Rexroth bitched that Leite remained a communist, though more of a dupe than an organizer.[128] With Lamantia and others, he complained about Leite's eclecticism, mixing mysticism, Theosophy, interests in various art movements—and Forteanism. Art and mysticism, they insisted, both required theoretical commitments.[129] Miller broke with Leite over his fecklessness, his inability to pay contributors.[130] While driving a taxi to make ends meet, Leite supposedly beat and robbed a fare, further distancing putative supporters.[131] Leite and Porter fell out, and *Circle* was quiescent for several years—then shocked back into existence by the charge that Bay Area bohemians were a cult of sex and anarchy.[132] But Leite and his mistress, Jodi Sccott, managed only one more issue; money was tight. In 1949, Leite sold daliel's to Sccott and her ex-husband.[133]

The imagination sought expanded vistas. Alcohol and drugs were one vehicle to explore these far frontiers. Drugs watermarked the avant-garde, another way bohemian artists stood against convention, mainstream institutions, part of the protest against stultifying artistic forms, capitalism, repressive sexual mores, censorship, technocracy, the bomb. Drugs worked in concert with Reichian Orgone

accumulators, with mysticism, surrealism, alternative religious practices, and Forteanism.[134] Marijuana was ubiquitous throughout the
islands of freedom. Other drugs circulated, too.[135] Inspired by earlier
bohemians, Lamantia discovered that peyote buttons could be purchased through seed catalogs; he started experimenting with the drug
in 1951.[136] About the same time, Leite's college roommate, Warren
d'Azevedo, returned from a stint in the merchant marine to study
anthropology at Berkeley.[137] Probably it was he who brokered a visit
by Leite, Lamantia, and Lamantia's wife Gogo to a Washoe peyote
ceremony in the Sierras. "The sky tasted like crystal star meat," Lamantia recorded.[138] Drugs accentuated the avant-garde focus on cultural renewal not via class revolution but by expanding individual
consciousness. They were dangerous, though, the price they exacted
often high. The Berkeley poet Jack Spicer died from alcoholism when
he was forty. Lamantia found his way to heroin and was lost for a long
time.[139] At some point, Leite mixed Dexedrine with peyote, suffered
a breakdown, and was admitted to the Napa State Hospital for psy-
chiatric care. Released, he cut his ties with the art community, went
to work as a teacher, his wife Nancy working for the school district.[140]

Even as the San Francisco Renaissance endured external attacks
and internal dissension, and its participants sometimes paid too
much to set their imaginations against modern determinisms, their
bohemian ideals spread to the culture at large. Out of the San Francisco Renaissance emerged the Beats. (William Gaddis also hung
out with the Beats.)[141] The GI Bill launched the Renaissance's ideals
deep into American culture. Hundreds of returning veterans used
their government benefits to attend the California School of Fine
Arts, carrying with them into their daily lives the Renaissance's central tenet: that imagination was an autonomous realm, a place where
one could take refuge against the military, government, corporations, and science; that this place promised peace and freedom. The
1940s and early 1950s set the stage for the counterculture.

Fort and Forteanism belonged to the Bay Area art scene. For a
while, Fort was in vogue among the Big Sur bohemians.[142] There
were Miller and Lamantia, of course, and Rexroth and Leite. As a

youth, Rexroth had been close to his maternal grandmother and shared with her what he called the "annoying habit" of second sight (annoying because his prescience always involved trivial matters). She told him stories of ghosts, monstrous births, the sea beast of Lake Erie, and stories about the family's tradition of horse whispering. According to Rexroth, she had tales of meteorological phenomena to "rival Fort." In his autobiography he wrote, "Her home in Elkhart seemed to be advantageously sited for the witnessing of inexplicable happenings in the heavens and I witnessed two of them myself." In the first case a spinning, sapphire blue fireball appeared in the attic, came downstairs, went out the front door, hit a bike, and exploded. In the second, a fish-shaped hole filled with dull red flames opened in the summer sky; the image lasted about fifteen minutes.[143] On at least a few occasions, Rexroth lectured the Libertarian Circle about matters Fortean.[144] He never explicated how Forteanism fit with his philosophy, but the affinity seems clear. Science, Rexroth concluded, like philosophy and all human opinions, was "a great work of art—man's construct over and against the ultimately unfathomable universe."[145] Beautiful, but also domineering. Fort helped resist it and all other determinisms, rooting his views in singular reports, the ground from which imagination grew, privileging the individual against the statistical constructs of modernity.

It was easy for Leite, always eclectic, to fit Fort into his worldview, and so his Forteanism was much more obvious. Like Miller, he joined the Fortean Society.[146] He introduced Fort to his son.[147] He swapped magazines with Thayer, *Circle* for *Doubt*, and sold both *Doubt* and *The Books of Charles Fort* at daliel's. (Along with San Francisco's Paul Elder & Co., daliel's was one of the key sources for Forteanism in the Bay Area.)[148] Leite, as well, advertised *The Books* in *Circle*: "If you like surrealism, if you read Blake, if Cagliostro intrigued you read The Books of Charles Fort."[149] (They retailed for a hefty $4.50, with the admonition, "*It costs a lot to be freed from stupidity!*")[150] But the connection between Forteans and Leite went beyond mutual admiration. He thought in Fortean terms. The very name *Circle* seemed to come from Fort's famous dictum "one measures a circle, beginning anywhere," a variation of which was the first clause in the magazine's opening manifesto.[151] Though unapologetically eclectic, Leite was, underneath it all, committed to a surrealistic aesthetic: surrealism, with its striking, confounding images, promised to awaken the masses, prime the pump of their imagination so that they could resist "authoritarian control."[152] Fort, as Lamantia and the surrealists knew, was a potent tool for such a project, his prose guaranteed to make the imagination so commodious as to render ridiculous the provincials who would censor Miller or dismiss the avant-garde as cultists.

Thayer was sympathetic to the San Francisco Renaissance, which accounts for his advertising in *Circle* and swapping magazines. Like the Bay Area avant-gardists, he was a pacifist and fancied himself an anarchist.[153] He had little interest in Reich, but Reich was but a hiccup in an otherwise cordial relationship.[154] Thayer read, advertised, and reprinted material in *Doubt* from the same anarchist magazines that inspired the Renaissance, supported the same COs, and admired the same virile arts, particularly the poetry of Pound, one of Leite's favorites.[155] He badly wanted other Forteans to develop their own forms of avant-garde arts in line with what was going on in the Bay Area.

Le Matin des Magiciens

As the San Francisco Renaissance gave birth to the Beats, surrealism returned to France and underwent its own transformation, though the parturition was much more difficult than what was happening in San Francisco and New York. Once more, Fort was part of the proceedings, central to the development out of surrealism of a movement called fantastic realism and its most important and popular creation, the book *Le Matin des Magiciens*. Fortean ideas were key to fantastic realism's understanding of power and how it moved through the world. Artists associated with the San Francisco Renaissance had also concerned themselves with power and its gathering into great institutions, but they had sought ways to oppose it rather

than theorizing how it worked, at least in their use of Fort. Forteanism for them was a means of expanding the imagination, the place
where they could stand against modern determinisms. Fort was a
tool for liberation. Partisans of fantastic realism thought Fort was
not a means of opposing the powers of government and the military,
corporations and science, but a way to access that power, to understand its passage through society, and to wield it themselves. Theirs
was a conspiratorial view. The powerful were hidden; they were not
the obvious institutions of modern life but secret cults. Only those
who nurtured the proper, Fortean perspective—not skeptical but
attuned to the fantastic—could see the true mechanisms of power
in the postwar world.

Elements of fantastic art, suppressed or cut off by the war and
Nazi occupation, returned to France in the late 1940s and early
1950s. Some of the fantastic was homegrown, a rebirth of the old
roman scientifique and magazines devoted to parapsychology that
built on a long history of French studies.[156] But much of the French
fantastic, both visual representations and the written word, relied
upon translations and importations. American parapsychological
studies of ESP formed the core of the French discipline. *Soucoupisme*
became the mode—UFO-spotting. Thayer observed:

> If you think the USA newswriters and cartoonists had fun with
> the "flying saucers" you should have a gander at the current
> press of La Belle France. wow! Once they took it up, they took
> it up. No issue of any daily home journal in France is complete
> these days without its photo of the latest peasant to interview a
> spaceman.[157]

American science fiction translations filled the bookselling market,
Asimov and Bradbury and Matheson and Russell and more. *Galaxy*,
clearest competitor to *Astounding*, started a French version in 1953;
so did Boucher's *Magazine of Fantasy and Science Fiction*, though
its French edition, *Fiction*, showed less fidelity to the original.[158]
Fort, too, was brought to France in those heady days. *The Book of the*

Damned was translated as *Le Livres des Damnes* in December 1955. *Fiction* carried a Fortean column and ran a translation of Miriam Allen De Ford's article on Fort to prepare the public for Fort's arrival. "Now that the barrier of language is down," Thayer said, "we expect the dissenting minorities of all Gaul to rally around Forteanism and forget forever the juvenile inanities of Sartre and company."[159]

Surrealists rode this wave of the fantastic. Breton was beguiled by parapsychology, wondered if surrealists could cultivate forms of telepathy or decipher the workings of everyday magic, by which he meant the odd coincidences of quotidian life. Surrealists remained committed to spiritualism and made clear their loyalty in the title of France's leading surrealist magazine of the time, *Médium*.[160] The French philosopher Jean-Paul Sartre dismissed surrealism as anti-art, painting that destroyed painting, literature that destroyed literature; his existential philosophy, in which humans were to take responsibility for their every act, valorized literature for its ability to resist violence. But the surrealists were opposed not so much to art as to its conventionality and sought to use aesthetic elements to reach the otherwise-hidden sublime.[161]

As had happened a decade before, Fort was admired as a writer and prophet of surrealism. A surrealist magazine ran a column on Fortean events. The surrealist Robert Benayoun translated Fort, and *Médium* published excerpts of *Le Livres des Damnes*. Fort exemplified the processes that so bothered Sartre, critiquing language, discovering in humanity's descriptions of the world slippages, contradictions, and odd conjunctions. Playing with these anomalies, he uncovered a deeper reality, one rule-bound but resistant to standard logic. He passed "from the everyday to the surreal," Benayoun said, "of which he is a native."[162] From the sky fell frogs, from the watery depths arose monsters. Dogs talked—or maybe they did not, and maybe only water fell from the sky and fish swam only in the sea. Fort's was an exercise in which the active mind contemplated, from many perspectives, weirdnesses that others ignored; from this musing he created marvelous new worlds, as rigorously materialistic as our own, but stranger. And then he destroyed them, only to suggest other, equally bizarre notions. By contrast, most of the science fiction being translated seemed, to the surrealists, banal. Almost certainly they would have agreed with Rexroth's assessment that it was "entertaining enough, but very much in the class of something to read while drinking yourself into oblivion": the genre was the true negation of art, with no promise of the sublime on the other side.[163]

At the center of the Venn diagram linking Fort, surrealism, and the fantastic were the authors Louis Pauwels and Jacques Bergier. Pauwels, a teacher-cum-journalist, struggled to make meaning of the world after the horrors of World War II, the casual destruction of property and life, the unbelievable inhumanity of Germany's death camps, the existential terror unleashed by the atomic bomb. He knew Breton, the surrealists; moved through the social circle of mystics; experimented with socialism, Hinduism. Through a surrealist daisy-chain, he met Bergier in 1954. A Jewish immigrant from Ukraine, Bergier had fought for *la Résistance*, been tortured, and survived a concentration camp. The two bonded over a love for science fiction, for surrealism (Bergier was on the editorial board of an important surrealist magazine), and for Fort. Pauwels tugboated *Le Livres* into print, the first book in a series he edited, Collection Lumiere Interdite (Forbidden Light Collection). Bergier wrote the Fortean column for *Fiction* and one of the three forewords to *Le Livres*. (Benayoun and Thayer wrote the other two.) He also persuaded Pauwels to reconsider his quest to apprehend the world and its mysteries. "I rearranged and orientated the various intellectual and spiritual experiences which I had exposed myself to," Pauwels said.[164] Over the next five years, they worked on a new philosophy and contemplated the strange ways that power moved through society, through history.

Fort, "one of our most cherished idols," was the linchpin, holding together their assemblage of surrealism, parapsychology, and science fiction.[165] Fort showed that science was a conspiracy against the real; the upholders of orthodoxy damned strange things with such vigor that most people never saw—never experienced—the

truly real. Fort, Pauwels and Bergier said, sought to counter the conspiracy, to open eyes to new possibilities, doing for science what the
surrealists had done for art. "A fresh start would be needed," they
interpreted Fort as saying, and they, building on his ideas, outlined
what the world might look like once the dross was cleared, even
as they acknowledged that they surely had many details wrong.[166]
Fort had suggested that the world might be controlled by "secret
societies"—his earth owners. Pauwels and Bergier took the idea literally.[167] In their view, humanity, over the vast course of history, had
repeatedly discovered the fundamental laws of the universe. Those
in the know formed sprawling organizations. Pauwels and Bergier
borrowed John W. Campbell's idea that intellectual elites had no
reason to hide, for they were assured that no one else in the world
could understand their actions, their talk. These were "open" secret
societies, Breton's Great Invisibles.[168] These elites were worldly people, living among the common mass of humanity but possessed of
extraordinary power, viewing the non-elect as puny-minded, of little
importance. Some of these societies, though, were mercenary, sold
themselves to nations, to governments. The Nazis had allied with
some of these powerhouses. The Soviets and the US had secret societies encysted within their enormous government bureaucracies.
Others sold themselves to India, Argentina, South Africa. Still, there
was hope that newly emerging secret societies could be turned to
good purposes. The vast knowledge accumulated by scientists since
the nineteenth century need not be so insular, need not hide reality,
but could be used to mutate humanity, move it to a higher plane,
as though Theosophically evolving. A new open secret society was
aborning, and humanity might yet benefit. Look at the nuclear
physicists who manipulated reality's most basic forces; consider Einstein, who was humanistic in orientation—precursor, perhaps, of a
time when the elite uplifted the benighted.

Pauwels and Bergier called their perspective fantastic realism.
They made literal surrealism:

But unlike [the surrealists] we were exploring not the regions of sleep and the subconscious but their very opposites: the regions of ultraconsciousness and the "awakened state." We call our point of view Fantastic Realism. It has nothing to do with the bizarre, the exotic, the merely picturesque. There was no attempt on our part to escape the times in which we live. We were not interested in the "outer suburbs" of reality: on the contrary we have tried to take up a position at its very hub. There alone, we believe, is the fantastic to be discovered—and not a fantastic leading to escapism but rather to a deeper participation in life.[169]

The surrealists denied this insistent child. Politics underwrote some of the disagreement. Pauwels and Bergier supported France's colonial war in Algeria, while surrealists opposed it. It was easy enough for critics to draw from this seed larger political meanings; Pauwels and Bergier, fascinated as they were by ruling elites, were accused of latent fascism. Chief among the opponents of fantastic realism was Benayoun, himself of North African descent. He found Pauwels and Bergier's writing sloppy. They had plagiarized him, plagiarized Fort. More troubling, he said, echoing George Leite's critics, was their undisciplined eclecticism, so different from the rigorous techniques of surrealism. Pauwels and Bergier, he said, "make a mystery of everything, of pregnancy, robots, yoga, mathematics, women, Bretons, youth, monkeys, the human body . . . [They] see everywhere genuine mutants, flying saucers." They are "stupefied by the banal, and being mystified, attribute it to a conspiracy from which they are upset at being excluded."[170] Fort moved from the mundane to the poetic, while Pauwels and Bergier subjugated fact to fascination with master races. Benayoun's foreword to the 1967 edition of *Le Livres des Damnes* lambasted them.[171] Breton, for his part, cut the idea of the "fantastic" out of surrealism. Once at the heart of the movement, a synonym for the marvelous, in 1962 the fantastic became its antonym, standing for everything surrealism opposed. At the same time, he ended surrealism's association with parapsychology.[172] Breton wanted no part of fantastic realism.

Nonetheless, Pauwels and Bergier persisted, their work becoming increasingly popular. They published their ideas as *Le Matin des Magiciens* in 1960. The book was translated as *The Dawn of Magic* for the British market in 1963 and *The Morning of the Magicians* for the American the following year. In less than a decade, it sold around a million copies. They started a magazine, *Planète*, whose circulation peaked at about a hundred thousand. There were Spanish, German, Portuguese, and Italian versions.[173] Among those published, reprinted, and celebrated in its pages were surrealists. Though Pauwels and Bergier were obsessed with conspiracies and the ways in which secret societies wielded power and shaped the history of the world, fear was not the overwhelming emotion provoked by their theorizing. Rather they replaced the "gloomy, tedious" sensibility of Sartre's existentialism with a dewy openness to possibility: the modern world was yet magical, enchanted. Anthropologist Mircea Eliade wrote:

> What was new and exhilarating for the French reader was the optimistic and holistic outlook which coupled science with esotericism and presented a living, fascinating, and mysterious cosmos, in which human life again became meaningful and promised an endless perfectibility. Man was no longer condemned to a rather dreary *condition humaine*; instead, he was called both to conquer his physical universe and to unravel the other, enigmatic universe revealed by the occultists and gnostics.[174]

On the Fortean Visual Arts

- Martha Visser't Hooft, *Fall of Related Objects (Tribute to Charles Fort)* (painting, 1948): Visser't Hooft, in Buffalo, though educated in Paris, belonged to the emerging abstract expressionist movement. Against a black background, four multicolored, legged (ciliated?), amorphous objects, bisected by black, seem to fall.

- Clay Spohn, *The Museum of Unknown and Little Known Objects* (exhibition, 1949): Spohn was a professor at the California School of Fine Arts in San Francisco during the height of the San Francisco Renaissance and, like Visser't Hooft, was inspired by abstract expressionism, as well as surrealism and Dada. His installation at the school, in association with the annual costume ball, displayed scrap metal, objects found in the trash, lint from vacuum cleaners—many of the items strung on wire. Spohn's work "helped transform the relationship many midcentury California artists had with the art object."[175] Although Spohn did not explicitly refer to Fort, it is impossible not to see his collection as a gathering of the damned, especially given how many of his artistic peers were inspired by Fort, talked of him, his books, and the Fortean Society.

- Clarence Laughlin, *Monsters from the Sea* (photograph, 1950): Ridged patterns in sand lead from camera to driftwood to ocean. Laughlin was a surrealist photographer; two years earlier, he'd bonded with the poet Myrna Loy over a love of Fort, "the William Blake of our time." Much of his photography incarnated Fortean ideals, capturing damned objects, the holes left by lost things, circles and spirals, anomalies of time, witchcraft, the pluripotency of nature.[176]

- Arrow MMoray, *Animals from Outer Space* (exhibition, 1957?): "MFS Moray [has] gone far beyond the line of duty, carrying several of the Fortean arts to new heights. She is the former Arrow Mackey, poet and publisher of little magazines, herbalist, and now taxidermist of space fauna. Her recipe for Fortean Bouquet for Split Pea Soup will be printed soon. At the mo we can only commend her exhibition of Animals from Outer Space which took over the San Lorenzo Public Library for the month of September. Seventeen species were represented, all authentic. At least YS can vouch for the two specimens sent to him, (1) the MLINK, adult male, and (2) the CHOLLYFORT. These two marvelous creations appear to be composed of whatever the artist puts her hand on—pipe-cleaners, bits of fur, cornflakes or whatever."[177]

Clarence Laughlin took this photograph about two years after Ivan Sanderson investigated strange tracks in Florida. But Laughlin was not interested in literal monsters or logical explanations; he was inspired by the surrealists and inclined more to the alogical, the illogical, the unlogical. Writing of monsters in *Lo!* Charles Fort said, "I am tired of the sensible explanations that are holding back new delusions." Laughlin's explicitly Fortean photography pressed beyond the sensible to new delusions. (Clarence John Laughlin Archive, Historic New Orleans Collection, acc. no. 1981.247.1.19.)

Beauty Is Difficult

While the use of Fort by modernist authors was an important way in which Forteanism related to literature in the 1940s and 1950s, it was not the only one. Thayer also inserted himself and the Fortean Society into debates over modernism and how it should be canonized. Throughout this period, artists like William Gaddis probed the limits of modernism, pointing toward future artistic movements; critics, meanwhile, grappled with the previous half century of avant-garde art, how to make sense of it, collate it in didactic collections, install it in galleries, teach it to a new generation. The idea that the avant-garde shared some quality or constellation of qualities that unified modernism was, of course, inherently nostalgic, coming "when the basic impetus of those movements was already moribund."[178] The Forteans—Thayer, Russell, Blish, and Lilith Lorraine—focused their efforts on one front in this battle: how to understand the poet Ezra Pound, and whether his poetry was worth celebrating or was too juvenile, too racist, too tainted by fascism, too technically jarring to warrant inclusion in the canon. His fate was emblematic of how modernism would be remembered and, therefore, how Forteanism would be recollected, for Forteanism had been born at the same time as the modernist avant-garde, its existence supported by the same social conditions. More than a contest over arcane literature, the fight was to save both Forteanism and modernism.

Pound had announced himself noisily in the 1910s and lived loudly for decades thereafter. He was convinced he'd discovered the key relationship between language and reality, a scientific analysis that "dared show modernity its face in an honest glass."[179] Certain of himself, his "violent gods," he seemed, as his friend the poet W. B. Yeats said, possessed of a "desire to personally insult the world."[180] Pound hectored and lectured Margaret Anderson and Harriet Monroe and wore out his welcome in *New Age* with attacks on English culture. Having left America for London, he then left London for Paris, and Paris for Italy, where he settled in 1924. He held himself as an *enfant terrible* even as he was less and less an *enfant*.[181] In his *Who's Who* entry, he listed his hobbies as fencing, tennis, and "searching the *Times* for evidence of almost incredible stupidity."[182] This bomb-throwing Pound, so much like Fort in his obstinance, was the writer whom Thayer discovered and admired for the rest of his life. Pound, noted Thayer, was dismissed as "a talented, angry, exhibitionistic small boy thumbing his nose at the world," which was all to the good: "Are there, then, other kinds of writers—other reasons for writing books."[183]

As he moved into economic propagandizing in the early 1930s, Pound remained as recalcitrant as ever. C. H. Douglas's social credit and the equally radical ideas of German economist Silvio Gesell were the world's best hope of overcoming the Depression that had descended upon it, Pound thought. He wrote books and essays on his ideas, tried to influence national leaders, kept up a large correspondence on economic topics. (Thayer was drawn into this network in the 1940s.) The scientifically precise poetry he'd invented was put to use explaining economic ideas in his Cantos, a lifelong cycle of poems that he'd started in 1915 and updated periodically.[184] After an initial rush of promise, though, social credit ideas gained little influence, petering out at the end of the 1930s, by which time Pound was resigned that economic reform would not come through mere advocacy. "Till 1940 we are INSIDE the nude eel," he said, using wordplay also employed by Iktomi: nude eel = New Deal.[185] Pound turned his attention to politics and made common cause with fascists, whom, he thought, seemed most likely to support the kind of economic reforms he favored. He contributed to fascist magazines in Britain and in Italy. His concern over finance transformed easily into a bitter anti-Semitism, the conviction that an international cabal of Jewish bankers conspired against reform and led the world to war so as to reap bloody profits. (Thayer imagined a similar conspiracy, but while Pound placed the blame wholly on Jews, Thayer regarded Jewish financiers as one group among many criminal operatives.) Pound's poetry became increasingly anti-Semitic, and in 1941, he began a series of radio broadcasts for Italy that supported Mussolini's fascists and attacked Jews as "filth," "kikes," accusing three score of orchestrating all of World War II with a handful of non-Jewish collaborationists.[186]

Pound was arrested by American forces in May 1945, as World War II in Europe came to an end. Still apologizing for Mussolini and Hitler, though denying anti-Semitism, Pound spent three weeks caged in a military camp. There, he started writing a new entry for his Cantos, apparently scribbling with pencil on toilet paper.[187] Returned to the US, he was involved in a long legal battle, eventually deemed insane, and installed at St. Elizabeth's hospital, in Washington, DC, where he would spend twelve years.[188] Although his freedom was reduced, Pound continued to write and, through the devices of his lawyer, Julien Cornell, was permitted to entertain visitors. He turned his room into a literary salon, receiving poetic pilgrims, issuing ex cathedra rulings on the state of art, the state of the world.[189]

Pound's institutionalization proved to Thayer and other Forteans that America was rotten.[190] Thayer repeatedly asked Cornell—previously, he'd been the Fortean Society's lawyer—for updates on the likelihood of release. He sent Pound letters.[191] He dithered over how strongly to condemn the shameful treatment of "the greatest poet ever to have the misfortune to be born in the United States."[192] Cornell advised Thayer to keep quiet so as to not upset the delicate arrangement, which he did grudgingly, nonetheless slipping surreptitious homages into *Doubt*, such as following Pound in calling newspapers "wypers."[193] Kathleen Thayer visited Pound, as did Tiffany; he encouraged other Forteans to do the same.[194] He connected with Pound's family, taxied them when they came to America, and encouraged a love for Fort in Pound's son-in-law, Prince Boris de Rachewiltz, helping spread Forteanism into Italy.[195]

Through the 1940s, Pound and Thayer shared a deepening bond. Thayer sent to St. Elizabeth's books, magazines, *Doubt*; offered money, food, even underwear.[196] Pound annotated his copies of *Doubt*, underlines and marginalia highlighting stories on economics, the perfidy of Catholics.[197] Thayer pumped him for information. The Duke of Bedford—Nazi sympathizer, social credit advocate—

was an Accepted Fellow of the Fortean Society, but Thayer found his politics "a bit foggy" and asked Pound for clarification.[198] He made a similar request about the poet George Sylvester Vierick, whose "principles" remained "imperceptible" to Thayer.[199] (Vierick was a poet and author, imprisoned during the war as a Nazi asset, part of the extended circle of fascists with whom Thayer kept company.)[200] Thayer remained allied with N. V. Dagg because of Dagg's connections with Pound.[201]

Pound reciprocated the collegiality, sending a clipping and a suggestion. Conduct a census of "tentative" or "potential" adults, he said, people who understood the basis of Pound's political philosophy: modern society's need for individuals uncowed by the state, unbowed by the economy. As he would say in another context, "The whole fight is for the conservation of the individual soul."[202] Pound also sent some aphorisms to run in *Doubt*. Thayer begged off publishing them, because, he said, they were insufficiently Fortean—a weak line, given his capacious definition of Forteanism; probably he was concerned about publicity.[203] As for the census, it would only embarrass: "I should like to be able to say that the entire roster of the Society would do for a start, and, alas, that isn't true."[204]

The poet and the Fortean also debated the purpose of art. Pound believed that transcendence through art was possible but difficult, requiring the kind of intense study and precise technique that characterized his works. Thayer, still influenced by the Nietzschean nihilism of his youth, dismissed the very notion of transcendence, of inherent beauty:

> The only arguments I have heard in defense of reverence adduce aesthetic values as a basis for art—a rose, a sunset, the female breast, a desert sky at midnight, a perfect fifth, and so on: but these are all arbitrary, relative, all "acquired characteristics." Wherefore revere—and what? . . . I feel stabs of ecstasy as I observe some of these phenomena (witness my long-time memory of and devotion to your Exile's Letter), but I think I am able to trace my pleasant pangs to experience—personal or racial. Even if

I could not, I should never bow. And not because I am proud or stubborn but because I see no reason for it. The miracle would be if there were no roses—and all the rest.[205]

The disagreement stuck with Pound, and he remembered Thayer's comments fifteen years later during an interview with *Paris Review*.[206]

As he corresponded with Thayer, Pound prepared for publication of the latest addition to his epic, to be called *The Pisan Cantos* in honor of where he'd been held as a prisoner of war. It is too much to say that Thayer or Forteanism played a significant role in the Cantos' construction, but also a mistake to ignore their influence, for Pound, in his encyclopedic way, digested his Fortean interactions and inserted them into the poem. One reference may not have been a reference at all, Thayer's imagination riding ahead of the facts. One was minor. And one, more speculatively, was closer to the core. The first supposed callout to the Fortean Society comes as Pound taunts the English, hoping for economic reform there:

> Now that there's room for doubt
>> And the bank may be the nation's
>> And the long years of patience
>> And labour's vacillations
> May have let the bacon come home.[207]

Thayer took the word "doubt" to be a hat-tip to his magazine, and an acknowledgment that several people associated with the society were among those working to free England's banks from the grip of financiers.[208] That reading is creative but built on too slender a reed, a lot of weight placed on a single word. The second reference, though, certainly concerned Thayer's organization. In one section of the poem, Pound muses on the passing of beautiful things, the mind's ability to remember them and all their associated imagery, at the same time wondering where those beautiful things go. Some might invoke reincarnation—what Pound called metempsychosis—

but he undercuts that possibility, citing no less an authority than the Theosophist G. R. S. Mead, who sarcastically poked at those who claimed to be reincarnations of great historical figures: "what have they done in the interval"—Pound asks—to have been born to such lower forms; it must've been something devilish. Having dispensed with metempsychosis, Pound writes, "and there are also the conjectures of the Fortean Society."[209] Perhaps in the vast corpus of *Doubt* lay some explanation for beauty's fate, something occult, something bizarre, some mechanism that explained how beauty was damned. Thayer was ecstatic at having his society recognized by Pound.

Immediately before his meditations on the passing away of luminous items, Pound writes, "Beauty is difficult." The phrase is repeated several times, attributed to the decadent artist Aubrey Beardsley (and his pallid imitator John Kettlewell). Pound was responding to the flux of his life, its decomposition into fragments—he'd lost control. But adrift though he was, Pound was not without resources, the ability to rescue the fragments, his memories, and through them touch the beautiful.[210] Beauty was difficult: difficult to create, to appreciate, and to recreate after its loss. Importantly, the phrase recurs immediately after the reference to the Fortean Society. There is no proof, but it is hard to resist the interpretation that Pound organized the lines with Thayer in mind—a third reference, then. He was talking back, answering Thayer's nihilistic philosophy with his own. The Fortean Society, in the form of its secretary, speculated that beauty was impossible, but Pound knew better. However hard to access, beauty was real, obdurate and difficult.

The Pisan Cantos was published in July 1948. The following February it won the inaugural Bollingen Prize, given to the best book of verse put out in the past two years. Pound's admirers had gamed the system, timing *The Pisan Cantos* to be eligible and assuring that most of the judges were supporters, thinking such recognition would be cause to have Pound freed.[211] Instead, the award provoked a controversy, one that engulfed the Fortean Society. America's artistic community was changing in fundamental ways, modernism and the avant-garde losing their edge, just as William Gaddis's novel *The Rec-*

ognitions suggested. Modernism's techniques were passé. The avant-garde's focus on individual liberation had opened the movement to co-optation by institutions. The government and staid bodies such as the Rockefeller Foundation and the Ford Foundation supported modern art as a weapon in the Cold War: proof of America's freedoms and their superiority to the tyranny of the communist Second World.[212] (Perhaps the Ford Foundation's openness to patronizing modern art was why Thayer applied to it for support.) The same language of individual fulfillment helped transform works of avant-gardism into consumer products—not an opposition to official culture, but a bibelot of the bourgeois, decor for corporate offices, the stuff of advertising: the poems, novels, paintings produced by artists who sought to liberate themselves could be sold to others, used as evidence that the owners had reached personal nirvana. Universities had been one of the institutions against which avant-garde artists defined themselves; there too, by the 1960s, an alliance had been forged, as colleges hired artists working in that tradition and offered classes on modern art. "The academic take-over," John Atkins called it.[213]

Pound's award came at the beginning of the avant-garde's end and so was especially fraught. His anti-Semitism, his defense of fascism— for many, these seemed enough to keep him out of polite company. Others objected that his poetry itself, jagged and without regard for formal conventions, made him unworthy of the prize.[214] But the protest was also about how modernism should be canonized. His award could be read as a sign that the boozy, brawling, explicit, political, and adventurous avant-garde, which had, for the better part of the century, insulted the middle orders, should be welcomed into the parlor. A confederation of conservative cultural criticism coalesced in the late 1940s and dominated the 1950s. Pound's opponents belonged to this movement. These critics repudiated avant-gardism, disdained modernist experimentation and fragmentation. A loosely allied group of anticommunists shaped the American poetry canon, shaved off its radical edges, simplified its history, all under the cloud of the Cold War, McCarthyism, the felt imminence of atomic war.[215] The battle absorbed, fractured the Fortean community.

Leading the movement to filter modernism of its so-called im-
purities was Lilith Lorraine, her close associate the science fiction
author Stanton Coblentz, and the historian and poet Peter Vierick (son of George Sylvester). Vierick initiated a new conservative
tradition, its politics opposed both to his father's authoritarianism
and to the utopianism of communism and liberalism. He wanted a
rebellion against modernism, itself now oppressively conventional:
conservatism was the true radicalism. This movement to reconstruct
modernism was never unified; rather various poets and commentators worked out their ideas publicly, in their writings and the collections they edited (though Lorraine and Coblentz did partner
around a particular vision of poetry). Within the cultural logic of
the Cold War and McCarthyism, there existed a narrow repertoire
of possible actions, most tending in one direction: the deradicaliza-
tion of modernism, the creation of a canon that was either apolitical
or tended toward conservatism. Even those, such as the editors of
Poetry magazine, who were sympathetic toward modernism but favored aesthetics over politics, willingly cut off poets who had once
been popular if they belonged to the wrong ideology or expressed
their political views too clearly. Pound, radical, overtly political, fit
poorly into the new canon. He was viewed with suspicion: his buzz-
saw rhythms not serving aesthetics but other, meaner gods.

Lorraine and Coblentz formed the League for Sanity in Poetry,
this before Pound's institutionalization, but his situation added
mordancy to the title. Their league worried over the fetishization of
modernism and progress, the turn away from eternal verities. *Different* offered articles sussing out the hidden communism in modern
poems. (A companion feature in Coblentz's 'zine *Wings* called "This
Is Not Poetry" showcased examples of how terrible modern verse
could be.) Lorraine assembled a mailing list of seventy-five thousand
names, cultivated conservative donors. Avalon, she announced, was
"dedicated to the development of a unified world culture, to the dis-
couragement of bewilderment, incoherence, cynicism and defeatism
in all the arts, and to the fearless analysis of those forms of mass de-
ception that menaced the peace of the world through the distortion

of truth."[216] Pound's award provided an excuse for her to attack him and the rot of modernism.

(Relatively) quiet about Pound's institutionalization, Thayer and other Forteans let loose against those who attacked his winning the Bollingen Prize. The science fiction author James Blish, who completed a master's thesis on Pound's poetry (and borrowed a Pound alias for his critical writings), insisted that *The Pisan Cantos* was not anti-Semitic. Pound used ugly Jewish stereotypes as symbols—metonymy—not as a description of reality. The focus should not be on the poem's supposed politics but on its aesthetic integrity.[217] That Pound had been institutionalized was evidence not of his insanity but of America's incapacity to withstand the acute insights of genius.[218] Thayer offered lukewarm support of *Different* when he first saw it, but subsequent attacks on Pound in its pages turned him, brought to light the same misogyny that motivated his and Russell's response to Judith L. Gee.[219] Russell, admitting he'd not read *The Pisan Cantos*, dismissed any extratextual complaints in an essay rejected by *Different* and published in *Doubt*, even as Thayer griped the piece was not nearly strong enough.[220] In the article, Russell reiterated comments he'd made to Gee, claiming anti-Semitism did not exist.[221] After the attacks on her, Thayer never again mentioned Lorraine, in correspondence or in *Doubt*.

Despite the strafing by her fellow Forteans, Lorraine proved victorious in the 1950s, or at least her vision of art did. The conservative critique did not have much of an effect on the San Francisco Renaissance or the Beats. Their description of the imagination as an autonomous realm where one could find refuge from what Lamantia called the "corny reasoning, sick politics" of "the stupid, synthetic half/life of postAtomic Bomb man" remained.[222] The tradition was vigorous, animating American culture into the 1960s, both mainstream and not. That was the exception, though.[223] As it was canonized, modernism had much of its radicalism sawn off, just as Lorraine, Coblentz, and Vierick had wanted. The energy that had launched Fort, as it had Pound, was dissipated. Finally freed from St. Elizabeth's in 1958, Pound fell into a depression. "You—find me—in fragments," he told an interviewer in 1960.[224] Thayer was antsy, concerned that the increasingly conservative literary scene left no room for Fort, and this at the same moment the society was reeling from changes in science fiction. In the late 1950s, Thayer suffered attacks from Art Castillo, then in Mexico, the literary critic Jack T. Jones, and journalist jack green, one of the leading voices in support of Gaddis's *The Recognitions* at a time when few defended the book.[225] They raked Thayer over the coals for what they saw as simpering contrarianism, while he saw these "New Scholiasts" offering only "abstractions . . . pitted against abstrusities." The world had changed since his youth, and there no longer seemed room for him, or the authors he loved:

> Without the humor, color, wit and wisdom of a Fort or a Joyce or a Pound, such stuff is nothing but suet, not worth the time it takes to spoon through it, less rewarding than working a crossword puzzle, more like "clearing" through Ron Hubbard or playing Tipperary on a comb.[226]

He worried for Forteanism's future, feared there might be none, the world so changed there was no longer room for it.

Caresse Crosby and Black Sun

In January 1948, Caresse Crosby, owner of the small but influential Black Sun Press and a patron of the arts, received a manuscript titled "Manifesto for Individual Secession into a World Community" by Rufus and Janice King. "I had in my hands the cornerstone of a new World Order," she thought. "I knew it would have no immediate political or financial success. The best I could promise was to get it into print in Paris in the spring, and to distribute copies."[227] After three decades of senseless war, tens of millions of dead, hundreds of millions suffering, the birth of an apocalyptic weapon—the world needed to be fundamentally changed, its way of being humanized. Crosby's was a variation of secularism, not skeptical of religion or supportive of science, but centering civilization on the value of the human being. Individual consciousness would be cultivated under this new system, partisans pledging allegiance not to corrupt nations but to humanity—a superior alternative to the new United Nations, then meeting in Paris. She spent the next twenty years using her considerable resources, working with humanists, secularists, and artists to support the formation of a world government.[228] Mondialization this movement was sometimes called—one world. It attracted a bevy of idealists. In May, Black Sun created the first world passport and found its first taker, Garry Davis, who renounced his American citizenship and harangued visitors to the UN offices in Paris. The Fortean Society supported the movement, a few of its members mondialists, among them Davis and R. Buckminster Fuller. Thayer became friends with Crosby, a skeptical ally.

Crosby had visceral reasons for her pacifism. The Great War had turned her first husband, father of her two children, into an alcoholic. Her second husband, Harry, was a veteran, too; his job collecting body parts from the killing fields warped him into an unsentimental nihilist. Caresse and Harry were both scions of wealthy families; Caresse (an adopted name) also had money from an invention she'd made while young, the brassiere. She and Harry established themselves in France during the 1920s, drank recklessly, took drugs, engaged in orgies and extramarital affairs, whiling away the hours until the date they'd appointed to kill themselves. Amid the debauchery, they ran Black Sun Press, publishing small, beautiful editions of their own work, of modernist experiments by the likes of James Joyce. Artists flitted into, out of their abode— Hemingway, expatriates damaged by the war. In 1929, Harry, suffering still from shell shock, killed a lover, then himself—a deep violation that sent Caresse into a spiral. She recovered, continued Black Sun, launched other publishing ventures, opened an art gallery in Virginia. She housed Henry Miller as he vagabonded, offered aid and comfort to Pound; it was likely through Pound that she met Thayer. In the 1950s, she took control of an Italian castle while traveling with Rexroth, hoping to make it the center of a world government, transforming it instead into an artist's colony. (Thayer tried but failed to see the castle on his trip to Italy.)

Even as Crosby moved into politics, she continued her patronage of the arts, notably supporting Charles Olson. Olson had been Pound's amanuensis at St. Elizabeth's, enduring his violent, racist rants in order to be mentored by "the great man," before, finally fed up, he engineered a break.[229] Returned from a trip to California, Olson told Pound that George Leite was not Aryan but Portuguese, provoking a "slew of racist commentary," some of it personal, that Olson used as pretext to end the relationship, though the separation was not clean, Olson forever responding to Pound.[230] Like William Gaddis, Olson was looking for something to invigorate the arts, a "post-modernism," he called it.[231] Pound had sought to condense energy, ideas, emotions into a single perfect image. Olson opted for the opposite, ridding the mind of its habit of classifying the world, releasing the energy Pound contained, his poetry propelling readers along lines organized "not by metrical feet but by the natural rhythms of thought, breath, and gesture," not by logic, syntax, or form but by each image ineluctably suggesting the next.[232] Against modernism's endlessly refracting mirror of self-fascinated-by-itself, Olson posited "the getting rid of the lyrical interference of the individual as ego."[233] In the 1950s, Olson became director of Black Mountain College, an experimental school in North Carolina with connections to the West Coast that nurtured a number of the Beats.[234] Black Sun published his (unintentionally Fortean-titled poem) Y & X, and Crosby supported him as he embarked on the work that would consume him the rest of his life—the Maximus poems, a response to Pound's Cantos, written according to Olson's own poetic principles.

Connection between the Fortean Society and Crosby started tentatively, Thayer approaching in 1950, she sending a clipping about a sunflower growing in a girl's lung, before a friendly alliance formed.[235] In 1953, they met face to face, subsequently dining together whenever Crosby was in New York.[236] Crosby distributed scores of copies of Doubt at peace conferences.[237] Through 1954, Thayer promoted Crosby's Citizens of the World, connected her with British Forteans he thought would be supportive (even Gee).[238] But for all his support, Thayer had his own ideas about the movement. World citizenship impressed him as regimented, too much like another form of government. Explaining mondialism to the society, he discounted the community-building in favor of its anarchic potential:

The individual is the all important unit. Each is his own universe. Living is the chiefest of the arts, and the aim of the organization . . . is

Charles Olson writing at Black Mountain College. Olson was influenced by Ezra Pound and seems to have had some familiarity with Fort—possibly picked up at Black Mountain College, maybe from Caresse Crosby. Olson sowed the seeds of a new cultural arrangement, exploding Pound's modernism into a postmodernism that, while indebted to Fortean skepticism of science, had less room for Forteanism, at least initially. (Courtesy of the Western Regional Archives, State Archives of North Carolina.)

> to see that the opportunity is not lacking anywhere on Earth for indi-
> viduals to realize their fullest creative potentials as universe builders
> and movers.[239]

As well as pacifism, Crosby's patronage of the arts linked her to the Fortean Society. In January 1953, Crosby paid her first dues and requested that Thayer also send *Doubt* to Charles Olson (and to architect Edward Stone and Andre Magnus, a French film publicist).[240] Subsequently, it seems, the society and Black Mountain College exchanged publications, as had been the case with Leite's *Circle*. Thumbing through *Black Mountain Review*, Thayer found the fall 1954 issue, featuring a Robert Duncan poem, a review of Pound's poetry, and Olson's meditation on the damned facts of history, relatively "exciting."[241] Fortean ideas circulated among Black Mountain students and alums, Fort's books appreciated by Jonathan Williams, R. Buckminster Fuller, and, though he tried to disguise the influence, William Burroughs.[242]

There's some evidence that Olson found support for his poetic practice, what he called "projective voice," in a dusty corner of Fortean thought, likely led there by *Doubt*. Reider T. Sherwin was a Norwegian immigrant who wrote eight volumes arguing that Norwegian travelers around the year 1000 taught their language to Native Americans—an encounter that forever altered the aboriginal languages of the New World.[243] The bulk of Sherwin's evidence was phonetic similarities. Sherwin became associated with the Fortean Society, which sold his books, evidence of pre-Columbian contact between Europe and North America being one of Thayer's pet topics. Olson collected all eight of Sherwin's volumes, wove references to them throughout the Maximus poems, which he worked on into the 1970s, when he died from complications of alcoholism.[244] For Sherwin, sound was the first clue to linguistic families—and for Olson sound was what helped connect one image to the next: "keep moving, keep in, speed, the nerves, their speed, the perceptions, theirs, the acts, the split second acts, the whole business, keep it moving as fast as you can, citizen."[245] For Sherwin, a word could be decomposed into an exact object; for Olson, the best poetic words were those that were the most concrete.

Olson's poetic practice was intended to clear away the debris left behind by modernism and make way for something new. Projective verse dispensed with the ego, so central to modernism, and with logical classification, which was what made Pound's poetry scientific: its ability to exactly specify and contain within words intense human emotion. Olson distrusted the very notion of "human," seeing it as a misleading category of analysis. His poetry—not unlike Fort's prose—turned the focus from humans and their strivings toward geology and astronomy, the movement of tectonic plates and the cycling of constellations. Things, the environment—these were more important to Olson than the passing ideas of humans. Postmodernism came to be associated with writers similarly inclined to question all categories, to see "history" not as a given but as something constructed by people for particular purposes.[246]

The idea of "humankind" was similarly a fiction, invented by scholars of the past and certain to be deconstructed in the future.[247] Artists began to explore the constructedness of these ideas, their instabilities. Their works, as Olson's poetry, displayed a "high tolerance for disorder," noted the scholar George F. Butterick. That tolerance was another attack on logic, classification, and categorization: "It is . . . intended to test our faith in the representative power of language."[248] Over the rest of the twentieth century, postmodernism would become increasingly dominant, elbowing aside (though not completely supplanting) modernism. For a certain segment of the artistic vanguard, it became commonplace to assert the impossibility of aspiring "to any unified representation of the world."[249]

In the end, it was Crosby's support of the arts that endured, her dreams of creating pacifist institutions unfulfilled. Bureaucratic hassles prevented her from setting up a center for her world government, and other events in the middle of the decade slowed her activism.[250] In August 1955, Crosby's son died of carbon monoxide poisoning; her daughter-in-law sickened but recovered. "We are stunned, devastated, speechless before the monstrous unfairness and injustice of a fate so hard," consoled Kathleen and Tiffany. "Where, in all classic tragedy, is a heroine to compare? O, dear, brave girl, our hearts ache for you."[251] The following year, Crosby's health declined; doctors told her to slow down. She was sixty-five. She recuperated, returned to her work. Thayer remained supportive, but *Doubt* less often reported on mondialism. He had been turned off by its worst excesses: yet another disenchantment. San Francisco, however, seemed intrigued by her ideas, and she found support among the elites, if not yet the masses.[252] In the 1960s, having outlasted the Fortean Society, she plugged along, preoccupied with the increasingly urgent questions of civil rights and her own legacy. Crosby donated her papers to Southern Illinois University (where Buckminster Fuller had found an academic home), hoping to endow a "threshold for peace"—not quite the center of a new world government, but a seed, perhaps to grow if given water and a bright sun.[253]

5

The Cosmic Aquarium

Why so desolate?
　　And why multiply
　　　　in phantasmagoria about fishes,
　　　　　　what disgusts you? Could
　　　　　　　　not all personal upheaval in
　　　　　　　　　　the name of freedom, be tabooed?

Is it Nineveh
　　and are you Jonah
　　　　in the sweltering east wind of your wishes?
　　　　　　I myself, have stood
　　　　　　　　there by the aquarium, looking
　　　　　　　　　　at the Statue of Liberty.
MARIANNE MOORE, "Is Your Town Nineveh?" 1916

In late 1950, Waveney Girvan wrote to Eric Frank Russell asking for help.[1] During the war, Girvan had been the director of a military contractor while in his off hours agitating for peace with Hitler. He wasn't a pacifist, rather an admirer of fascism, if not quite a fascist himself. Deeming communism the greater threat, he was concerned that the financiers who involved Britain in the war would make off with bags of money once Germany was defeated but fail to prevent the nation from falling into socialism.[2] His actions earned him an investigation by the government. After the war, Girvan shook off suspicions that he was a potential traitor, left his job with the military contractor, and moved into far-right publishing ventures, some of which, funded by the Duke of Bedford, employed fascists. At the time he wrote to Russell he was with Carroll & Nicholson.[3] The house was putting out a book on flying saucers, mostly, it seems, at his behest. He'd read *Tomorrow* for its political articles, and that had led him to an interest in the occult, the extraterrestrial, the Fortean.[4] The problem was, he'd learned that the astronomer royal planned a devastating review of the book. Girvan had arranged for *The Riddle of the Flying Saucers* to be serialized in the *Sunday Herald* but was worried that would not stir up enough support. Could Russell organize Forteans to present a public defense of the book, he wondered.

Although Girvan did not convince the Fortean Society to join his fight, he was right to think Forteans were allies. Forteanism was fundamental to what John Keel called the flying saucer subculture.[5] Long before UFOs caught the world's attention, Fort had written about strange objects in the sky, speculated about their provenance. The Fortean Society continued in this vein, Thayer gathering both reports of atmospheric anomalies and members inclined to wonder about them in ways unlike scientists and government officials. Fort and Forteanism expanded the imagination of those studying the anomalies, transforming unusual phenomena into new views of the universe and humanity's place in it. By applying Fortean formulae, students of flying saucers thought they could expose structural features of the cosmos and explain the forces that organized reality. They could see how power was arranged, who was in control of the world. The fit between the flying saucer subculture and Forteanism, though, was not perfect. As with science fictioneers and avant-garde artists, UFOlogists did not restrict their imaginations, descriptions, and explanations to strict Forteanism but mixed it with fundamentalism, New Agery, and conspiracism. Flying saucers were variously understood to be scientific objects, negations of science, and the fulfillment of modernity: the final reconciliation of religion and science, for good or ill. These speculations, and the phenomena that prompted them, fractured the Fortean community. Many members decided that flying saucers were a proper subject for the society, even subsuming all Forteanism to the study of UFOs. But Thayer despised both the topic and the subculture. And so the members and their secretary jockeyed for control of the society and of Forteanism.

New Lands

Charles Fort worked not just at the beginnings of science fiction and the transition from decadence to avant-garde modernism, but smack in the middle of a popular genre. The early decades of the twentieth century saw a plethora of stories about the upper atmosphere, the region just beyond the reach of crewed balloons or airplanes. With the coming of machines capable of reaching those places, in the years around World War II, these tales would move deeper into space. But for several decades, the focus was nearer earth: "the realm of the air," as the meteorologist Charles Fitzhugh Talman called it.[6] Maurice Renard's *Le Péril Bleu* belonged to this class of narratives, as did its English versions. There were stories of air serpents and batlike creatures, of strange races inhabiting sky islands.[7] In 1913, as Fort struggled with what would become "X," Arthur Conan Doyle published "The Horror of the Heights," about an aviator investigating pilots who'd died trying to set new altitude records. Forty thousand feet above the earth, the hero discovers, are jungles populated by creatures, snakelike or resembling jellyfish or with beaks and tentacles like squid: as though the upper air were an ocean and pilots making their way through a kelp forest.[8] Fort's description of a Super-Sargasso Sea, in which could be found the flotsam and jetsam that occasionally rain down and, perhaps as well, the monstrous owners of the earth, meshed neatly with this convention.

Doyle's tale, like many scientific romances, was presented as truth, written as if taken from a recovered diary fragment. Still, readers were presumed to understand it was fiction; it was published in a magazine of fiction, after all. Fort was more intent than his genre peers to blur fact and fiction. He worked with the same fictional tropes but presented his ideas without the trappings of fiction.

Digging through newspapers, scientific journals, and magazines, he uncovered genuine reports of strange objects in the sky, which he then offered as evidence for his philosophizing and what seemed to be genuine scientific speculations. Among his finds were what would later be called airship waves. From California, late 1896, came accounts of weird aircraft, big, seemingly heavy and powerful, first seen above Sacramento. Over the next several months, newspapers across the country reported similar objects maneuvering above other cities.[9] There was another flap around 1909 and another in 1913, this time in Europe.[10] "Extra-mundane voyagers," Fort suggested in *New Lands*.[11] The early twentieth century, he implied, was not so different from 1492, when residents of North America had watched three craft approaching over the horizon. Indigenous wise men might have dismissed their lights as the glow emitted by rotting fish, a feature of their culture's superstitions. But we knew better now.

Fort's playing at the boundary between fiction and nonfiction reflected a general cultural confusion: how could one make sense of the upper atmosphere and the spaces just beyond it? Were the romances of Doyle and others pure imagination, or did they point toward the possible? Talman, the meteorologist, dismissed Fort and anyone else who failed to distinguish scientific romance from fact. He laid claim to expertise about all "variety of phenomena and appearances in the atmosphere." Such things fell under the technical definition of "meteor," not just shooting stars, he said, and he was a scientist who vouchsafed the truth about such matters.[12] There might be fish falls and frog rains, he acknowledged, but they could be explained by tornadoes and waterspouts.[13] As for those who imagined life in the atmosphere or on near planets—ridiculous. His first published writing, in 1900, debunked fancies about Martian civilizations; more than forty years later, he scoffed at Thayer's offering him membership in the Fortean Society.[14] He wanted nothing to do with such rubbish. But Fortean speculation on extra-mundane life could not be so easily dismissed. Just as *The Book of the Damned* was published, in December 1919, Guglielmo Marconi, inventor of the radio, claimed to have detected signals from Mars.[15] Fort's publisher wasted no time in connecting his book to Marconi's report.[16] Perhaps Fort was not a crank but a visionary. "It is a bold man who says 'Impossible' these days," editorialized one journalist.[17] A few years later, the nations of the world collaborated to silence terrestrial radio communications for thirty-six hours to see if, with Mars at its closest to earth, signals from the red planet could be discerned. Something was detected, though no consensus was reached about what the "dots and dashes" meant, if anything.[18]

Talk of celestial seas, atmospheric jungles, and Martian civilizations was astounding but also cozy: the solar system was small and connected; easily traveled, as Fort wrote in *New Lands*, traversed in a matter of days. Pay no attention to the likes of Talman, Fort said, and his cosmological theories about the vastness of the atmosphere and space. Meteorology was, with astronomy, the epitome of hubristic science. Back in the 1770s, French peasants insisted that they'd seen stones fall from the sky, illuminated and hot, but savants, insisting that the earth was unconnected to the rest of the cosmos, dismissed the reports as popular delusions; the stones had not fallen but had been hit by lightning. And so, for a century, meteors were damned, until scientists finally admitted that, yes, stones fall from the heavens.[19]

Who could trust meteorologists, their arrogant explanations? The universe was not an unimaginable fastness, rather a tidy space that a stone could cross, as if thrown by a poltergeist, some "sly afreet," in the words of one Fortean.[20] Clark Ashton Smith once said that he meant his fiction to take place beyond the "human aquarium."[21] Fortean anomalies provided a means of escaping the claustrophobic earth and its rules. But Fort's ideas could push in the other direction too, shrinking the cosmos to human dimensions—a cosmic aquarium, with earthlings living on its aqueous bottom. Luminous objects in the sky were as fishes in the sea, and there were many. In late 1944, Allied pilots reported seeing oval things darting about them; "foo fighters," they came to be called. Not quite a year and a half later, observers in Nordic countries told of "ghost rockets" streaking across the heavens. Hardly more than a year after that, the world was introduced to flying saucers.

What to make of the anomalous sightings was as difficult, as controversial a question in the 1940s as it had been in decades past. Kenneth Rexroth thought them projections of subconscious mechanisms onto the screen of the firmament: reflections of the self, not real objects.[22] This view was ratified by the psychologist Carl Jung.[23] "The starry vault of heaven is in truth the open book of cosmic projection, in which are reflected the mythologems, i.e., the archetypes," he wrote.[24] Other commentators also explained accounts of atmospheric anomalies as psychological, though in a different sense: they expressed anxieties about the modern world and the death that had rained from above in both world wars, worries that grew almost unbearable with the atomic bombing of Japan and the brinksmanship of the Cold War.[25] Flying saucers did not exist, but people worried that they did as a displacement of their true fears. Fort's (and other's) argument that these things should be taken literally, as visitors from some other place, opened new interpretive possibilities. Perhaps the question to be asked was not the modernist one, about how these reports illuminated the self, its many aspects, rather one about the relation between the self and reality. Perhaps Charles Olson was right: one should attend to how the self is shaped by external forces, how stories are objects themselves, constructed according to patterns. Perhaps the focus should be not on how the mind works but on the universe of objects—a return not to realism and its concern with the attributes of quotidian life but to wonder about how reality itself is built.[26] Perhaps UFOs represented a way through the modernist impasse.[27]

Or perhaps these visitors proved correct yet another science fiction prophecy. Science fictioneers had accurately predicted atomic weaponry and, against the scorn of scientists, held on to the idea that rocketry was the science of the future. Now, it seemed, they had been right to obsess over objects in the high atmosphere and space voyagers. They had discovered yet another piece of history's motor. If humanity could but gain control of these things, who knew what powers might follow? Humanity might soon learn the secret of interstellar travel, be able to move freely through, even conquer, the cosmic aquarium. Or perhaps the powers might be not material and scientific but spiritual. The visitors might be Theosophy's Ascended Masters. Perhaps they portended a New Age, escorts to the next stage of evolution. Already there was a tradition among Theosophists of claiming contact with inhabitants of other planets. In 1935, Guy Ballard had told of going inside Mt. Shasta, that Theosophical hotspot, where he met a dozen Venusians that teleported—in Fortean fashion—before him. They played music, broadcast scenes of their home world, and informed him that earth would see dark times before an inevitable era of peace. Ballard and his wife Edna created a group, I AM, to spread the word.[28] Maurice Doreal's Brotherhood of the White Temple also conflated extra-mundane visitors with Ascended Masters and published pamphlets explicating occult interpretations of flying saucers.

There were other interpretations, too, a bevy of them.[29] The lights in the sky were natural phenomena, Venus or the moon or St. Elmo's fire or refracted lights. They were delusions, hoaxes, lies. They were experimental craft being tested by a terrestrial nation. They were not technological at all but organic. Maybe, others said, they were qualities of the atmosphere, ripples in the ocean of air that humans would understand once they reached high enough. Or perhaps, as Jung said, there was something psychological about them, some operation of the human mind that was worthy of study. They might be beings but not intelligent ones, as science fictioneers and Theosophists assumed, rather driven by instinct or impulse. Gerald Heard, whose book Carroll & Nicholson was publishing (a British Fort, Girvan said), claimed flying saucers were driven by beelike creatures from Mars. Still others challenged the Theosophical conclusion that such craft brought good news. They might be signs of the apocalypse. Fortean Norman Markham versified:

It's later than we thought!
Across the shrunken world a shadow creeps
And prayers are powerless against
The frightful fear that curdles in its path.[30]

The Fortean Society included partisans of all these perspectives—and more, too, coupling flying saucers to durable Fortean categories. Atmospheric anomalies were connected to teleportation, falling objects, mysterious booms, and bullets from nowhere (what Thayer called ballisterics); to disappearances, appearances, and monstrous creatures; to the blindness of astronomers and the duplicity of government officials. Earthquakes and oceanic anomalies could likewise be linked to the supposed crafts. These could be analyzed rigorously, as Norman Markham tried to do by mapping Fortean events against the movement of the planets, or discussed with a wink and a nudge—an ironic appreciation of the marvelous. Walter H. Kerr, a stolid MFS, joked that Forteans knew poltergeists—jinn, genies, afreets, what-have-you—were behind all the phenomena, cosmic tricksters, even as astronomers prevaricated and governments accused each other of rattling sabers.[31]

The Fortean Society was preadapted to the coming of flying saucers, ghost rockets, and foo fighters, prepared with explanations and categories of analysis. The flying saucer subculture would always be broader than the Fortean community, and an adversarial relationship would develop between it and Thayer. Nonetheless, the society was one of the sources of the subculture, and Forteans, broadly speaking, proponents of taking the objects seriously.[32] Fort's ideas could serve as a vessel to contain the phenomena, although not without controversy. Some would insist even Fort was too cramped a thinker and would shatter his theories to make room for the strangeness they saw.

Trail to Empire

The end of World War II, thought Robert Lee Farnsworth, promised the beginning of an American empire that would stretch into the cosmos. Nuclear power had destroyed two cities and brought imperial Japan to its knees; now it could be harnessed for peaceful purposes, used to fuel rockets to the moon and warm astronauts when they arrived.[33] Within a few years, he predicted in the late 1940s, earth's lone satellite would be the "Chicago of the solar system," a crossroads of trade.[34] Farnsworth had dreamed of this moment for several years.[35] Born in 1909, he was raised in a Chicago suburb, Glen Ellyn, the son of a real estate agent and village elder. He married Evelyn Swanson and they had two children, a son and a daughter. Farnsworth worked for a rubber company, then Pennsylvania Oil (in Chicago). In 1942, he founded the United States Rocket Society.[36] There'd be a race for the moon, he said. Whichever country reached it first would gain a military advantage, a cultural renaissance, unimaginable scientific advances. From the satellite, cartographers would map the earth's surface, meteorologists its weather, with uncanny precision; artists, freed from the constraints of gravity, would pioneer new techniques. The moon would be a jumping-off point for further explorations. In 1944, Farnsworth made news, petitioning the government for permission to homestead Luna.[37] He published a pamphlet, *Rockets, New Trail to Empire*.[38] He proselytized to science fiction fans in *Startling Stories* and reached out to rocket enthusiasts in *Popular Science*.[39]

Farnsworth's bumptious chauvinism could grate. He was technically proficient, engineers said; at least, his designs were reasonable. But his aims were abhorrent, said others. The science fiction writer Marion Zimmer Bradley accused him of lengthening America's adolescence. Rather than confront the problems they had created, war and pollution and monstrous scientific Frankensteins, men lit out for the frontier, only to create new problems. In the past, this had led to the extinction of the bison, the genocide of Native Americans. No one knew what horrors would follow the rockets, but that there would be horrors could not be doubted. Ignore the siren-call of the stars, she insisted, ignore Farnsworth, and remain on earth, solving the planet's problems, growing up before going out.[40] Members of the British Interplanetary Society were also turned off by Farnsworth. They wanted to reclaim rockets from the military and sketched cooperative, international plans for space exploration.[41] Arthur C. Clarke penned a bitter attack on Farnsworth's pamphlet, bemoaning his jingoism and water-carrying for American business. He feared Farnsworth's use of "all the most blatant tricks of journalistic advertising" would turn astronautics into a "laughingstock."[42] P. E. Cleator, cofounder of the BIS, went after Farnsworth in the *Journal of the British Interplanetary Society*. Farnsworth's response in *Rockets*, his own magazine, did nothing to mollify his critics.

Amid his self-promotion, Farnsworth buddied up to the Forteans. He came to know the fanzine editor and Fortean Donn P. Brazier, sending him material, even an autographed self-portrait, impressing Brazier that he was well connected to the military-industrial complex.[43] He purchased several copies of the omnibus edition of Fort's books in 1945, about the same time he joined the Fortean Society.[44] As he did with Brazier, Farnsworth swapped publications with Thayer and contributed material.[45] His Forteanism was broad; Farnsworth sent the society clippings on balls of fire, dead fish washed ashore by the ton, a baby strangled to death by a necklace, and cryptozoological reports. He shared with Thayer a distrust of the government, particularly Civilian Defense and the ways schools brainwashed students into accepting this limitation on their liberty.[46] And he became the force behind Chapter Three, the Fortean Society's Chicago branch.[47]

It was difficult, though, keeping on the right side of the Forteans while still appealing to (some) science fiction fans and maintaining credibility, if not good will, among those with a technical interest in rockets. Science fiction fans quibbled that Farnsworth's publications paid too little attention to their genre, and cost too much besides.[48] After an initial flush of enthusiasm, Thayer, as was his wont, turned on Farnsworth, finding him boringly besotted by the moon and too conventional. Farnsworth, Thayer griped, bought into any scientific pronouncement about rockets and space because it was in his financial interest to do so—just another grifter, then. Farnsworth's championing of nuclear power was particularly galling. "We wish that this compromise with the Great Atom Fraud were not necessary, but if Farnsworth wants $350,000 to build his rocket he must assassinate his conscience to get it."[49]

Both science fictioneers and engineers, though, considered him not prosaic but irritatingly Fortean. By 1946, complained fan Harry Warner, his "science had become adulterated with mysticism and Forteanism."[50] Farnsworth's magazine *Rockets* had a column on Forteana, and he was exchanging data with another Fortean, R. DeWitt Miller. Was the United States Rocket Society meant as a serious provocation promoting space exploration or a nose-thumbing joke at the scientific establishment? It wasn't always clear. Rocket enthusiasts wanted to separate rocketry from crankery, to make it a real science, and Farnsworth was hurting the cause. In a 1947 editorial for *Rockets* that was picked up by the United Press, Farnsworth surmised that the moon's craters were relics of a pre-historic atomic war between alien combatants. And perhaps those Lunarians came to earth, after that war, and were somehow involved with now-lost continents, such as Atlantis. A trip to the moon, the editorial said, could solve these ancient mysteries.[51] When flying saucers were first reported, Farnsworth was, not surprisingly, among those saying they were visitors from Mars.[52]

The Coming of the Saucers

On the afternoon of 24 June 1947, Kenneth Arnold flew his plane across the state of Washington on a business trip. Arnold sold and installed firefighting equipment across the West. Around three o'clock, he was interrupted by several flashes of light. Looking for the source, he noticed a line of objects, nine in all, that he could not identify: not birds, not a plane. Over the years, he would give different descriptions of their shape, their motion, but he was sure, based on his observations, measurements, and calculations, that they were large, each at least a hundred feet long, and fast, traveling well over a thousand miles per hour, which exceeded the capabilities of any plane then built. After landing, he reported what he'd seen, and the story made its way to the press: a businessman, an experienced pilot, he was a reliable witness.[53]

It took time for the story to become a sensation and for the national media to settle on a small repertoire of descriptions for such observations, categorizing the mysterious objects as flying disks, or saucers, based on Arnold's comment that the round forms looked "like a saucer would if you skipped it across the water."[54] Along the way, Arnold's report, and the many that followed, were attached to Fort, Forteanism, and the Fortean Society. Arnold could not decide what he'd seen—military craft of unique design or something stranger. Fort—and science fiction more generally, both materialistic and Theosophical—provided an interpretive framework. Fort had not been wedded to his "extra-mundane" thesis any more than his other suppositions, but the idea that the objects originated somewhere other than earth became, probably, the dominant position, though there was less agreement as to their nature. The objects were transformed into technological crafts (disks, flying saucers), or into creatures, or into elements of a drama that filled the entire cosmic aquarium; ideas about those nine luminous shapes in the sky rearranged humanity's relationship to the cosmos. Forteanism expanded the imaginative possibilities of Arnold's disks, spinning a few unusual images seen by a pilot into a reason to turn all of reality upside down. Perhaps they promised communication with other planets, or with unimaginable places. Perhaps humanity was indeed owned, or soon to be saved.

By early July, Arnold's story had spread around the world. American newspapers and magazines were filled with discussions of flying saucers, additional reports of disks, quotes from those who took Arnold's report seriously and those who thought it a grand joke. On 8 July, newspapers ran an Associated Press column by Hal Boyle that mocked the craze as an enormous flapdoodle. Martians were on a scavenger hunt, he joshed. One Martian craft had captured him—as he stumbled out of a New York bar.[55] It wasn't an auspicious date for *pshawing* disks, though. The day before, another businessman, this one in Texas, claimed to have found a disk stamped with a return address for the Army Air Force.[56] Also on 8 July came word from Roswell, New Mexico, that a crashed flying saucer had been captured. Subsequent developments only deepened suspicion and concern: in Texas, the businessman admitted he was pulling a prank, but the military's refusal to comment on the disk's existence made some speculate that the air force was hiding its involvement. In New Mexico, a military spokesman announced that the 509th Operations

Group had recovered a disk on a nearby ranch—a report later altered to say that the object was a weather balloon, dampening interest in the story for a long time.[57] But flying saucers—those remained a mainstay of the American press. More than a decade after Arnold's report, Thayer, surveying the continued interest, said, "It's hard to tell *Amazing* from the *Times* these days."[58]

From the beginning, the flying saucer phenomenon was connected to Fort, thanks in large part to Robert L. Farnsworth. Farnsworth was already known to Chicago-area journalists as a space expert; either he reached out to them in the days after articles about Arnold's experiences hit the papers, or they to him. In any case, he let them know about Fort. The first article linking flying saucers to Fort was written by an unnamed Chicago-based AP reporter, who called *The Book of the Damned* "a rare book in Chicago's Newberry Library," then listed a host of previous accounts of saucers he'd cribbed from Fort, suggesting the vast imaginative possibilities that inhered in flying saucers.[59] That story appeared in newspapers on 7 July. The next day an article by Claire Cox recounted an interview with Farnsworth. (Cox had previously written about his rocket advocacy.)[60] Farnsworth praised Fort as an essential source for understanding current flying saucer stories. Three days later, an article in the *Chicago Tribune* by Marcia Winn—who'd previously covered the Fortean Society's peace activism for the conservative paper, which was pacifist and isolationist in outlook—stated, echoing Fort, that whatever the objects were, they had been seen for a long time and in many places, so there was no cause to blame Russia or rush to war over fears that they were some heinous weapon.[61] Farnsworth made the same point in 1952, publicly pleading with President Harry Truman not to shoot down the saucers.[62] They were no threat, rather tourists who had visited repeatedly.

Around the time Farnsworth talked with Chicago journalists, Thayer also spent a Saturday fielding calls from reporters and broadcasters and answering telegrams.[63] He granted all who asked permission to quote Fort at any length. By the middle of July, journalists had not only discovered *Lo!* but upgraded Fort from the author of a rare book to an old friend. "Charlie," one writer called him.[64] Thayer devoted an entire issue, *Doubt* 19 in October 1947, to the subject, compiling and digesting hundreds of news stories. "The entire disk issue seemed a waste of time to me, but the membership expected it," he told Russell.[65] In 1948, Bernard Newman published *The Flying Saucer*, the first book to use the phrase in its title. *The Flying Saucer* was a thriller in which scientists colluded to achieve world peace by faking an alien invasion. At one point, Newman mentioned Forteans, quoted from *Doubt* 19, and gave the Society's address, prompting an inundation of queries.[66] Thayer was flabbergasted at the attention. "Now we are being taught in high schools. This is terrible."[67] New flying saucer groups were formed; magazines, amateur and professional, were published.[68] Waveney Girvan dedicated himself to the subject as editor of *Flying Saucer Review* and author of *Flying Saucers and Common Sense*, which included a chapter on Fort.[69] Kenneth Arnold went on to report more sightings, joined by thousands of witnesses through the 1950s and 1960s. He seems to have discovered Fort around 1950.[70] Two years later, he teamed up with Ray Palmer to put out *The Coming of the Saucers*. There's some evidence, though not definitive, that Arnold joined the Fortean Society at about this time and was one of the hundreds who sent Thayer clippings about disks each month.[71] In 1967, Ted Bloecher, a physicist at the University of Arizona, wrote an analysis of the 1947 events, choosing a quote from Fort as the epigraph:

> The little harlots will caper, and freaks will distract attention, and the clowns will break the rhythm of the whole with their buffooneries—but the solidity of the procession as a whole: the impressiveness of things that pass and pass, and keep on and keep on and keep on coming.[72]

Fort afforded the heft of history to flying saucer reports—no phantasmagoria could persist for a century—and also enlarged the imagination, giving those interested in the subject freedom to construct meaning from the reports. Even among those inclined to skepticism there was room for clowning, the flexing of the ironic imagination. David Bascom, an Oakland advertiser, told that city's *Tribune* that he'd seen lights in the sky, just as Arnold had, only they looked less like saucers than gravy boats.[73] (Thayer, his senses apparently dulled by reading through so many reports, included Bascom's description in his summary for *Doubt* 19, not recognizing it as a joke.) The following year, Bascom wrote the *Tribune* to contradict its reporting on the moon, which was obviously "misinformed" by "completely erroneous information" from science and astronomy texts. It was essential, he said, that Americans "establish bases on the moon before some other nation or planet gets the jump on us." But his methods were far more imaginative than those of Farnsworth and other rocketeers. The moon, he explained, was composed of green cheese, its craters made by nibbling rats; the lunar atmosphere was rich in Cheddaroxide, which could easily be converted into oxygen via a "common cheese-gas converter" added to any gas mask. As his authority for these claims, he cited Fort.[74]

Also circulating in the early days of the flying saucer craze was the suspicion that the flap was in fact the creation of Bascom's peers in ad agencies. Marcia Winn opened her column by jesting, first, that the disks were being thrown by a champion collegiate discus team and, second, that a soft drink company was somehow behind the sightings. The piece's title, "Pie in the Sky?" harkened back to Fredric Brown's "Pi in the Sky," which had run in *Thrilling Wonder Stories* a few years before. In that story, stars were rearranged in the night sky for the banal purpose of selling soap.[75] The pulp writer Kathleen Ludwick connected these musings about advertising directly to Fort. Like Bascom, she lived in Oakland and had read the *Tribune*'s mealy-mouthed editorial on the saucers, which conceded only that the topic was fun. The editorial also made passing reference to Forteans.[76] Ludwick wrote into the paper, excited to see her people called out. "Well, shiver my timbers! Was I astounded, amazed to read the reference to Charles Fort, the Apostle of Doubt, the High Priest of Skepticism, in the *Tribune*!" she began. Ludwick was skeptical of

Cover of *Doubt*, no. 25 (Summer 1949). Forteans had always been suspicious of official pronouncements about objects that seemed to come from space. At first, Thayer thought sightings of flying saucers were evidence of astronomers' blindness, but as reports continued into the 1950s, he came to suspect an official conspiracy: the people were being lied to.

both astronomers and flying saucers. Like Fort, she'd concluded that the universe was smaller than scientists said. She doubted the saucers were weapons, thought it more likely that they were an advertising stunt, like the blimp that flew over Oakland selling gasoline.[77] She quoted Fort, "One looks up and sees, instead, an illuminated representation of a can of spaghetti in tomato sauce, in the sky."[78]

Although deflating, Ludwick's interpretation overlapped with those that offered more substantive theorizing, for in invoking the actions of advertisers, she conceded that the lights seen in the sky were real, not the reflection of Venus through beer goggles or the humbug of practical jokers. Flying saucers were part of the universe's furniture set. The question was, What kind of objects were they? Were flying saucers no different than the commercials Thayer wrote? Or were they something stranger? Fort was as useful to those who took the disks as an indication of reality's strangeness as to those saw in the saucers only the dispiriting mechanics of corporate capitalism. John Keel remembered that as a seventeen-year-old writer and science fiction fan, "I assumed, after reading Fort . . . they must be spaceships. Fort didn't really come right out and advocate the ET thesis but he said there was something there and that it had been around for a long time."[79] Ivan Sanderson was also beguiled by the things—"things" his catch-all for unusual phenomena, his way of insisting they were tangible, not spiritual.[80] He developed a theory, not unlike that ventured winkingly by Eric Frank Russell and seriously by Norman Markham, that the saucers were connected to the oceans. Perhaps, Sanderson said, the unaccountable aerial anomalies being reported around the globe originated not on other planets but from the watery depths. Perhaps there were submarine civilizations yet unknown.[81] In either case, flying saucers made weird the aquarium in which humanity lived; Keel, Sanderson, and a host of others sketched a large drama in which humans played only a part. Cozy as the universe was, its corners had not yet been fully comprehended by scientists and conventional thinkers. It would take the imagination unleashed by Fort to fully fathom reality.

Or perhaps the saucers were not machines at all but organisms, like Russell's Vitons, creatures that swim in the ether. Fort could be read to support this contention as well. John Philip Bessor, a paranormal researcher who'd given up investigating "ghosties" to concentrate on saucers, claimed "these aerial objects represent a highly attenuated form of intelligent 'animal' life of extra-terrestrial origin—possibly stratospheric or ionospheric."[82] Perhaps these creatures, similar to those of Doyle's air jungles, moved by teleportation as well as undulation. "Strange, luminous creatures inhabit the depths of our seas," Bessor told readers of *Life*. "Why not similar creatures of highly rarefied matter in the heights of our heavens, and as diverse in size and shape as living things on earth?"[83] Farnsworth offered a similar explanation for the saucers, at least provisionally. It was possible, he said, that life had evolved on Venus that could fly by use of electric currents, "such as a sting ray fish which has an electric charge." These "fish from Venus" might then have traveled through space and hovered above the earth because they could not survive the planet's atmosphere.[84] A onetime astronomer (and a friend of Sanderson), Morris K. Jessup, postulated that the region between the earth and moon might be inhabited by a race of beings that collected humans and animals for experiments, like Fort's Ambrose collector with a more catholic diet, occasionally discarding remnants as garbage that falls to the earth.[85] This hypothesis would make sense of mysterious disappearances, bizarre rains, and aerial peculiarities.[86] Perhaps humans were indeed property, the fished-for prey at the bottom of a space ocean.

These imaginings placed humanity in a fundamentally new relation to the cosmos, a mixture of the disenchanting and the wonderful. Since time immemorial, objects in the sky had been interpreted as signs, tokens of fate and divine will. Over the previous two centuries, scientific investigation seemed to challenge this assumption, reducing the universe and its elements to dead materials following mechanistic laws. Einsteinian physics allowed incredibly precise predictions about planetary motion, the structure of space-time. Mastery of natural forces made possible the launching of human crafts into the skies; no longer, then, the province only of the gods. Some

Fort-inspired musings on flying saucers synthesized these two views: the objects of space could be characterized by scientific equations; they yielded themselves to mechanistic analyses. Nonetheless, they influenced human life in ways that remained imperceptible but might soon be as legible as Venus's orbit or the precession of Mercury's perihelion.[87] In 1948, a fiery bolide shot across the Midwestern skies, shocking many observers, but not Fortean Norman Markham. He noted that the meteor's appearance might be tied to the moon's position relative to the earth. Perhaps there was a lunar civilization, like the ones he thought existed on Venus and Mars. Extra-mundane beings came to earth when travel was easiest, at what would have been called—in days past, using a different connotation—propitious moments. Markham explained his idea to the US Army, which passed his letter on to Lincoln La Paz, a meteor expert consulting on disk matters for the military. La Paz thought Markham's conclusions too "fantastic" to be worthy of consideration but admitted that he had compiled a number of anomalous happenings that could not be easily explained.[88] Some UFO historians wonder if Markham's method of correlating reports with the position of nearby planets influenced the military's procedures, for the army and air force, too, compared sightings with the periodicity of heavenly bodies, as though life on earth might be influenced by the planets, a conjunction of the astronomical and the astrological.

The coming of the saucers allowed others to reimagine history, and humanity's role in the cosmic drama. An engineer and "born Fortean," Frederick Hehr saw an entire squadron of disks in June 1953 performing maneuvers over the Pacific Ocean near Santa Monica, California, for ten minutes.[89] They left him with no doubt that they were material crafts controlled by intelligent beings, capable of actions far beyond anything humans had invented. Hehr had another source of information, too: Venusians, the first of whom he reportedly met when he was a thirteen-year-old in Germany, back in 1903.[90] They told him of earth's true history, which had started half a billion years before, with a war between two races, one of which seems to have been the Venusians. The two groups battled for galactic supremacy; Venusians, following the universe's general inclination toward peace and cooperation, protected humanity from the depredations of the evil beings, preserving humans against several attacks that nearly eradicated earlier advanced civilizations—Atlantis among them—and shepherding the world through two world wars—which were caused by the Venusians' enemies, some of whom had taken human form. (Hehr thought the British elite and Jewish financiers were connected to the plot in some way.) The arrival of the disks presaged a third and final war, an epic battle after which humanity would have a chance to attain true peace and earn the privilege of interplanetary flight.[91] Hehr's story was not unlike others circulating among Forteans and the flying saucer community. The disks that Arnold had seen skipping like stones through the atmospheric ocean were not merely objects, were more than lights in the sky. They were the visible parts of a cosmic architecture that could then be seen in its entirety only by nonscientific means, among them the imagination.

Visitors from the Void

While newspapers blared headlines about flying disks, newsstands displayed the June 1947 issue of *Amazing Stories*. Its cover showed a sleek, speeding red car, under the gaze of gargoyles and in the sights of someone aiming a ray gun. The magazine sold for a quarter. Inside was Vincent Gaddis's Fortean article "Visitors from the Void," which contemplated the arrival of strange craft from beyond the earth.[92] Neither Gaddis nor his editor, Ray Palmer, had reason to be surprised by the coming of the flying saucers. Their imaginations were not expanded by the strange lights in the sky for they had already fantasized about them, already constructed entire mythologies. Rather, the UFOs were proof of their theories. The coming of the saucers confirmed that they had figured out the universe's hidden structure, its contours and dimensions. Nor were they the only ones in this position. The Fortean R. DeWitt Miller, the New Age thinker N. Meade Layne, and a few others were similarly situated, finding the disks a revelation, yes, but also a confirmation—proof that what they had imagined was indeed true. Unlike science fiction-eers and avant-garde artists, flying saucer enthusiasts did not think humanity could use its knowledge of the universe's once-hidden machinery to control the future, nor did they believe humans could resist the dominating institutions. The only choice wasn't a choice at all but surrender: accept that the universe moved according to the rules of strange mechanisms, which had brought humanity to its current state; reality would do what it will. The question, rather, was how to describe these hidden structures, these dominating institu-tions. Like an X-ray, the saucers illuminated the universe's skeleton, but what could be said of the bones? Were they material? Interdi-mensional? Spiritual?

There were many Forteans and Fortean allies who saw the sau-cers, to varying degrees, as material things, visitors from outer space. This was a Fortean view, in a strict sense, the forces that held the uni-verse together multifarious and material. Donald Keyhoe, a former pulp writer and US marine, argued the case first in the men's adven-ture magazine *True*, then in his 1952 book *The Flying Saucers Are Real*, which name-checked Fort. R. DeWitt Miller also saw the disks as physical, though he preserved a spiritual realm within his larger metaphysics. Miller had been interested in Fortean topics since he was a child and made a name for himself writing Fortean columns for *Coronet* magazine in the 1930s and early 1940s: "Forgotten Mys-teries," "Forgotten Experiments," "Your Other Life," "Not of Our Species." He circulated through southern California's metaphysical community and went on to write *Forgotten Mysteries*, a Fortean compilation that, to its detriment, avoided Fort's playful theoriz-ing.[93] After Arnold's sighting, Miller wrote an article on flying sau-cers that was sent over the wires and published on that miraculous day, 8 July 1947. He noted that there had been, over the last century and a half, hundreds of accounts like Arnold's. Miller thought there were three likely explanations: either the military was experiment-ing with new weapons systems, the objects were crafts visiting from other planets, or they were interdimensional travelers. "Something very queer is going on," he concluded.[94]

As he continued to study the phenomenon, Miller became convinced that the second of his hypotheses was the correct one. The objects had been seen for too long and over too wide a field for them to be experimental weapons. And while some evidence supported the interdimensional notion, Miller ultimately rejected the idea because he drew a bright line between material and spiritual realms. The spiritual domain was ineffable, the source of auras and the will toward evolution, the place where ventured the ghosts of the dead. Some adepts could access this realm via astral projection.[95] But saucers were stubbornly tangible. Vincent Gaddis's ideas were materialistic like Miller's—unsurprising, since Gaddis had been Miller's researcher until his own writing career steadied in the 1940s; his life steadied a few years later when Miller introduced Gaddis to the mystery writer Margaret P. Rea, who was researching a classic Fortean legend about an enigmatic woman who spoke an indecipherable language. After a whirlwind romance, the two wed in Clearwater, Florida, in July 1947.[96] (The event was announced in *Doubt*.) After his marriage, Gaddis stopped writing about Deros and Teros, instead mixing the Shaver Mystery, which he contemplated in the pages of *Amazing Stories*, with Fort and his Christianity into an explanation of flying saucers. God imbued the universe, he said, and humans were nodes of energy in a vast field. Still, the universe could be cleaved, as Miller said, into noumenal and phenomenal realms, which met at only a few points, what he called vortical windows. Child prodigies reached through these portals, which accounted for their wild talents, as did fire walkers and accident victims who had lost part of their brains yet functioned normally.[97] The things in the sky, seen by Arnold and hundreds of others, belonged to the physical universe. They were material—specifically, colloidal animals existing in deep space, beyond the region through which Sputnik traveled. Like Vitons, they fed on human energy, and they were ubiquitous now because there was so much to consume. In strikingly Fortean vocabulary, Gaddis explained, "There are strange fish in the sea above us. And we are soft, two-legged creatures dwelling on the bottom."[98]

Another of Gaddis's mentors in Forteanism saw the saucers differently, not as material entities belonging to the known universe, rather as interdimensional visitors; not in strictly Fortean terms but in New Age ones more open to spiritual theories. Newton Meade Layne (Newton usually cut to its initial) was another member of southern California's metaphysical community and a Fortean.[99] After studying Theosophy and psychic research and reading *Doubt*, he started his own magazine in 1945. *Round Robin* was a mimeographed collection of reports on all manner of occult, psychic, and Fortean events modeled explicitly on *Doubt* and Miller's *Coronet* columns.[100] Initially, Layne wrote and edited most of the articles himself from his home in San Diego, but in time he gathered a roster of regular contributors, many of them Forteans, including Vincent Gaddis. *Round Robin*, Layne said, was not interested in rehashing debates over the reality of spiritual communication, the afterlife, teleportation, and other such subjects. He accepted them as certain and based his studies on their existence. This he termed the "New Realism" or "Higher Realism."[101] Recognizing the existence of the occult and the psychic offered a plan of research: How do we understand human psychology when humans are alive? How does that understanding translate to dead humans? How do we help the newly dead? In 1946, Layne formed what he called the Borderland Sciences Research Association to study these areas more formally.

Combining Theosophy with the theories of maverick physicist and Fortean icon Albert Crehore, Layne argued that space was not empty, not a void, but filled with an intangible ether. This was a throwback to scientific theories disproven by Einsteinian physics:

> Space? Space is not nothing. Space is stuff, is matter. What kind of matter? The matter which makes up ether. More dense—not less, but more dense by far than the rarefactions of our world. The matter of the Etheric world![102]

The denser realm was the realm of the spirits, of the dead, of Theosophy's Ascended Masters. These were beings of great but not infinite knowledge; they were subject to the same psychological quirks as physical humans and so their pronouncements, provided through mediums, needed interpretation.[103] Higher Realism was different from spiritualism. That was a religion, accepting the words of mediums as holy writ, while Layne's practice was scientific, a dispassionate study of the etheric realm, which, while it held the discarnate intelligences, was still material.[104]

Layne applied his metaphysics to the flying saucers. When he saw reports of Arnold's disks, he knew exactly what they were.[105] Etherians. "They are NOT delusions, fantasy, or lies. They are not constructed by any foreign government, or by our own. They do not come from the depths of the sea."[106] The flying saucers and their inhabitants came from another plane of existence, a denser one, and reached our own world only by changing their vibrations, which altered their density. To avoid connotations of science fiction, he renamed the things *lokas*, a word he'd picked up from his readings in Theosophy that meant place or location. "Etheria is here—if we know what here means! Along-side, inside, outside our world."[107] The saucers were not visitors from the void; their coming was more like what would happen if Breton's Grand Invisibles manifested themselves.

Sallying so far from standard science, Miller, Gaddis, Layne, and the many others who improvised on their theories had to erect novel evidentiary systems. What proof was there of etheric realms? What proof saucers traveled through the void of space or the "vast net," as Layne had it, of interpenetrating dimensions?[108] Miller arrived at his hypothesis by way of a baggy syllogism: the saucers had been seen too often to be dismissed as delusion. Scientists said ether did not exist. The spiritual realm was an aspirational state for humanity and would not send messengers. Ergo, the saucers must be from space.[109] Vincent Gaddis, like Fort, was an assembler of clippings. As evidence for his ideas, he could point to thousands of newspaper reports—which could easily be gathered when he went to work as a journalist for Indiana newspapers—glued together by references to theorists of various sorts. The science fiction author Rog Phillips, who had some scientific training, argued in *Amazing Stories* and *Round Robin* that ether did exist, contrary to what modern scientists said; earlier physicists had been correct.[110] Gaddis leaned on these ideas. He also borrowed from the sociologist Pitirim Sorokin, who contended that cultures evolve from spiritualism to idealism to materialism.[111] Gaddis reversed this progression, saying that his collection of facts proved humanity was moving away from the sensate to an era of spiritualism, in which imagination would have free rein. Unconfined by useless conventions, those blessed to live in this time would recognize the existence of ether and the energy that traveled through it. Flying saucers, he noted, were often seen in the same place—hotspots—which was evidence that they followed lines of etheric energy. Frederick Hehr made the same point.

Layne turned to the tools of spiritualism for proof of his theories. In 1945, he met the San Diego medium Mark Probert. In weekly seances over the course of three years, Probert introduced Layne to fifteen members of the "Inner Circle," including the ancient Chinese philosopher Lao-Tzu.[112] The Inner Circle tutored Layne on the structure of reality, its etheric nature, and the role played by the disks, even before Arnold's sighting. On the evening of 9 October 1946, a bulletish, winged structure appeared in the San Diego sky for an hour and a half. It had two red lights but was otherwise dark, moving at varying speed and raking the ground with a light. At Layne's suggestion, Probert established telepathic contact with the thing, learning that it was called Kareeta. Driven by an advanced people, Kareeta was constructed of balsa wood coated with an alloy and powered by a small electric motor. The visitors were afraid to land but open to meeting with a committee of scientists.[113] Layne finessed this description to fit with his own thesis about ether—as Madame Blavatsky had shown, Theosophical mediumship required interpretation, not mere stenography—and applied it eight months later to the flying saucers. He was humble enough as a Fortean to admit that he might be wrong; or at least that his ideas could be guaranteed no more than one could guarantee Tuesday would follow Monday. But mediumistic methods had been refined and made rigorous over a century of research; they accorded with the lived experience of regular people and so could not be easily dismissed. "It's a point of view, and coherent and possible and probably interesting to everybody with a smattering of esoteric knowledge—which leaves most of our intelligentsia out of it, but happily lets in about fifty million lesser folk of this Pilgrims' Pride."[114] Flying saucers illuminated, for the open-minded, the universe's true structure.

High Strangeness

In the late 1960s, the theorizing of Layne, Gaddis, Miller, and their ilk, and the reports on which it was based, began to be gathered under the umbrella "high strangeness," a phrase introduced by J. Allen Hynek. A Yale-trained astronomer, Hynek acted as an advisor to the US Air Force's investigation of UFOs starting in the late 1940s. His skepticism slowly gave way to cautious acceptance that UFOs were more than misidentifications, hoaxes, imaginings. Sober, scientific, Hynek was the respectable center of UFOlogy, focused on mechanics, engineering, quantification.[115] At some point, he introduced the practice of plotting witness credibility against what he called "strangeness." Likely, he borrowed the term from atomic physics, where strangeness is a property associated with the decay of subatomic particles.[116] With this notion of strangeness, Hynek meant to capture "a measure of the difficulty of fitting, by scientifically trained persons, the contents of a report to a highly likely physical explanation."[117] The more inexplicable the event, the higher the strangeness index—the more worthy of consideration. Strange reports by credible witnesses could not easily be dismissed as lies, as misidentification of some known phenomena—that was why Arnold's initial account was so troubling to the mainstream: a rock-ribbed Republican and stalwart businessman witnessing the inexplicable. Repeated exposure to reports with a high strangeness index, offered in good faith by credible witnesses, eroded Hynek's skepticism, pushed him toward the conviction that UFOs warranted genuine investigation. First uses of the term hewed closely to Hynek's definition. Over time, however, "high strangeness" came to stand for weird ideas generally; it was used liberally in the 1970s to describe all

manner of phantasmagoria, "a term of art for particularly intense
and bizarre experiences—especially anomalous experiences associated with paranormal phenomena, occult practices, synchronicities,
and psychedelics."[118]

Preceding the term were the examples. The 1940s and especially
the 1950s saw the rise of highly strange theories, built on Fortean
foundations, especially (although not exclusively) regarding UFOs.
As Fort had decades earlier, members of the flying saucer subculture
profligately blended fact, fiction, and philosophical speculation,
creating trippy rearrangements of reality. These were not ludic inventions like Fort's, however, but serious theories. They meant to
explain the ways that power flowed through space and time, how it
pooled in the hands of some and ran through the fingers of most.
These highly strange theories ranged from a forms of fundamentalism to an anxious conspiracism, proponents certain that earth was
controlled by some being or cabal but disagreeing as to whether the
ruler of the cosmic aquarium operated through mundane or spiritual means. They clashed, as well, in their prophetic vision of the
dawning new age, some promising a period of peace and prosperity,
perhaps after initial challenges, others a dire future. The political scientist Michael Barkun calls these prophecies "Improvisational Mil-
lennialism."[119] Those heralding the coming era eclectically blended
scientific detritus, pulp stories, and modern worries into visionary
revelations.

A Polish-American Theosophist living in southern California
was among the most conspicuous of these prophets. George Adamski claimed that in November 1952 he had met Venusians in the
California desert, they having landed their saucer to see him and
telepathically communicate. The following year, his manuscript
describing the encounter was packaged with another, by Desmond
Leslie, and sold under the title *The Flying Saucers Have Landed*. A
sensation, the book was the beginning of Adamski's career as a contactee, and the inspiration for dozens to follow, the beginning of the
modern contactee movement.[120] In the manner of Theosophists dating back to Madame Blavatsky and continued by Guy Ballard, Ad-

amski cast the Venusians as Ascended Masters; they possessed occult knowledge, which they passed on to humanity, as though they were messengers of God. Indeed, the Venusians told Adamski he stood in a long line of contactees, the latest of many deputized emissaries; Jesus Christ himself had been an envoy. These Space Brothers, as they would later be characterized, brought both a warning and good news: atomic war might destroy the planet, but earth could yet save itself if it followed what amounted to the universe's divine will and gave up war. Adamski's fundamentalist announcement freely blended Theosophy, mediumship, and science fiction tropes into a vision of the eschaton, giving humanity the choice between imminent apocalypse and entering, quite literally, the kingdom of heaven.

Initially, contactees were a small, controversial clique within the broader flying saucer and Fortean movements, but eventually they became dominant. So prevalent were those who worked in the tradition that Hynek classified them separately in his studies, "the Adamski-type" standing for those who saw flying saucers as chariots of earth's saviors.[121] Hehr's purported experiences were clearly elaborations of Adamski's, blended with the teachings of L. Ron Hubbard, whose Scientology liberally blended space opera, religion, and Theosophy. Long John Nebel hosted a steady stream of contactees on his radio show.[122] Waveney Girvan also did his part to launch contactees into the spotlight. He was the one who got Adamski's manuscript into print, and the magazine he edited, *Flying Saucer Review*, made contactees a central concern.[123] Contactees and their supporters claimed to have discovered not only the hidden parts of reality but the true power behind all of history.

Ray Palmer, mixing the lore of contactees with the mythology of Deros and Teros, also contemplated the movement of power in a Fortean universe, though his thought tended toward the conspiratorial and a bleaker view of earth's fate than those who trumpeted the coming of the Space Brothers. *Fate* magazine debuted in the spring of 1948, a year after Arnold's sighting, and featured an article by him in its first issue; the cover was an illustration of his encounter. Although focused on flying disks, *Fate* freely mixed in stories of Fortean interest, including a feature on Fort himself in the third issue. An article in *Amazing Stories* from the same month, by Marx Kaye (a house pseudonym), was titled "Fortean Aspects of the Flying Disks."[124] The author argued that Fort's Super-Sargasso Sea existed and was, just as Fort supposed, the source of strange rains as well as the flying saucers, attracted to earth by the explosion of atomic weapons. Whether beneficent or martial, the saucers should be welcomed, the article concluded, for they would force humanity to overcome petty national disputes and unite against the unknown. Thayer noted *Fate*'s appearance on newsstands and doffed his cap to the opening editorial's "high moral tone," hoping the "magazine lives up to it."[125]

Under Palmer's editorship, *Fate* became the preeminent Fortean publication, adding to standard Forteana a large dollop of parapsychology and spiritualism, combining a Fortean focus on mundane and multifarious mysteries with excitement about myriad spiritual forces that might interfere with or shape human lives—ghosts and reincarnation and ineffable realms. "National Geographic for explorers of the anomalous or weird," said Palmer's biographer Fred Nadis.[126] Palmer gathered a set of regular writers, what another biographer called an "inner circle," to ensure a steady supply of material—exactly what Thayer desired for *Doubt* but failed to bring about. Vincent Gaddis was a frequent contributor under his own name and various pseudonyms.[127] Saucers were a perfect topic for *Fate*, an enticing enigma that suggested a panoply of solutions—spiritual, material, Theosophical, esoteric, occult, scientific, Fortean—but that could not be resolved, a perpetual motion machine for articles.

Shortly before he left *Fate*, Palmer started another magazine, *Mystic*, which published articles deemed too radical for *Fate*.[128] *Mystic* considered flying saucers too, but also N. Meade Layne's psychic Mark Probert, who was frequently discussed in the magazine as a messenger of higher truths. *Mystic*, as well, resurrected the Shaver Mystery. In Palmer's hands, the Deros and Teros were transformed into aliens—connected, therefore, to flying saucers—engaged in a spiritual war for the soul of humanity. Some powerful group, though, obscured the truth; Palmer was called to pierce the veil of secrecy and reveal how power moved through the universe, how hidden groups manipulated reality.[129] He flogged the point in subsequent publications. Not that Palmer's ideas were consistent or easily summarized. He had little regard for coherence and so flitted easily between views of the disks as interdimensional, material, and spiritual; Deros and Teros were equally fungible. What was clear through the squall of ink was that mid-twentieth-century humanity was on the verge of a sea change, swimming in a cosmos far weirder than any single mind could comprehend. To capture the breadth of its strangeness required leaving behind logic and forgoing the divide between fiction and fact. Only from this position could one understand that humanity was small, not nearly as powerful as was presumed, subject to forces controlled by more powerful beings.

Others, wading through the same waters as Palmer, ventured deeper into conspiracy and strangeness as they hunted for earth's owners. Morris K. Jessup, the former astronomer, wrote four books on flying saucers in the mid-1950s as his life unraveled—divorce, remarriage, depression, suicide.[130] His first, *The Case for the UFO* (1955), argued that an ancient civilization, Atlantis or Mu, invented spaceflight and colonized the region between the earth and the moon—the "Up Above" of the old scientific romances. A subsequent book, billed as the first in an annual series, attempted to establish him as a clearinghouse for UFO reports. The series never came to fruition, but the failure did not stop him. Jessup kept on, arguing that he was revealing earth's true history and that the history was Fortean. He found Fortean elements in the Bible, cited *Doubt*, explained organic rain: effluvia of the sublunar civilization. The history was also Theosophical: Jessup thought he was improving Blavatsky, using more rigorous methods. Flying saucers heralded the final reconciliation of science and religion. The Bible predicted a time when the "Son of Man" returned to earth, and that time was now—the saucers, the civilization that existed between earth and its satellite was literally the offspring of humankind. To meet them required forgoing the false promise of rocketry, instead following the ancestors by learning to manipulate gravity. Only then could humans (once more) escape the planet and meet their destiny.

In early 1957, Jessup received an invitation to the Office of Naval Research. The ONR had obtained a copy of his first book annotated in three ink colors, three different voices. They read like government officials and an extraterrestrial. A couple of people in the ONR privately printed a small edition of the annotated book. Jessup immediately recognized the handwriting. It was that of Carl Allen (who also went by Carlos Allende). Allen had been writing Jessup for some time about naval research undertaken during the war that had successfully turned a ship invisible. Jessup became increasingly paranoid, certain that he was the victim of some baffling conspiracy. He was also depressed, visibly so when he visited Ivan Sanderson at the end of 1958. On 20 April 1959, he was found dead in his car, having poisoned himself with the vehicle's exhaust. His passing only furthered his legend. Although the facts about Allen's life and the navy's research were easily verified, a story arose that the navy had, truly, made a ship disappear in 1943, learned the trick of levitation, and housed an alien—all secretly. Jessup had gotten too close to the truth, whispered some, and had been killed.[131] Fortean things, including flying saucers, were evidence that the earth was ruled by a secret cabal, its power hidden but unchecked.

Robert Ernst Dickhoff was a prophet in the same tradition as Adamski, Palmer, and Jessup, announcing that he understood power, who wielded it and why, and that those who followed him could understand it too and save themselves from its harrowing effects. Dickhoff had reinvented himself as a spiritual teacher at the end of World War II, leaving behind his job as a printer, a flirtation with communism, and a wife, but not his status as stateless; he'd jumped ship back in 1927, when he left behind his German homeland.[132] Dickhoff synthesized Theosophy, pulp fiction, and Fort into a heady space opera that he elaborated in a series of pamphlets.[133] Over the next decade and a half he improvised his theology, arriving at the idea that Martians had established a utopia on earth in the distant past, which had been destroyed by Venusian snake people, who slithered through the

Jacket of Robert Ernst Dickhoff's *Homecoming of the Martians*, designed by Albert Roger and Salwi Art Service. The book represents the synthesis of science fictional, Theosophical, and Fortean themes that animated much of the UFOlogical community. Dickhoff, a member of the Fortean Society, opens his book with an encomium to Fort and his famous phrase "I think we are fished for." As he saw it, *Homecoming of the Martians* carried on where Fort left off, thumbing its nose at "cultish" scientists and feeding hungry minds.

planet's subsequent history, a shadowy force behind governments, whispering the secret of atomic weaponry to those who approached the apple tree. The Martians, hiding in underground caverns, rallied behind esoteric Buddhism: a light leading humanity back to bliss. Dickhoff was a teacher of that form of Buddhism. It was in the mid-1950s that his statelessness was resolved—deportation proceedings were discontinued—and history approached its final stage. His 1957 booklet *Homecoming of the Martians: An Encyclopaedic Work on Flying Saucers* tells how UFOs herald the final battle between Martians—defenders of all that is good—and Venusians—who plan to devour all of humanity in the Gobi Desert. *Homecoming* opens with a poem by Lilith Lorraine and is dedicated to Charles Fort, "an uncrowned philosopher king."[134]

On 15 July 1968, Dickhoff printed the last of his pamphlets, *Behold … the Venus Garuda*. By this point, he had fully absorbed Fort, having read his complete corpus in 1964 and having been involved with the Fortean Society before its demise. *Behold* was thoroughly Fortean in structure, weaving news clippings into an alternative cosmology; it was also Fortean in approach, arguing that humans are property. In this text, Venus, while still home to the snake people, is also a way station for the Hindu demigod Garuda, an eaglelike monster that feeds upon humans. Dickhoff blends fact and fiction so thoroughly they cannot be separated: his was not an ironic imagination, but one fully immersed in a world of its own invention. This was the prophet's exhortation to resist the demonic—Garuda, the snake people, Castillo's Cerberus—and return to a society guided by the principles of morality. The only way to neutralize the power of evil is to embrace that of good, but to do so one must first understand the true operations of power, which are revealed by contemplation of Fortean things.

Through the 1950s, such improvisations were common, stranger on the margins, certainly, but reaching into the mainstream too. Perhaps these were superficial manifestations of deeper cultural anxieties about atomic weapons, the dangers of science, and the possibility of another war, one that humanity would not survive. Whatever their ultimate cause, this high strangeness represented an enchanted version of modernity in which monsters were real, used modern weapons, and were responsible for Fortean anomalies. The earth had its owners, power residing in places unknown to the masses. This was, indeed, Fort's era of the hyphen. Fiction telescoped into nonfiction, reality into pulp, religion into science, fear into wonder.

Thing from Another World

An envoy from the stars arrived at the Parisian United Nations offices in the summer of 1947. He wore monkish robes, carried an astronomical chart, and offered a plan for peace. Julian Parr, a Fortean and part of Britain's embassy in Germany, noticed the short blurb about the visitor in a German paper, translated the article, and sent it to Eric Frank Russell.[135]

Had an alien arrived?

They doubted it but relished the Fortean gag, chortling when a bureaucrat declared that there was no need for the peace plan, as though UN meetings, long tables, and paperwork could be more useful than an ET at averting World War III.

Then in December 1948, a man wearing "an old army blanket like a flowing toga," a dirty face—the same man?—tried to lecture the UN General Assembly. An astrologer—sans papers, sans even a name, briefcase carrying only a towel, soap—he called himself a "citizen of the universe."[136]

In between, in May 1948, Garry Davis, taking up Caresse Crosby's challenge, renounced his American citizenship, claiming to be a citizen of the world. The move left him stateless, made international travel difficult, and inspired others. He too stormed the UN General Assembly, in November 1948, demanding an end to nations, which were the cause of war.

Davis had seen war's devastation from a godlike position, flying in a bomber during World War II. He'd felt its vicious claws: his brother, who joined the navy, died. And he'd been transformed from callow youth of prosperous privilege into an idealist. End nations, he insisted; institute world government. He spoke before crowds of thousands that autumn. The surrealists, Breton, the novelists André Gide and Albert Camus, and others came together to support his movement.[137]

Davis found validation of his ideas in a short story by Eric Frank Russell. (A longtime science fiction fan, Davis had read Fort's *Lo!* when it was serialized in *Astounding*.) "And Then There Were None" concerned a planet of anarchic pacifists who cannot be enticed into a universal alliance; rather, they convert emissaries of the alliance to their way of life.[138]

Davis joined the Fortean Society, attended a dinner with Thayer and others.[139]

The dirty-faced, blanketed astrologer evoked the Greek philosopher Diogenes the Cynic, who went about in ratty clothes, calling himself a citizen of the world. That was Davis's title, too, though in the flood of mail he received (as many as four hundred letters a day) others saw him in more exalted terms: "I think you are Christ come back to earth."[140]

(And why not, Thayer asked of others calling themselves thus: who are we to say? On what authority?)[141]

The language was easily translated into metaphysical, Theosophical terms.[142] An advertisement that ran in American papers in the fall of 1948 asked, Do you know that you can "extend your consciousness so that you can become a citizen of the universe?"[143] Was humanity poised to enter its next stage of evolution? Was Davis, like the Space Brothers, beckoning earthlings toward a better future?

On 6 April 1951, Christian Nyby's film *The Thing from Another World* (really, the director was also the producer, Howard Hawks) opens in American theaters. Based on a science fiction story by John Campbell (writing under the pseudonym Don A. Stuart), revised for the screen, according to the credits, by Charles Lederer (in reality, Ben Hecht did much uncredited writing), the movie tells the story of a scientific expedition to investigate a meteor strike at the North Pole. The meteor, it transpires, is a flying saucer, accidentally destroyed by the scientists; the pilot turns out to be adversarial, a shapeshifter. The expedition battles the invader—except for one scientist who, like the physicists employed by the Manhattan Project, places knowledge above human life. He defends the creature. Very Fortean: science and scientists cannot be trusted.[144] This was exactly opposite the moral in Campbell's story, which had the scientists working together against the alien.[145]

This alien was no space brother. This future, one of suspicion about motive, identity.

Life magazine suspected that Garry Davis's "cult," as it was called, was not only naive—he "a corny, carrot-topped young American in Paris"— but dangerous, a weakening of national resolve that invited communist invasion.[146]

Through the mid-1950s, Davis ping-ponged from one European country to the next, arrested and sent on.[147] His exploits were reported in the press so often that Thayer worried to mondialist and patron of the avant-garde Caresse Crosby that Davis was becoming the face of the one-world movement. Davis was ridiculed as hypocritical—demanding the right to live in his own country even as he rejected the notion of nations—a Don Quixote tilting at windmills while atomic bombs were poised to drop.[148] Crosby admitted to Thayer, "*L'affaire Davis* has become quite a problem," particularly his refusal to join her organization, instead promoting competing institutions, his own flags, his own passports, as though to make the movement about himself, not peace, fostering antagonism among people who decried antagonism.[149]

Thayer and his wife watched in wide-eyed horror as Davis was interviewed on Mike Wallace's news program. "Kathleen almost cried to witness your ideals and my sympathies so badly abused. I expected nothing less. Although it may be impossible—or impracticable—for you to disavow any connection with his brand of world citizenship, in my opinion it is an enormous fallacy to tolerate his incompetence and irresponsibility for the sake of the press he commands."[150] Thayer had identified Davis as a phony shortly after he'd had dinner with him; *Doubt* stopped reporting on his activities immediately after that night.[151]

In a rare display of hope, Thayer tried to buck Crosby's spirits: "This is not intended as advocacy of complaisance [sic] on your part or mine, but do accept my assurance that 'your ideals and my sympathies' . . . are inevitable of realization, whether we live to see it or not. What we are trying to tell them today will BE, eventually. Nothing can stop it, not even Garry Davis."[152]

In the meantime, however, in the real world of late-1950s America, Davis had become a useful idiot to the powers-that-be, Thayer thought, promoted as the face of mondialism so that it could be ridiculed. The "world citizen" was not Diogenes, with his lantern, searching the world for one honest man; not Christ, bearer of good news; not even evil. At base, he was naught but a "clown."[153]

The United States of Dreamland

Flying saucers ballasted the Fortean community for the rest of *Doubt*'s run, and Tiffany Thayer was not happy about that fact, coming in time to wonder if the reports were part of a new conspiracy and Forteans, as they had been with Sputnik, dupes or, worse, coconspirators. Hundreds of clippings about flying saucers inundated the society's post office box at Grand Central Station each month—306 between July and October 1950; 281 between June and October 1952—as did air force reports and books—so many books!—by Arnold and Adamski and Heard and Newman and Keyhoe and H. T. Wilkins and Frank Scully; and magazine articles from *Life* and *Harper's* and *True* and *Argosy* and the *Saturday Evening Post* and *Reader's Digest* and *Fate*.[154] Thayer received letters, often long and abstruse in their reasoning. Some of what clogged the mailbox was technical and dry, some Theosophical space opera, some only tenuously connected to disks at all. Some MFS offered objective evidence for their theories, many more speculation, and others hard-to-credit stories about communication with Venusians. Some was Fortean, some New Age or fundamentalist or conspiratorial. Thayer tried to sequester discussion of flying saucers in brief columns, often decoupling contributor names from the reports, perhaps in an attempt to deprive members of any sense of satisfaction in seeing their name appended to a "datum" (as Thayer called the clippings) or perhaps because he was simply overwhelmed by the amount of material he had to digest. Avoidance was impossible, though. By 1950, nine out of ten Americans had heard of flying saucers, and Fort was the prophet of their coming.[155] Thayer was repeatedly browbeaten into doing large spreads, including by Eric Frank Russell.[156] In 1947, 1949, 1953, and 1957, issues of *Doubt* offered extensive coverage of UFOs.

Members were not pleased by Thayer's foot-dragging. Norman Markham sent in long letters elaborating on his theories—"You should see his bulging folders in the archives!!!" Thayer told Don Bloch in 1949—but Thayer printed only a fraction and responded to less.[157] "Such communications as passed between Thayer and myself were mostly on my side, he being a very laconic fellow in his letter writing," Markham said.[158] John Philip Bessor was pointed. "Thayer, frankly, is NOT doing the Society one iota of good. I can tell you that. [Parapsychology researcher] Harry Price agreed with me 100%. So does Vincent H. Gaddis."[159] Gaddis's complaints expanded beyond Thayer's reluctance to fully consider flying saucers: he thought that Thayer wasn't much of a Fortean at all (and admitted that he himself fell short, as he couldn't help but have beliefs). Thayer's "dogmatic atheism" and attraction to "ridiculous" ideas—calling the atomic bomb a hoax!—ruined *Doubt*.[160] It was the same complaint made by Chapter Two's George F. Haas and Robert Barbour Johnson: Thayer simply ignored good Fortean data if, for some reason, it didn't conform to his preconceived notions. And so, out of obstinacy, he was missing what might be the most important story in world history.

Over time, Thayer's belligerence toward the flying saucer subculture became more apparent. Thayer had initially welcomed N. Meade Layne's *Round Robin*, as he had *Fate*, as an expansion of Forteanism. "There's meat on that Robin," he said in response to very warm regards sent by Layne to the society. Layne had written:

We think [the Fortean Society] is doing an admirable piece of work. We're not so much interested, personally, in its social and political iconoclasm, tho others may find it worthwhile. But the effort to salvage the "damned" facts, hold them up for all to see and for the confusion of all orthodoxies, is a matter of very great importance.[161]

The following year, Thayer reversed course and mocked Layne, his organization, the Kareeta ballyhoo. He'd decided that *Round Robin* was a spiritualist magazine and Probert's conversation with the Etherians no more real than the seances Thayer had trawled through back in the 1930s—entertaining enough but obviously fake. He'd had enough of the matter.[162] Thayer, Layne replied, was confused, with no idea of the differences between spiritualism and spiritism— between science and religion. Spiritism could not be ignored, since it uncovered facts "as startling as anything in the *Book of the Damned*, including 'explanation' which can at least be used as hypotheses and points of departure."[163] Nor did Thayer recognize just how well-supported the San Diego sighting had been. Thirty-five people claimed to have seen the atmospheric anomaly, a far sight better than most Fortean reports Thayer chose to highlight.

Most importantly, Layne said, Thayer's dismissal of the Kareeta story showed just how insulated Forteanism was becoming under his direction. Thayer had become so bedazzled by minor matters he could not see what was essential. "Strange as it may seem," Layne said sarcastically, he valued the evidence adduced by Probert and those like him as "more worthwhile than columns devoted to fire balls, fish impaled on telegraph poles, and mysterious rumblings within the earth. Fortean data are certainly disturbing and import- ant, but just how to turn them to account remains unsolved. Apart from saying Oh my Gawd! most of the Forteans seem to make very little progress." Nonetheless, Layne remained a contributing mem- ber of the Fortean Society deep into the 1950s. But he also would not concede. The society was useful as a corrective to the hypocrisy and arrogance of science and, as he thought of it now, as an adjunct to his magazine, rather than the other way around, since *Round Robin* was so stuffed it could not print a quarter of the Forteana that was submitted to it. Just don't go to *Doubt* looking for answers, he warned members of his Borderland Sciences Research Association.

"The Forteans haven't any explanations, of course, and probably wouldn't admit it if they did."[164]

Thayer turned on *Fate* as soon as he realized the editor was "Palmer, the Astounding man who gave Shaver to a gaping world."[165] He'd been thrown off the scent initially by Palmer's using the pseudonym Robert N. Webster on the masthead and editorials. Thayer's irritation was doubled by the third issue's article on Fort. "This 'author' rewrote selected paragraphs from the Introduction to the Books and called it an article," Thayer told Don Bloch. "I wish some of them would dig up something new."[166]

A few years after Arnold reported those nine disks over Washington, advertisements started appearing in Palmer's *Mystic* and *Flying Saucers* for a shampoo called "Turn-ers," sold by Guy L. Turner, out of Arnold's home state of Idaho. Arnold had introduced him to the product, Palmer attested, and it worked: cured his itch and dandruff, darkened his hair, resolved his wife's rash—and it was manly: "Ken's no sissy and doesn't put perfume on his hair."[167] Hawking snake oil was a little on-the-nose, confirming for Thayer that the saucer craze was itself just advertising, a way to sell publications, panaceas, a nefarious ideology. Thayer wanted to cleanse the Fortean Society of any connection to *Fate* and Palmer. In 1950, Russell resold an article from *Tomorrow* to *Fate*. "I just hope the piece you sold him does not mention the Society. We will never willingly be named in FATE."[168] Palmer's magazine would go on to popularity and a long life, while Thayer's curdled, heading toward death. But Thayer was unbowed. He said in 1953, "I am now killing every man woman or child who says 'saucer' to me."[169]

Thayer was disappointed by the poor quality of the books published on disks, even if they mentioned Fort approvingly. "Shit is shit," he said.[170] Waveney Girvan was trying to make *The Books* more widely available in the UK, but that didn't mean his own products were any good. Thayer received Heard's *The Riddle of the Flying Saucers*—the book Girvan had wanted Forteans to rally around and defend—Girvan's *Flying Saucers and Common Sense*, and other books put out by Girvan's publishing house, "and the stench is something awful."[171] He was aghast at Frank Scully's 1951 *Behind the Flying Saucers*. Scully was an MFS and an acquaintance of sorts.[172] In 1949, Scully had reported in his usual column for *Variety* magazine about secret government research on flying saucers, their use of magnetic propulsion, and the fact that one had crashed, with extraterrestrial bodies recovered. A few weeks' work, and he expanded the columns into a book-length manuscript, which he sold to Henry Holt. The book seemed to promise positive press for Fort and the society. Girvan saw Scully's book as an excuse to bring Fort to the world's attention again, perhaps justifying a British edition of the Fort omnibus.[173] The Fortean writer Donald Whitacre told fellow residents of Ohio that *Behind the Flying Saucers* was "one of the greatest works ever to hit the press." He read the book's chapter on Fort "with great pride."[174]

Thayer thought "the kindest way would be to ignore this little item," as repayment for how positively Scully had written of Fort and the Fortean Society. But he couldn't quite do that, opting instead to "not let go with both barrels" in his review for *Doubt*. He contented himself with suggesting that Scully was himself part of a vast conspiracy, something organized by the government. Probably Henry Holt was a useful idiot in the conspiracy, but Thayer was still disappointed that the publisher of *The Books of Charles Fort* should be associated with this sham. By the end of his review, Thayer quit trying to be polite and concluded, "Frankly, the book is a waste of your time," but, in deference to the Fortean value of open-mindedness, the society still offered it for sale, at $2.75.[175] (As it happened, later investigation by *True* magazine revealed that the sources for Scully's magazine column and book were not government agents passing on classified information but con men.)[176] Thayer was less circumspect about Dickhoff. His syncretization of Buddhism, Theosophy, science fiction, and Forteanism, unsupported by any facts, may have been alluring to those with an esoteric bent, but it offered nothing for Thayer. "A curio of curios for the saucer addict," he said of Dickhoff's *Homecoming of the Martians*.[177]

The flying saucer reports continued through the mid- and into the late 1950s, and Thayer became increasingly irritable; his interest in the Fortean Society lapsed as his contempt for its members grew. He heard from Frederick Hehr about communicating with Venusians, and Hehr's scuffle with Layne over what the Kareeta was: the very definition of a tempest in a teapot, he implied in *Doubt*.[178] Alan F. Wilson, a science fiction fan from Cleveland, sent Thayer and Russell a succession of rambling, inscrutable letters synthesizing space opera, Dianetics, UFOlogy, physics, and Fort. Thayer rolled his eyes and disregarded them, encouraging Russell to do the same.[179] When Russell could not resist responding, Thayer merely shook his head. "What you-all don't know is how many Alan F Wilsons this land of the free and home of the brave can produce in a bumper season. You get only a random few. Me? Wowee!—to put it mildly."[180] He was the victim of his own desire to encourage Fortean thinking in the wide world. Indeed, the spreading of the Fortean imagination only muddied Fortean matters; flying saucers were ruining everything. "Discs, disks, saucers, Svenskarockets, balloons, meteors and all flying sky lights are now so thoroughly mixed up by all reporters that sound evidence has become extinct," he complained in 1948.[181] For a time, Thayer thought he might have developed "a technic" for "categorically" siloing the many reports, and even encouraged members to send in more clippings.[182] "As a former reporter, former publicity man, advertising man and sometime master of word-twistery, [I feel] rather competent to take the syntax apart."[183] But nothing ever came of his method.

What he supplied, instead, was what science fiction writer L. Sprague de Camp called, borrowing a term from Robert Heinlein, a "devil theory."[184] Or, rather, two devil theories, for Thayer's initial understanding of the conspiracy behind the flying saucer reports changed after 1947, his new view then reinforced over the years through selective reading. In both formulations, the problem was that the press, in concert with the world's governments, had created a false reality in which most people lived—not him, of course, clear-sighted, open-eyed, awake in the United States of Dreamland.[185] What changed was Thayer's understanding of the conspiracy's aims, and the conspiracists relation to the disks. Originally, he thought the press opposed to saucers, resentful that everyday people should take a stand against science, like those French peasants in possession of meteors. A small elite controlled the media, which credited some things as true, discredited others as false. The press, he believed, had hidden that the attack on Pearl Harbor was coordinated with the Roosevelt administration and had invented the myths of the atom and of the atomic bomb. Now, people were reporting lights and objects in the sky that defied scientific explanation, and so they had to be ridiculed, deemed hysterical, in order to protect the dream against reality.

By late 1950 or very early 1951, Thayer revised his opinion: far from threatening mainstream powers, flying saucers were pawns in their game of global control. He offered this idea, at first, with a Fortean sense of proportion, but soon enough gave in to certainty. How could stories about disks continue to appear in the papers, more than three years after Arnold's sighting, accompanied by appreciative comments about Fort, if they truly threatened the powers-that-be? For some reason, the powerful wanted newspapers to fill their pages with articles on UFOs. "The only agency known which could command the space and time, the brains, paper and ink, devoted to this topic since it began is the Office of Strategic Services in Washington, D. C.," he said, referring to the bureau that would become the Central Intelligence Agency. "There the world's future is planned for at least three generations to come. There is decided what the newspapers are to print when. There is decided every question of peace or war or life or death for millions of people—strategically." Government officials wanted to muddle all the atmospheric anomalies into a single mass called flying saucers for their own degenerate reasons. "One suspects that the purpose is to keep the public in a state of confusion, but always looking up—high up into the sky where 'Peace on Earth' is no longer to be seen."[186]

The point of all this plotting, Thayer thought, was to justify defense spending.[187] Think of the sources behind Scully's book, the constant commentary from the army and air force, the involvement of ex-marine Donald Keyhoe in the flying saucer subculture. "About 90%" of saucer reports were provided by Pentagon press agents, he said, and even the brass's scoffing was a way to keep the story going. Corkscrewing the narrative of Bernard Newman's novel ever so slightly, Thayer concluded that the reports had helped avert World War III, which he had once presumed inevitable. The Soviets and the Allies had agreed to abstain from military engagement and use "interplanetary defense" as "the new boondoggle."[188] Billions of taxpayer dollars would be funneled into this new battle—as in his own Perpetual Peace Program, taxpayers were pickpocketed to keep the fat cats fat. Infuriating, but "it beats killing 18-year-old kids. Even the Senators must like it better."[189]

Russell did not appreciate Thayer's amended devil theory. He was not content to dissolve the flying saucer phenomenon into government chicanery or cloak-and-dagger pantomimes between journalists and spies. The disagreement was another reason for the rift between the two Forteans in the mid-1950s. Apparently, Russell wrote to complain about Thayer's spread on the disks in *Doubt* 40, sent out in April 1953. Thayer responded with condescension. "You should re-read the lead paragraph," he said. "The phenomenon dealt with were not objects seen, but newspaper items. . . . Take that slow and easy so you're sure you see the difference."[190] Thayer was writing media criticism where Russell wanted him to deal with the anomalies. The difference in opinion became starker when Thayer argued that flying saucers, like science fiction, were a prologue to the more odious Sputnik conspiracy. Stories about disks being tracked by radar served to gauge the public's willingness to believe that objects could be tracked through the high atmosphere and near space.[191] If the masses were convinced that aerial crafts could be followed by machines on the ground, that lights in the atmosphere could be explained as technological objects moving through space, then they could be hoodwinked into believing media reports that a human-made satellite had been launched, was being monitored, and could even be seen by the naked eye at particular times in particular places. Arnold and Keyhoe had been part of a setup, like Robert Spencer

Carr's science fiction story in the *Saturday Evening Post*. Russell
didn't like Thayer's conclusions, but Thayer insisted, even if more
in sadness than anger. He was getting Perpetual Peace, though at the
price of widespread imbecility. He was willing to pay the cost "as
long as it's the other fellow's brain they wash and not my own."[192]
Thayer would not pledge his allegiance to the United States of America, nor to the United States of Dreamland.

John Keel and the Disneyland of the Gods

He was still a Fortean, would always be a Fortean, but on UFOs, John
Keel decided, Charles Fort was wrong. It had taken him the better part
of two decades to sort the matter out. Keel would become, perhaps, the
best-known Fortean writer on UFOlogical topics during the last third
of the twentieth century, combining elements of Fort's thought, Layne's,
Palmer's, and Dickhoff's. From this mixture would emerge a theory that
looked like a throwback—an unembarrassed demonology—but that was
nonetheless thoroughly modern.

Born in western New York in 1930, Keel (or Kiehle, in official documents) had something of the bohemian to him. He moved to Greenwich
Village when still in his teens, wrote poetry and comic book stories, and
ran in science fiction circles—he'd started putting out his own 'zine
while still living in Perry, New York. Drafted into the military, he was
assigned to the Armed Forces Network, for which he wrote scripts, and
he continued to work on the fringes of the entertainment industry after
leaving the military. In the mid- to late 1950s, he traveled through Asia,
his exploits the basis first for articles in the men's adventure magazines,
then for the 1957 book *Jadoo*, which also traded on his fascination with
stage magic.[193]

Keel read Fort just a few years before Kenneth Arnold discovered flying saucers to the world. He saw a UFO himself in 1952. In these years,
Keel ingratiated himself into Fortean networks.[194] He read Ray Palmer's
Amazing Stories and N. Meade Layne's *Round Robin*. He clipped reports of
the strange in newspapers. He called himself a Fortean. The designation

stayed with him throughout his life. Between 1969 and 1974, he irregularly put out a Fortean newsletter called *Anomaly*. In the 1980s he ran the New York Fortean Society, printing its newsletter. *Jadoo* was followed by a series of books on paranormal topics, through all of which ran Fortean topics, Fort's name, Fort's mysteries. The Fortean subject that most beguiled him—the one around which all the others orbited—was flying saucers.

By 1966, Keel gave up on the idea that flying saucers carried aliens from some other planet, the so-called extraterrestrial hypothesis. He saw flying saucers as something weirder. UFOs were beings from another dimension. At first, he said, like Layne, that they moved through the ether, but he would dispense with this idea too, restless in his search for a new language by which he could comprehend the phenomenon. Not UFOs. Not ETs. Not Etherians. Ultra-Terrestrials, he called them, stating that they hailed from the Super-Spectrum, the "upper frequencies of the electromagnetic spectrum." As bundles of energy, Ultra-Terrestrials traveled along this spectrum. But whatever the exact verbiage—Etherians vs. Ultra-Terrestrials, ether vs. Super-Spectrum—Keel's borrowing from Layne was obvious. As was his use of Palmer and Shaver, for he did not see these visitors as benign. They were Deros, manipulating humans to achieve their own depraved ends.[195]

Keel called the period 1966–1967—when he left Fortean ideas about UFOs behind—the "year of the Garuda," invoking the same bird-god as Robert Ernst Dickhoff. Why a bird-god? Because, in Keel's estimation, the Ultra-Terrestrials had no essential form. They took on mythological appearances as needed—angels, demons, gods, machines, whatever was necessary to bring off their plans. They were shapeshifters. These earth-owning beings created Fortean events and left evidence for outrageous legends, such as vampires, in order to confuse humanity and lead it astray.[196] "UFOlogy," he said, "is just another name for demonology," and the Ultra-Terrestrials were demons.[197] Familiars of these beings took the image of humans, Men in Black, who paved the way for their coming while obfuscating their intentions.[198]

In 1988, Keel published *Disneyland of the Gods*. The title chapter referenced Fort's famous dictum, "I think we're property," while suggesting

Keel's own explanation of the universe's structure. In time immemorial, gods "conned" ancient peoples into constructing a network of monuments to serve as guideposts, transforming the planet into a Disneyland, a place the Ultra-Terrestrials could come to tour, to entertain themselves at humanity's expense.[199] Ancient people had disappeared at the hands of these gods, and now, as modern people unraveled the secrets of the visiting Ultra-Terrestrials, Keel wondered whether humanity would be put to work constructing new guy-wires for the gods—or be made to disappear? Perhaps humanity had served its purpose and was to be swallowed up by the Garuda, the Deros, the demons of old. Apocalypse awaited, in one form or another.

A modern Jonah, then, Keel—at first mistaken about the word, the world, then warning of the cosmic aquarium's imminent destruction by vengeful gods. But although these were the old gods, Keel's vision was not a rejection of modernity. The old world had passed away—this was a new one, enchanted, and threatened. Keel's prophetic voice proved influential. Jerome Clark, a onetime acolyte turned dissenter wrote:

> Keel has been more widely read, and it is largely through him that ufologists and Forteans, or at least some of them, have plunged into the thickets of occultism and obscurantism, into a realm where words like elemental and superspectrum and ultraterrestrial and transmogrification are actually supposed to mean something. Into, in other words, a domain of incoherent theory and dubious data and, finally, numbing irrelevance. If Keel were a humorist like Charles Fort rather than a windmill-tilter like Tiffany Thayer, one could smile and shrug it off as an ongoing, offbeat joke. No Fortean, to my knowledge, has ever championed Fort's sky islands or Ambrose-collectors, knowing that Fort wasn't championing them, either. But Keel is deadly, gloomily, blusteringly, spittle-spewingly in earnest. Though usually politer and calmer about it, so are the legions of acolytes who since then have dropped a ton of Keelist doctrine on all our heads.[200]

6

Future History

Forest of branching selves, my various masks, my
    serio-comic souls, my antique, half-remembered
    egos: are ye that?
And here I now stand peering tensely curious into the
    crater of Eternity, seeking out Demiurge there
    in the depths,
BENJAMIN DECASSERES, "Beyond Sense," 1917

In midsummer 1959, Tiffany Thayer wondered where Eric Frank Russell's head was at.[1] Back in February, Russell had written Thayer admitting that he hadn't been at his best, and Thayer had looked forward to a revival of the Fortean Society, the two old dragons breathing fire again, wreaking havoc on this fallen world.[2] Thayer had thrown himself back into society work. He'd been indexing volume 1 of *Doubt*, and offered to bind the index with the first twenty-seven numbers of the magazine and sell it to members for cost.[3] He also volunteered to study one of the categories into which he'd divided Forteana and wrote an essay summarizing the faultiness of computers.

Thayer had long thought computer technology overblown, massive "electronic brains" not even able to calculate as quickly as a trained human using an abacus. The machines were yet another boondoggle, which he lumped with Adamski, the government pro-

motion of fallout shelters, and lies about space exploration. A poem
by MFS B. Goldstein, he thought, summarized the matter:

> These figures astronomical
> Make me QUITE gastronomical-
> ly ILL.
> Like U.F.O.s they climb the sky
> To soar vertiginously high.
> Who'll pay the bill???
> WHY
> You—
> and I—
> WE Will!!!![4]

After that February letter, however, Thayer hadn't heard from Russell again. Maybe Russell was quiet because Thayer had circled back
to the topic of Sputnik, arguing now that official claims about artificial satellites marked the third step the US government had taken
on its path to assuming godhood, after the bombing of Hiroshima
and the execution of Ethel Rosenberg, "the first woman ever elec-
trocuted for spying in Peace time." Maybe Russell was irritated by
Thayer's inconsistency, highlighting the day "when our dear, benig-
nant Uncle dropped the first atom bomb" while simultaneously calling atomic weapons hoaxes.[5] Perhaps he was reminded of all Thayer's annoying eccentricities. Or perhaps he really was tired. Whatever
the cause, Russell remained silent, and Thayer was bothered. He said
so in a letter dated 1 August. A week later, he and his wife went to
Nantucket for a vacation, where they bicycled, rowed, sailed, and
walked.[6] Flat and treeless, Nantucket doesn't strike one as paradise,
Thayer said. "The only thing about it is that it is Nantucket."[7]

There's no record of what he thought about on that trip, what
he and Kathleen discussed, but it seems reasonable he would have
contemplated the society that took so much of his time, where he
had been and where his remaining energy and enthusiasm would
allow him to go. Thayer had just put to bed the sixty-first issue of

Doubt, embarking on what would be the magazine's fourth decade. It was a good time to sum up—better than he could have known. For on that island, the Fortean Society would come to its end, as would a Fortean era. The twenty-ninth Fortean year would be the last. Fort would continue to be read; Forteans would march on, and new ones be born or made; Forteanism would, in the 1970s, see an efflorescence that outshone the golden age of Thayer's project in the late 1940s and early 1950s. But there would also be significant ruptures. The Forteanism that proliferated after 1959 served not Fort's strange, orthogenetic gods but new deities; it grew in a different soil, nourished by cultural conditions different from those that had given rise to Fort and sustained the society. The full scope of the new Forteanism, its continuities and discontinuities with the Age of Doubt, is a story for another day, but it can be outlined, the better to define the era that preceded it and preceded the death of Tiffany Thayer.

Earth's Shifting Crust

Thayer's intention since he came to the public eye in the early 1920s had been to shock. He didn't want to save the world, wasn't sure it was worth saving, but he wanted to send the shiver of an earthquake through its crust. Freeing minds, that was his desire, and changing the way the universe was seen, revealing it as both more wonderful than science allowed and more shot through with corruption than anyone could imagine. His strange geological theories made manifest this intention. The earth, he said, grew, in the process changing shape, from cube to sphere and back again, ad infinitum. While Thayer only rarely presented this idea in the pages of *Doubt*, he clearly carried it with him, if only in the back of his mind. Earth mysteries were a mainstay of his Fortean Society, and important. In the first issue of the magazine, he published a map showing everywhere on the planet then being visited by a scientific expedition, an index of topics he planned to discuss in later issues. In the very last issue of *Doubt*, he instructed a member who wanted to sharpen his Fortean eye to avoid weather-related topics and, instead, scan the papers for articles on earthquakes and islands that had disappeared (as well as reckless spending by the government).[8]

There was, however, at the heart of Thayer's practice, a contradiction. He both desired to shock and sought approbation. Thayer encouraged dissent and cultivated relations with those who advocated fringe ideas. He connected the Fortean Society with flat-earthers and those intrigued by Atlantis, among them Egerton Sykes, a British intelligence officer who created the largest library on Atlantis.[9] Thayer was not, however, quite as skeptical as he made out, or, rather, he struggled with what his skepticism meant and the relation to the world it required. Thayer disdained expertise. Introducing his piece on the locations of explorers throughout the world, he wrote, "Science is a lucrative and honorable field walled all about with a mystic nomenclature intended to obscure. The Latin of the piscatologist and the cosines and square roots abstrused by Einstein and the rest are merely NO TRESPASSING placards to keep mortals out." At the same time, though, he yearned for official recognition. Nothing thrilled him so much as a scientist taking Fort seriously. He told Sykes:

> Forteans generally are cordial to all your specializations, with fingers crossed, alert on the one hand not to embrace the Faith, and on the other, not to be beguiled by the authoritative, scientific approach of some of your authors.[10]

The contradictory impulses, though, weren't as he described them to Sykes. They were not balanced but cyclical. Thayer dismissed scientific pronouncements; then he dreamed of validation; next, he scorned any recognition Fortean ideas received from the mainstream, before craving, once more, official recognition—an endless circulation of intense emotion that drove him from delight to acidity and back again.[11] Alternative understandings of the earth, he sometimes asserted, needed to be taken seriously. The right one might have saved Amelia Earhart and Fred Noonan. Then another time he'd insist they be dismissed as ridiculous, craven attempts to create new sciences, to imprison minds for lucre and glory.

These alternating urges brought Thayer to praise Dave Kelley. Kelley joined the Fortean Society in the late 1940s, when he was studying anthropology at Harvard University on the GI Bill, specializing in Mesoamerica; he remained a member at least until the mid-1950s, though his interests in Fortean matters outlived both Thayer and the society.[12] A polymath—he was also a noted genealogist and had a more-than-amateur interest in astronomy—Kelley connected to Forteanism in several ways. Probably the most obscure link ran through the poet Charles Olson. Kelley studied Mayan hieroglyphics, which played the same role in Olson's poetic practice as Chinese in Pound's: an exemplar of what poetry should be. Mayan writing instantiated the propulsive force Olson wanted to bring to English: the origin of speech, according to Mayan myths, was the sound of nature, of animals. Theirs was a language without human categories. "O, they were hot for the world they lived in, these Maya, hot to get it down the way it was—the way it is, my fellow citizens."[13] Kelley approached Mayan hieroglyphics more academically, but also controversially, as was his wont. He argued that Mayan writing was phonetic, not pictographic, and in this conclusion he was broadly correct, an insight that would open the door to new historical understandings of the ancient Mayans.[14]

Kelley's other controversial positions were less academically productive. Over the course of his life, he returned repeatedly to studying calendrical systems, trying to differentiate those offered by Mesoamerican and Polynesian societies from European ones. This classificatory project was connected to a broader question: how cultural ideas traveled. At the time he was working, most anthropologists accepted that apparently similar ideas found in far-flung societies had come into being through parallel evolution. Kelley resurrected an older idea, that similar ideas instead spread from one place to the next. Diffusionism, the idea was called. Kelley had presented a diffusionist hypothesis in his PhD thesis, arguing that similarities in Mayan and Polynesian societies could be accounted for

by transoceanic contact. Other Forteans had developed similarly diffusionist ideas. That all of world culture might derive from a single
source was, as well, central to the claims of Atlantean scholars and
Theosophists, who said, *inter alia*, that the presence of pyramids in
Egypt and Latin America could be explained by both places having
learned construction techniques from the now-extinct Atlanteans.

Another expositor of earth mysteries connected to the Fortean
Society (though more obliquely than Kelley) was Charles Hutchins
Hapgood. Born to bohemian royalty—his mother was the writer
Neith Boyce, his father Hutchins Hapgood, whose book on his open
marriage was on Boni & Liveright's fall 1919 list alongside *The Book
of the Damned*—Charles was raised in Westchester County, New
York, and Provincetown, Massachusetts, by his mother, along with
three other siblings, Neith finding that her gender circumscribed her
bohemian prospects. (Hutchins continued his freewheeling ways after the birth of his children.)[15] Charles attended Harvard, receiving
a bachelor's and a master's degree; the Great Depression forced him
to quit his studies before he could finish his doctorate. For most of
the 1930s, he lived with his cousin, Elizabeth, an artist, and worked
as the executive secretary of the fine arts commission and director
of a community center, both in Provincetown, where he became
close friends with Frederick Hammett, the scientist and Fortean,
who was suffering badly from tuberculosis. (Hapgood named his
first child Frederick in honor of Hammett.) After serving in military intelligence during the war, Hapgood took a job teaching at
Keystone College, in Pennsylvania, soon moving to Springfield College in Massachusetts, where he stayed until 1952, when he left for
Keene State College in New Hampshire. He ended his career at New
England College, also in New Hampshire.[16] His specialty was the
history of science. In 1949, he was stumped by a student who asked
him about the lost continent of Mu, a Theosophical favorite, and
thereafter began structuring his classes around the investigation of
another missing civilization, Atlantis.[17]

The research opened Hapgood's eyes to new ways of looking
at the world and its history. He was not the only one so disposed

in the 1950s. There was Kelley, with his renascent diffusionism. And there was the psychiatrist Immanuel Velikovsky, whose *Worlds in Collision*, published in 1950, argued that mythologies encoded true stories of major catastrophes—specifically, comets that, striking earth during humanity's prehistory, altered the earth's orbit and axis. Velikovsky's book received no attention from Thayer—though it inspired a cover illustration by Castillo—but it became a public sensation.[18] Hapgood came across another proponent of maverick paleontology, Hugh Auchincloss Brown, an electrical engineer who argued that the accumulation of ice at the poles caused the earth to tilt every five thousand years or so, at which time the mantle slipped and the continents were rearranged. (Another shift was imminent, he said in 1948, and advised that atomic bombs be used to break up the polar caps to prevent catastrophe.)[19]

In time, Hapgood was drawn to other alternate histories, excited by strange carvings supposedly discovered in Acámbro, Mexico, which some said showed humanlike and dinosaurlike figures, suggesting that humans and giant reptiles had walked the earth together.[20] Hapgood built on Brown's idea, gathering evidence for regular slippages in the earth's land masses. The last one, in 9500 BCE, had moved the North Pole from Hudson Bay to its current position and started the process by which the South Pole was covered in ice (it was naked at the time of the dislocation). He cultivated some important supporters: Waldemar Kaempffert, one of the founders of science journalism and a longtime foe of both the Fortean Society (although he was friends with Thayer) and Fort, having refused to publish *The Book of the Damned* when he was editor of *Popular Science Monthly*; George Sarton, the founder of the history of science; and Albert Einstein. Einstein knew Velikovsky, having worked with him in the formation of the state of Israel, but he preferred Hapgood's theory. The idea of continental drift, around since the early twentieth century, had not yet taken hold, and Hapgood's theory made sense of the shape of continents and the similarities of some plant and animal species across the vast oceans, and provided a mechanism for how continents could move. Einstein wrote the fore-word to Hapgood's 1958 book *Earth's Shifting Crust*.

Hapgood also tapped into the network of Forteans. Although he almost certainly knew of Fort and the Forteans via Hammett, he seems to have been an indifferent MFS, rather reaching out to Ivan Sanderson and attracting the notice of Frederick Hehr—both ambiguous supporters. They urged him to incorporate flying saucers into his theories, which seemed a bridge too far for Hapgood. Moreover, Sanderson's own speculative writing on earth's prehistory threatened to poison the subject for mainstream editors and publishers, ruining Hapgood's chances of being taken seriously.[21] At the same time, Sanderson warned him off the Acámbro figures; he'd been to Mexico, seen them, and determined they were hoaxes.[22] (Inevitably, *Fate*, with Ray Palmer still at the helm, ran a story about the figurines: "Did Man Tame the Dinosaur?" Archeologist Charles DiPeso said no; he too had studied the sculptures and, noting how complete they were, how clean, lacking the patina of age, dismissed them as tourist tchotchkes.)[23] Sanderson advised Hapgood not to publish in the men's magazine *Argosy*, as it was too déclassé.[24] (Likely this reflected Sanderson's professional position more than a genuine assessment of the market; at the time he was writing for *Argosy*'s competitor *True* but would soon enough move to *Argosy*.) When Hapgood asked Sanderson to sponsor his application for a Guggenheim fellowship, Sanderson begged off. He knew his place in the scientific community, that he was dismissed as fringe, that any public support he offered could undermine Hapgood's quest for scientific respectability.[25]

Instead, Sanderson worked behind the scenes. He was very involved in the development and refinement of Hapgood's arguments and tried unsuccessfully to interest his European friends in the book.[26] Hapgood took the lack of scientific recognition stoically, imagining himself another Charles Darwin, offering a revolutionary idea to a hostile public, which would later acknowledge his theory as obviously correct.[27] Likely, Sanderson connected Hapgood with his own literary agent. As the book was being readied, Sanderson let a secretary from the publisher visit his New Jersey compound and copy the mailing list he'd accumulated over the past decade, people who would be interested in the kind of unusual theory being proposed by Hapgood.[28] Then, once *Earth's Shifting Crust* appeared in bookstores, Sanderson arranged to put Hapgood on Long John Nebel's radio program; he was a frequent guest himself.[29] He also tried to convince his rich patron, a Texas oilman, to support Hapgood, though he could not pull it off.[30] Hapgood squeezed out a few articles on the book as a means of advertising; one he sold to the *Saturday Evening Post*, another appeared in *Coronet*. A child-friendly edition of the book, *Great Mysteries of the Earth*, came out in 1960.[31]

Still, Hapgood was unfulfilled and resentful of the scientists who either ignored him or labeled him a crank. (Even the book reviewer in *Astounding* called the book pseudoscientific bunk.)[32] He took another run at a PhD, starting a dissertation that anatomized "the dogmatic science of the 19th Century and its further degradation in the 20th Century," but he never finished.[33] He had divorced shortly before his book came out and was too poor, he told Sanderson, to remarry, unhealed by the psychotherapy to which he subjected himself or the chiropractic he chose simply because the American Medical Association opposed the practice.[34] His three mainstream champions, Kaempffert, Sarton, and Einstein, all died during the book's writing. Sanderson advised him to take up yoga, Hehr to resign himself: World War III would soon start, after which the world would be renewed and more accepting of heterodox ideas.[35] Eventually, Hapgood did calm himself and return to work, studying an old map that he thought showed Antarctica before it was covered in ice, the foundation for his next book, *Maps of the Ancient Sea Kings*, which argued for a globe-spanning civilization in 7000 BC as the source of all later cultures. He even claimed to have found the location of Atlantis, although he would not specify it.[36] Over the years, he continued playing with these ideas, revising *Earth's Shifting Crust*, investigating other evidence that implied earth's history was not as taught. Despite Sanderson's admonitions, he also published on the Acámbro figures.[37]

Thayer was ecstatic when he received a copy of *Earth's Shifting Crust* not long after it was published, probably from Sanderson, who recommended it highly. Thayer had designated *Doubt* issues 1 through 27 as volume 1, and 28 through 57 as volume 2, restarting the pagination each time. The announcement of Hapgood's book came in the very first issue of the third volume, an auspicious new beginning. "It comes within hailing distance of the Thayer Theory which has been mentioned here before, but never fully expounded," he said in *Doubt* 58. Hapgood's only mistake was to assume that the earth did not change size. Once that fact was accounted for, Hapgood's theory was completely reconcilable with Thayer's, and the basis of a whole new vision of the world—"of an Earth that has grown from the size of a mustard seed and is still growing." Buy the book, he told his readers, and if it could not be procured, he would make an exception to the Society's getting out of the book selling business and provide it for the princely sum of $6.50. In the meantime:

> YS has already begun an almost page by page commentary on this work, demonstrating the relations of this enormous quantity of material to the larger (and *smaller!*) concept, without distortion of any of the facts and documentation presented.[38]

Thayer admitted his annotations would take a long time to finish, and it is quite possible he was still working on them ten months later when he vacationed in Nantucket.

Vile Vortices

In 1914, the year Ambrose Bierce went missing in Mexico, the poet George Sterling imagined a cosmic vampire, driven by Satan's thirst, supping on the earth's vitality:

> A shape that clutched the deviating earth
> And checked its headlong flight and held it fast,
> Draining the bitter oceans one by one.[39]

Concern that lives might be swallowed by an unnatural fiend or fiends was a thread running through the Fortean body, like an artery:

Charles Fort: I think there's a vast, black vampire that sometimes broods over this earth and other bodies.[40]

The Fortean George Christian Bump pressed on Thayer a novel from 1853, C. W. Webber's *Spiritual Vampirism*, which suggested vampires were real but misunderstood. Not lappers of blood, they fed on a so-called odic force that was present in everyone to varying degrees. Those overflowing with the odic force, described as something like electricity, insensible but vital, those were the great men of history. Those deficient in odic energy, usually women or nonwhites, formed a secret cabal, organized to feed on the odic energies of the more endowed. In Webber's novel, every sort of reformist group—feminists, abolitionists, spiritualists—belonged to the parasitic conspiracy of vampires. The vampiric cult, Bump said, was Fort's true earth-owner, organized to drain away the life force of every-day people to sustain "the supreme vampire which holds humanity in thralldom." It appears he read Webber's novel not as fiction but as a description of reality.[41]

1964: Vincent Gaddis summarized research on disappearing ships for the men's adventure magazine *Argosy*, amplified it in a book a year on. One "particular slice of the world" had "destroyed hundreds of ships and planes without a trace." He specified the region: "Draw a line from Florida to Bermuda, another from Bermuda to Puerto Rico, and a third line back to Florida through the Bahamas. Within this area, known as the 'Bermuda Triangle,' most of the total vanishments have occurred."[42]

Long before he became a Fortean author, Ivan Sanderson had a Fortean experience. Traveling the world as a collector of natural history specimens, he made his way through an African jungle. There he was attacked by a giant bat: "Then I let out a shout also and instantly bobbed down under the water, because, coming straight at me only a few feet above the water was a black thing the size of an eagle. I had only a glimpse of its face, yet that was quite sufficient, for its lower jaw hung open and bore a semicircle of pointed white teeth set about their own width apart from each other."[43]

Charles Fort: Rabies in vampire bats, reported from the island of Trinidad. Or a jungle at night—darkness and dankness, tangle and murk—and little white streaks that are purities in the dark—pure, white froths on the bloody mouths of flying bats—or that there is nothing that is beautiful and white, aglow against tangle and dark, that is not symbolized by froth on a vampire's mouth.[44]

George Sterling was San Francisco's King of Bohemia. Tutored by Bierce, he created a bohemian community in Carmel after the 1906 quake, where gathered Jack London, Mary Austin, Upton Sinclair and his socialist retinue. Harry Leon Wilson, having made his fortune writing plays with Booth Tarkington, settled there. In 1912, aged forty-seven, Wilson married seventeen-year-old Helen Cook.[45] Bierce visited but was disgusted at Sterling: reports of opium use, cheating on his wife, a suicide on the commune.[46] The Carmelites indulged in the pleasures of a religious life loosened by secularity, experimenting with sundry esoteric religions. In 1921, Sterling introduced Theodore Dreiser and John Cowper Powys to Albert Abrams, the doctor with the mysterious box that could diagnose and cure any illness.[47]

Charles Fort: I note that it is ten minutes past nine in the morning. At ten minutes past nine, tonight, if I think of this matter—and can reach a pencil, without having to get up from my chair—though sometimes I can scrawl a little with the burnt end of a match—I shall probably make a note to strike out those rabid bats, with froth on their bloody mouths. I shall be prim and austere, all played out, after my labors of the day, and with my horse powers stabled for the night. My better self is ascendant when my energy is low. The best literary standards are affronted by those sensational bats.[48]

Gaddis's term was met with approbation in the men's magazines, the phrase "Bermuda Triangle" repeated, again, once more, again.

But it was not the only mysterious, aqueous danger. Sanderson argued that UFOs seen entering the sea were not from another world but returning home, the craft of earth's invisible, submarine residents. The

George Sterling in February 1926, not quite nine months before his death. Sterling, a poet of the fantastic, was one of the seeds from which the Fortean community grew—connected to Bierce, Wilson, Dreiser, and Mencken, and mentor to Clark Ashton Smith. His poem "Lilith" may very well have inspired Lilith Lorraine to adopt that name. His imagery influenced writers throughout Forteanism's twentieth-century career. (Johan Hagemeyer Photograph Collection, © The Regents of the University of California, Bancroft Library, University of California, Berkeley. Creative Commons Attribution 4.0 International License.)

Deros gathered not in caverns but in the world's oceans.[49] Not content with a single vortex swallowing planes and ships, Sanderson identified twelve, arrayed along the thirtieth parallels, north and south. Vile vortices, he called them, the product of vampiric cultures.[50]

> Charles Fort: I don't know whether I am of a cruel and bloodthirsty disposition, or not. Most likely I am, but not more so than any other historian. Or, conforming to the conditions of our existence, I am amiable-bloodthirsty. In my desire for vampires, which is not in the least a queer desire, inasmuch as I have a theory that there are vampires . . .[51]

Sterling mentored Clark Ashton Smith, imparting his decadent sensibilities to the teenage poet. But even at the time, Sterling was falling out of fashion. He'd been accepted into Margaret Anderson's *Little Review* until she decided his decadent fantasias were old-fashioned. Smith refashioned Sterling's ideas into a cosmic vision for *Weird Tales*, sending the magazine several stories of vampires. In late 1926, a few months before Kenneth and Andrée Rexroth reached the city, Sterling swallowed cyanide after H. L. Mencken did not visit him at his apartment in San Francisco's Bohemian Club. Said Benjamin DeCasseres, "George committed suicide at fifty-seven. The perfect age for poets to pass. He took a quick curtain instead of a slow one."[52]

In 1974, Charles Berlitz published *The Bermuda Triangle*, a bestseller that further nestled the Fortean anomaly into popular consciousness. Just off the coast of the world's most powerful nation, or perhaps, strung around the world's oceans—it proposed—was a malevolent maelstrom that bent space and time, that sucked the vital blood of capitalism into its depths. The scholar Brett Neilson writes, "A brief examination of the myth's infectious spread through 1970s information networks reveals continual cross-hatching between media, genres, languages, and technologies."[53]

> Charles Fort: We have data that suggest the existence of vampires . . .
> but the brazen and serialized—sometimes murderous, but some-

times petty—assaults upon men and women are of a different order, and seem to me to be the work of imaginative criminals, stabbing people to make mystery, and to make a stir. I feel that I can under-stand their motives, because once upon a time I was a boy. One time, when I was a boy, I caught a lot of flies. There was nothing criminal, nor of the malicious, in what I did, this time, but it seems to give me an understanding of the "phantom" stabbers and snipers. I painted the backs of the flies red, and turned them loose. There was an imag-inative pleasure in thinking of flies, so bearing my mark, attracting attention, causing people to wonder, spreading far, appearing in dis-tant places, so marked by me.[54]

A year after Berlitz, Lawrence Kusche published his investigation of the Bermuda Triangle. Kusche was one of the first fellows of the Commit-tee for the Scientific Investigation of Claims of the Paranormal, which had institutionalized Martin Gardner's style of skepticism. The book carefully examined the data underlying the legend. No story here, he said. Ships and planes disappeared in the Triangle no more often than in any other area of the earth; the so-called mysterious vanishments, upon inspection, often proved explicable. Not that Kusche's book ended the legend. "As the decade wore on," Neilson notes, "the stories were recycled for children's books. . . . But most important for the myth's expanding notoriety was its reworking in B-movies and television documentaries."[55]

Inherent in Bump's cosmology was a negative fundamentalism— there was a singular spiritual force that controlled human life, one that opposed humanity, a kind of Satan. Thus, it is unsurprising that later in life he became a fundamentalist Christian. It took only the flick of a tog-gle, turning the owner of the earth from a repugnant ruler to a magiste-rial one. Bump divorced, remarried, took a degree in mathematics from Stanford. He settled in southern California, wrangled with the Board of Equalization, which handled the state's taxes. He and his wife wrote "ob-ject" on their tax forms and refused to pay any money to the state. They'd had no income, they said, because the only legitimate currency was gold or silver, or a certificate redeemable for those metals, and having received neither they owed no taxes.[56] Like Webber in *Spiritual Vampirism*, Bump railed against reformers.[57] "America is a Christian nation and people; and when we cease to be Christian, we shall cease to be either a nation or a people. Christianity is based on the Bible."[58]

> Charles Fort: Dear me—once upon a time, I enjoyed a sense of amusement and superiority toward "cranks." And now here am I, a "crank," myself. Like most writers, I have the moralist somewhere in my composition, and here I warn—take care, oh reader, with whom you are amused, unless you enjoy laughing at yourself.[59]

The Death of Tiffany Thayer

By the time Thayer reached Nantucket, he needed a break. He and Kathleen had planned a trip to Greece the previous year but had to scrap it for work.[60] The holidays that year had not been a respite, and even small mercies were hard to come by. Once he could see the East River from his home in the Sutton Place neighborhood, but his view was now blocked by taller buildings, giant "egg crates."[61] Given his beliefs, he probably hadn't seen a doctor. "All our foods are so contaminated that I am on a straight gin diet," he told Russell, perhaps a joke, perhaps not.[62] He smoked, no surprise—way back when, T. Swann Harding had pooh-poohed the idea that cigarettes were coffin nails, and it is unlikely subsequent evidence would have convinced Thayer otherwise.[63] The vigorous exercise he undertook on the island would have been very different from his activities in New York City, writing ad copy, working on the next installment of *Mona Lisa*, handling the correspondence of the Fortean Society, putting out its magazine four times a year, and making a life with Kathleen.

On Saturday, 22 August, Tiffany and Kathleen bicycled to the water's edge to watch the final day of the Flying Dutchman Nationals, a series of yacht races.[64] Harry Sindle of New Jersey had won the previous four championships and, only a few days before, had been announced as the US representative in the Pan American Games, but he came in a disappointing third this year, losing to a husband-and-wife team, Jack and Pat Diane from Florida.[65] The Thayers returned to their room at the Holiday Inn that night, prepared to return to New York City after the weekend. Sunday morning, Kathleen attended church while Tiffany went to the hotel's restaurant for breakfast. At 8:15, fellow diners reported, he fell to the floor. A local doctor said the heart attack probably killed him in less than a minute—there was nothing anyone could do. Kathleen, hearing the news, "collapsed" in church, overcome by grief and shock. Tiffany Ellsworth Thayer was fifty-seven.[66]

Word of Thayer's death made the papers that Monday. He'd reaffirmed his notoriety as a writer of mainstream smut only three years before, and many were those who remembered his drugstore novels. "America's foremost fiction exploiter of the boudoir and the libido," noted the *New York Daily News*, quoting some advertising puffery from earlier in Thayer's career.[67] Kathleen was beset by "countless problems."[68] There seems to have been some confusion over what to do with Tiffany's body. The press reported that he would be returned to the city, but he was laid to rest at Prospect Hill Cemetery in Nantucket.[69] Thayer was a rich man, with cash assets totaling more than $100,000, plus a separate apartment to house his thousands of books and the society's archives, but he died intestate. Eventually, his estate was divided. His second wife, Tanagra, received a little over $10,000; his mother, $25,000; Kathleen, $83,000 and his property.[70]

Russell fell into a depression. He'd been extricating himself from the society; only a year before, Thayer had reprinted the Fortean Society brochure, removing Russell's address to stem the flow of unsolicited correspondence.[71] He seems to have thought that he was too aloof to be saddened, but he surprised himself; he'd lost a constant in his life, having spent twenty years helping run the society, reading the newspapers with Thayer's eye. Perhaps he felt lonely. His daughter, his only child, had moved out of the country. His best friend, Fred Shroyer, lived in Los Angeles and had run Thayer down after *Mona Lisa* flopped: "Jesus Christ, you'd think the guy discovered sex! And that thesaurus erudition he flaunts . . ."[72] Two months after reading of Thayer's death in the papers, Shroyer reached out to Russell, conflicted about Thayer's legacy:

> Sure as hell sorry about Thayer. It was in the papers here. I liked the man immensely, though I disagreed with him in the matter of his attacks on science. The big battle is still science versus religion, and until that one has been won, we've got to keep our ranks closed. Anyway, I lay laurels on his barrow. And as the Romans had it, "May the earth rest lightly on you"—Tiffany.[73]

Others were more sympathetic, though even more tardy in their condolences. "I am grieved . . . to learn of the death of Tiffany Thayer, for I know how you will miss him," I. O. Evans told Russell in January 1960. "I too know what it is to lose old friends."[74]

Russell confided his sadness to John W. Campbell, who intellectualized the problem, turned to science fiction for answers. Maybe Heinlein could explain Russell's unexpected flood of emotions: "The trouble is the lack of People." Heinlein had suspected, according to Campbell, that "most of the 'people' in the world weren't real—just some sort of automatons put in the scene to fill in the background for the real people." Campbell thought he was right. Thayer (like Russell, like Campbell himself) was a real person. Russell's depression, Campbell diagnosed, was "the usual reaction of one of the People that another of the all-too-rare People had left the world. The deep and heavy feeling of 'Who, now, is left to carry on . . . ?' The answer is, of course, that when one of the People dies, there never is anyone to carry on; the thing that makes the People, People, is simply that they're unique individuals, and being unique, have no exact replacement."[75] Campbell's elitist assessment may or may not have comforted Russell, but it did touch on a question that troubled him: How could the Fortean Society go on? At its best, Forteanism fluttered in the heart of every true individual, but it needed an institutional home. Who could manage that? Was Thayer so exceptional as to be irreplaceable? Russell had assured the British Fortean George Peacock that he'd let him know what would happen next, but even a year later the society's fate remained unclear, Russell silent on the matter, perhaps because he was himself confused. Peacock seemed to sigh: "I must assume its future is still shrouded in doubt."[76]

Kathleen was as uncertain as Russell about how to proceed. At first, she thought she'd continue on her own, an idea she contemplated through the end of 1959, past the time the next issue of *Doubt* should have appeared. She told Don Bloch, "I worked so closely with Tiffany for eighteen years I have high hopes (sometimes very low) of trying to carry on the work of the Society."[77] In the end, though, she opted against taking on Thayer's burden. She consulted with Aaron Sussman, the only surviving original Founder, and they decided that it was best to close shop.[78] She assured Russell her decision was a sign of respect:

> No, the Fortean Society has never been and never could be an embarrassment to me. You must surely realize that only Tiffany could have kept it functioning. No other human being I have ever known could have or would have worked so hard for such an unrewarding cause. Therefore, the Society had to be terminated.[79]

In April 1964, Kathleen remarried.[80] She claimed to own no membership lists, no files from the society, and would not say where Fort's notes had gone. (They ended up where most had been born, at the New York Public Library.)[81] She had sold off Thayer's library back in 1960 and presumably destroyed the Fortean Society archives at some point, or perhaps they were thrown out after her own death.[82] The letters Thayer had filed and whatever other material he'd accumulated over his nearly three-decade career as a Fortean were lost.

Russell came around to the idea that the Fortean Society was best left buried, and if there was to be any revival that he would not take part.[83] Aged fifty-four, he'd suffered his own heart attack the same year Thayer died, which slowed him down and perhaps was the reason he was so tardy answering Thayer's call to rejuvenate the society. According to the Fortean and science fiction fan Roy Lavender, Russell received letters encouraging him to restart the organization, with him at its helm, but he "pour[ed] cold water on the idea, mainly because some 80% of the membership was in the U.S."[84]

In the 1960s, Russell reconnected with Sydney Birchby, a science fiction fan who'd run in the same circles as Russell back in the 1930s and 1940s; both were Forteans.[85] The postwar years had been hard on Birchby. He drifted from the science fiction community, as it had become too focused on parties, and life became a grind, especially as his wife suffered a bad back, which prevented her from working.[86] But then he'd fallen in with another old member of Britain's science fiction community, Harold Chibbett, an occultist and Fortean, who'd published his own Fortean 'zine before the war. Chibbett resurrected the *Newsletter*, as it was called, in the late 1960s. Birchby was revived; he eschewed UFOs and was bored by discussion of the afterlife (he'd already convinced himself that the spirit continued after death), but he'd nursed an interest in the esoteric for years, with no one but his wife to discuss it with, and now he was welcomed into a community of like-minded individuals. "If it does nothing else, it keeps me from having rust on the brain."[87]

Birchby collected Fortean clippings and encouraged Russell to join him and Chibbett, to get the band back together, but Russell was done with all things Fortean, as he was with most of science fiction. He begged off Birchby's entreaties (at some point he did seem to accede and join, but put little to no effort into the matter) and retired from writing, releasing only a handful of new works in the 1960s. Doubleday approached him about doing a biography of Fort, but he demurred—and Kathleen Thayer likely would not have helped, since at one point she claimed that her husband had been working on his own biography, based on the papers that she had inherited, and she would not want to see any competitors.[88] Unlike Kathleen with Thayer's personal papers, Russell preserved his archive. Eric Frank Russell died in 1978, and all his correspondence on Fortean matters, plus boxes of other papers related to the science fiction community, were given by his daughter to the University of Liverpool fifteen years later.

Tiffany Thayer was mostly forgotten after his death: a footnote.

As Kathleen Thayer Eliasberg, his widow renewed the copyright of a couple of his books shortly after her marriage, but all the novels were out of print. *Doubt* mostly disappeared into private collections, difficult to track down for anyone who hoped to read it or continue Thayer's project.[89] His name could be found in science fiction reference books and encyclopedias of the strange. Some of his papers, but nothing personal or connected to the Fortean Society, wended their way to the New York Public Library. James Blish privately recalled him as a generous mentor who had encouraged his writing when Blish was in high school. Mostly, though, Thayer was remembered as an object of scorn. Critical comments by F. Scott Fitzgerald and Dorothy Parker were repeated—remarks on his vulgarity and inexplicable popularity.[90] Arthur Schlesinger Jr., the historian and public intellectual, in recounting his time at a prep school in the early 1930s, offered as a kind of eulogy an assessment of Thayer's place in the literary memory of the late twentieth century:

> There was much clandestine reading of books frowned on by the authorities. Adolescent boys were inevitably preoccupied with sex, and bawdy novels passed from hand to hand until they were dog-eared. Favored authors were Tiffany Thayer (*Thirteen Men* and *Thirteen Women*) and Thorne Smith (whose *Topper* inspired the popular Topper movie series with Constance Bennett and Cary Grant). How ribald and (in the term of the time) risqué these books seemed in 1932! How tame they seem today![91]

But Memory?

In an undated letter, a single line only, Roy Lavender asked Russell, "Does the Fortean Society continue in any form but memory?"[92] Russell's response is lost, but the answer was clear: no. The society was, indeed, but memory. The finish was brought about by Tiffany Thayer's death, the death of the Founders, the decisions of Kathleen Thayer, the disappearance of Russell's enthusiasm, but not only by these. Had none of them occurred, the society would still have been headed for either extinction, revolutionary change, or irrelevance. Thayer recognized the impending breakup and sought new sources of support, new Founders.[93] He knew his conspiratorial view of Sputnik and flying saucers put him at odds with many Forteans. "I have walked way out on a limb, and I'm glad of it," he told Russell in 1958, hardly more than a year before his death. "No other course was open to me even if I 'believed'—which I don't."[94] In the late 1950s, the cultural conditions that had given rise to Fort and sustained the Fortean Society changed in fundamental ways. And so Forteanism, too, was bound to change.

Science fiction recovered from its late-1950s decline by reinventing itself, proclaiming a "New Wave." A generation of authors looked at what had preceded them and, seeing the genre played out, moved away from pulp conventions, forgoing space opera, a focus on the hard sciences, and stories of ideas to focus on new narratives, the softer sciences of psychology and philosophy and sociology, and aesthetic innovation, borrowing especially from postmodernist literature.[95] Although some science fiction authors continued to explore Fortean themes—notably Stanisław Lem and Lionel Fanthorpe—there was less room in this reoriented science fiction for Fortean topics, or rather, the methods by which Forteana were worked into the tales changed.[96]

The transformation of the modernist avant-garde and the rise of UFOlogical high strangeness also altered the ways that Forteanism related to experimental literature and the investigation of flying saucers. Modernism, at least those forms associated with Forteanism, had been intertwined with science. Surrealists saw themselves as investigating the workings of the mind, for instance. But postmodernism had a more fraught relationship with science, and so Forteanism was of less interest.[97] *Morning of the Magicians*, which grew out of surrealism, was intensely popular in the 1960s and 1970s, but the development of Higher Realism was increasingly cut off from literary trends. In a not dissimilar way, UFOlogy and its parent Theosophy also became increasingly insular in the 1960s and 1970s, esoteric descriptions of reality that were "foggy" compared to the specificity of science.[98] Fortean ideas persisted in the UFOlogical community, but their meanings became ever more abstruse.

Forteanism, however, continued. Thayer's records are lost, but it seems reasonable to conclude that, at the Society's peak, there were several thousand MFS, plus unaffiliated Forteans, and copycat organizations such as N. Meade Layne's Borderlands Science Research Association, Ray Palmer's *Fate*, Donn Brazier's Frontier Society, and, later, Sanderson's Society for the Investigation of the Unexplained. These Forteans did not disappear. They gathered, young and old, created new organizations, produced new magazines and movies and TV shows and, in time, websites. There were radio programs and books. Fort had set the stage for a new genre of writing, Fortean compilations by Fortean authors who collected all manner of unaccountable events and then accounted for them with outrageous hypotheses—what Jeffrey Kripal calls "science mysticism."[99] Vast new edifices were created that were supposed to change our understanding of the universe and humanity's place in it. These were not usually pure Forteanism, rather a mixture of Forteanism, conspiracism, New Agery, and fundamentalism. Nor were they usually inspired solely by Fort, rather combining his data and theories and fragmentary remnants of the society with the notions of other Fortean writers, like Ivan Sanderson and Vincent Gaddis, and whole swaths of other fields: religion, Theosophical, New Age, and Christian; science, mainstream and heterodox; pulp tales, horrific and science fictional; UFOlogy; and alternate histories.

The Fortean community was diverse and dynamic, but there was a common, if not universal underlying pattern: a distrust of science and the pronouncements of officialdom; an attempt to see the world in heterodox terms, mingling what was deemed false with what was possible and what was to create new visions of reality. Futile, perhaps, these gestures. Forteans did not stop, or even slow, the growth of scientific, national, industrial bureaucracies. Their promulgation of mystical, pseudoscientific, and contrary ideas, meant playfully or not, failed to reshape mainstream thought. But, perhaps, "ambitions were never the most interesting thing about" Forteanism, as political scientist Henry Farrell said of anarchists, first cousins to Forteans. Rather, what is striking about this history are the unexpected connections among disparate groups who used Fort to expand their imaginative vistas, to reveal the hidden structure of reality, and to map the movement of power. Who used Fort, in short, to think to new worlds.

Future History

In the mid-1960s, a pair of brothers set out to give American Forteanism a new institutional home. They would succeed, but the connection between what became the International Fortean Organization and the Fortean Society was never simple: there were continuities but also significant ruptures. That was true of all the Forteanism that followed Thayer's death and sprawled through the 1960s and 1970s. Forteanism did not die; rather, there were improvisations, reworkings of Fort and the legacy of the Fortean Society to fit a new era, and a brief examination of them brings into sharper focus what came before. The *INFO Journal*, as the new organization's publication was titled, joined *Fate* and Ivan Sanderson's *Pursuit*. Charles Hapgood continued to write, as did Vincent Gaddis, and a new generation of Fortean writers grew up around them, some of whom took up Dave Kelley's diffusionism. Fort's books continued to circulate, and new magazines came into being, notably *Fortean Times*. But this Forteanism was supported by conditions different from those that had underwritten Thayer's society, both intellectually and materially.

The brothers were Ron and Paul Willis, who'd come out of Missouri, moved to the Washington, DC, area. Their father, Eben, had been a science fiction fan and seemed to pass that on, along with a willingness to try anything—risking failure the sign of a vital life.[100] Ron, the older of the two, served in the army in the early 1950s; he may or may not have contributed to the society in its glory days—it's hard to be sure—but the name Willis did appear in *Doubt* when Ron would have been in his late teens. In the 1960s, the brothers contacted Eric Frank Russell and other erstwhile members of the society whom they could track down: Roy Lavender, Norman Markham, Don Bloch.[101] The first issue of *INFO Journal* appeared in 1966, and shortly thereafter the Willises purchased their own printing equipment and started the Goblin Press, which put out not only *INFO Journal* but a magazine of weird tales. *Anubis* was proudly Fortean, publishing stories and poems by Markham, reprinting Robert Barbour Johnson's tirade against Thayer, and running an article by Vincent Gaddis. Still, it wasn't clear if the Willises meant INFO to be a continuation of the Fortean Society or something different. Spread the word "that the FS is back in business and 'still kicking,'" Ron told Lavender. But a few weeks later, Paul said that there was no connection between the two Fortean groups.[102] He had no interest in old Fortean Society membership lists since INFO planned to generate its own audience, in part by advertising in *Anubis*. A few years on, however, he asked after just such a list and tried to make connections with previous MFS.[103] Chapter Two was resurrected and associated with INFO, giving Johnson and George Haas a chance to resume discussing the subject that had gotten them tossed from the society, the Stanford apports (as well as Haas's new obsession, Bigfoot).[104] *Anubis* lasted only four issues, but, despite Ron's early death in 1975, *INFO Journal* ran for years and grew into its name, finding subscribers in Australia, Canada, New Zealand, and England.

Hapgood was one of about a dozen INFO "research associates" (along with Haas and Gaddis). He still wrote on earth mysteries and communicated with the editors of *Fate*, but also moved into spiritualism. Around the time the Willis brothers put together INFO, Hapgood met the psychic Elwood Babbitt. Hapgood and his cousin Elizabeth started attending seances with Babbit, Charles recording and transcribing the sessions. Babbitt introduced the two to Jesus, Vishnu, and Mark Twain, and reacquainted Charles with Albert Einstein and Frederick Hammett. Elizabeth helped establish a school in Babbitt's honor; the trio was also connected to the Brotherhood of the Spirit, a commune in Massachusetts that claimed the earth was on the verge of cataclysmic changes that would usher in a spiritual revolution.[105] Charles published two books of Babbitt's sessions, *Voices of the Spirit* (1975) and *Talks with Christ* (1981). Hapgood died in 1982 after being hit by a car. Despite having the vast realms of the spirit to contemplate and being able to definitively evaluate Fortean ideas, Hammett's comments to Hapgood were inconsequential. The discarnate Hammett said he was no longer troubled by tuberculosis and offered Hapgood advice on his personal problems. Hammett also helped him write *The Path of the Pole*.[106]

Meanwhile, Hapgood and Dave Kelley's contemplations of earth mysteries inspired a new generation. Erich von Daniken's massive bestseller *Chariots of the Gods*, which claimed that aliens—earth's owners—initiated human civilization, cited Hapgood. Decades later, Graham Hancock's *Fingerprints of the Gods*, also very popular, found comfort in Hapgood's writings. Here was an accredited professor, with approval from Einstein himself, who studied history and argued for advanced ancient civilizations—how big of a step, then, to accept that aliens started those civilizations? Theorists of Atlantis, proponents of diffusionism, champions of the idea that earth was once home to an advanced global civilization, its wisdom lost or hidden, looked to Hapgood's writings as backing their contentions. Kelley's oeuvre was cited to support similar suppositions. Hapgood's work on the Acámbaro figurines was taken up by young earth creationists as proof that dinosaurs and humans existed at the same moment, before the Noachian flood.

Through the 1970s, a plethora of Fortean books flooded the market, authored by a generation of writers building on Fort, on Russell's *Great World Mysteries*, on R. DeWitt Miller. Vincent Gaddis continued writing in this vein into the 1990s. Damon Knight, a science fiction author and onetime member of the Futurians, was tapped to write Fort's biography after Russell declined. *Charles Fort: Prophet of the Unexplained* was published in 1968. Henry Holt put out *The Books of Charles Fort* until paperback editions appeared, first from the science fiction publisher Ace, then, with an introduction by Knight, from Dover (as *The Complete Books of Charles Fort*, 1974).[107] Loren Coleman and Jerome Clark became celebrated Fortean authors, both in tandem and individually; Clark had been associated with Ivan Sanderson's Society for the Investigation of the Unexplained and Coleman was another of INFO's research associates. The British realtor John Michell became popular after discovering earth mysteries in the mid-1960s, integrating them with UFOs and Atlantis. He argued, in ways not dissimilar from Hapgood and Kelley, that earth had once been traversed by a technologically and spiritually advanced race that had constructed monuments to channel its magnetic forces; their knowledge had been lost, he wrote, but would soon be rediscovered, ushering in a new age of spiritual awakening. His book on these matters was "a smash countercultural success."[108]

Another research associate on INFO's masthead was Robert J. M. Rickard—the only one associated with the "British Isles," as the *Journal* phrased it. Rickard was a science fiction fan and came across mentions of Fort in Campbell's *Astounding* and Russell's *Great World Mysteries*. Inspired, he started clipping anomalous reports from the papers. As he moved deeper into the British science fiction community, he came across some of the older Forteans, among them Harold Chibbett and Sydney Birchby. Chibbett was sending out his *Newsletter* at the time—less a 'zine than a chain letter, with recipients adding to the missive as it was passed around. Rickard was put on the list, with about forty others. By the early 1970s, Chibbett was slowing from the indignities of age, with Birchby picking up the slack. Moved by what he was reading in the *Newsletter* and *INFO Journal*, Rickard started his own publication, gathering about him a small cadre of supporters including John Michell. *The News*, "A Miscellany of Fortean Curiosities," appeared in 1973.[109] It grew slowly over the years, eventually transforming into the *Fortean Times*, which became, with *Fate*, the premiere Fortean publication of the twentieth century's second half. As *Doubt* once did, *Fortean Times* encompassed a wide pageant of Forteanism, more cynical, lighthearted, and materialistic than *Fate* but working the same beat.

All these improvisations, though, appeared in conditions significantly different from those that had allowed the Fortean Society to continue for so long. Many of the differences were material. Forteanism had spread mostly in a handful of specialized publications, plus a larger number of science fiction magazines and some books, both pulpish and modernist. In the 1960s and 1970s, Forteanism was still a print phenomenon, presented in journals, newsletters, and the books of Fortean authors. The market had changed, though, with science fiction magazines now less important than men's adventure magazines such as *True* and *Argosy*. These pitched Fortean events, even those invented out of whole cloth, as unambiguously true—not interesting ideas to mull over but facts.[110] There were other media now too, movies and television shows. Arthur C. Clarke lent his name to some, and to related books. Cryptozoology, UFOlogy, and earth and sea mysteries were common subjects of documentaries that played in theaters, large and small, and appeared on television.[111] Like the men's magazines, these mixed fact and fiction but presented the result as unvarnished truth. Following in Ivan Sanderson's footsteps, they tried to transform the fantastical and science fictional into the true. That there was a market for these products seems undeniable, given how popular many of the books, magazines, movies, and television programs were. At least partially, the success of these cultural products can be explained by the spread of ideas associated with the San Francisco Renaissance, which had given birth to the Beats, who, in turn, had spawned the counterculture movement of the 1960s.[112] Imagination was the last refuge of the individual, and these Fortean stories were a way to use the imagination to construct a different, less dispiriting world using the fictions of the previous generation. These new worlds were not understood ironically, however, but as fundamentally true.

The Passing of Pisces

The fervor of Forteana in the 1960s, combined with the increasing prominence of astrology, the birth of new spiritualities, the fantastic popularity of *Morning of the Magicians*—all of this the mainstream anxiously labeled "the occult explosion."[113] Counterculture utopians and hippies talked of the coming Age of Aquarius, based on an astrological notion that earth was moving from the control of the zodiacal sign Pisces, the fish, to that of Aquarius, the water-bearer—a change that would bring a new, more liberating age into being. Skeptics in the tradition of Martin Gardner fretted over this talk, seeing a recrudescence of superstition, an attack on rationality—an antimodernist retreat into ridiculous superstitions. Occultists and New Ageists threatened the authority of scientists. The skeptics paid little heed to the ways in which the world had never been disenchanted, the ways modernity had always been filled with wonder—New Agery, fundamentalism, conspiracism, and Forteanism being not separate from but constitutive of the modern world.

Charles Olson, the poet, wrote of the coming of a new age, which he also saw in astrological terms, in a vocabulary that would have made sense to Charles Hapgood had he read *The Maximus Poems*, which were, after all, a consideration of history from the perspective of Gloucester, Massachusetts, not far from Hapgood's old home in Provincetown. Western civilization had been moving northwest since the time of ancient Greece, Olson said, just as the earth's crust migrated and the continents rearranged themselves. America, where Western civilization had landed, was now at a crux, as was the earth in general.[114] "LEAP onto / the LAND, the AQUARIAN / TIME," he wrote.[115] His prophetic vision was pure Forteanism:

> swimming in schools, the lower atmosphere
> gelatinizing from their traffic, cities like cigarettes,
> country side
> sacrificed to their eating time & space up like fish
> fish-people   their own UFOs the end of Pisces
> could be the end of the species   Nature herself
> left to each flying object passing in a mucus
> surrounding the earth.[116]

Or, if the pivot did not lead to the end of the human species, then the Age of Aquarius, which would bring about "an archaic/postmodern Age of Graffiti."[117]

Olson associated the postmodern age with particular objects, the kind of material changes that forced Forteanism to adapt to new conditions: video, the computer.[118] "Postmodernism" also accrued meanings beyond what Olson intended; it became a term of art for cultural commentators and academics, applied to the works of William Gaddis and William Burroughs, some of the New Wave science fiction authors, a whole host of others. In one sense merely a demarcation of literary periods, "postmodern" also referred to an array of authorial techniques, although like modernism, postmodernism was a complicated, often contradictory movement. There are, however, some broad agreements about its outlines. If modernism aimed at a precise, scientific understanding of the processes of human thought and apprehension of the world, then postmodernism saw science as a repertoire of ideas with which to play, but not as a practice to which to aspire. Science, like all other systems of thought, was unstable—the crust on which human beings stood was always shifting; indeed, even the notion of the "human," the "individual," was suspect. There was no ultimate truth, rather any ideas about human life were constructed from remnants of the past, assembled according to arbitrary rules and the whims of power. As the scholar David Harvey notes, postmodernists insisted that "universal and eternal truths, if they exist at all, cannot be specified."[119] By contrast, earlier versions of Forteanism had, for the most part, maintained that there was truth but that it was obscured. Fort's recovery of the damned was an attempt to create a more accurate understanding of reality (even if that understanding allowed him to always be contrarian and argue against any version of truth). Postmodernism rejected the very notion of ultimate truth.[120]

Forteanism of the 1960s and the 1970s was deeply entangled with these questions. Jerome Clarke and Loren Coleman argued that explanations of Fortean events and UFOs were not true, in a scientific sense, but were nonetheless real. Witnesses to these extraordinary phenomena were experiencing an altered state of consciousness in which they were in touch with primal needs that could be expressed in terms of Jungian archetypes. After the experience came rationalizations, attempts to impose logic on what were otherwise literally indescribable incidents.[121] In 1977, John Michell and Bob Rickard similarly tried to finesse the problem of truth. "As phenomenalists," they wrote, "we accept everything; we believe nothing absolutely." Their phenomenalism, as they called their philosophy, was rooted in the writings of Fort, whom they acknowledged as "the father" of their ideas. But, in the manner of many sons, they were spending quickly what they had inherited. Fort indeed seemed to believe nothing absolutely and prophesied the era of the hyphen, but he was ultimately grounded in a sense of the truth: there were damned facts and heavenly ones, good sources and bad ones, reports to be considered and those to be doubted. Michell and Rickard simply dispensed with the notion of truth and falsity altogether. They wrote:

> Inevitably, with the modern confusion of all creeds and in reaction to the oppressive certainties of the great nineteenth-century theory-mongers, phenomenalism is becoming more generally appreciated for its inclusive approach to reality. With it comes a new and welcome tolerance. Policemen can spot pumas in Surrey or chase UFOs in Devon and still remain policemen. People mysteriously transported through the air are no longer burnt; sometimes they are even believed. Scientists view, and write papers on, such products of extra-scientific reality as fire-walking and spoon bending. Humanists, freed from such once obligatory concepts as spirits or the primacy of matter, study the divining rod and survival after death through the neutral evidence of their associated phenomena. There are even some psychiatrists who will listen uncensoriously to the impossible experiences of their patients.[122]

There was no ultimate source of truth, no final arbiter.

UFOs and aliens, among Fortean phenomena, especially raised problems for the notion of truth in the years after Thayer's death. The political theorist Jodi Dean wrote, "The alien seduces us into a critical reassessment of our criteria for truth: How do we determine what real is? Why do we believe? The claim to truth and its challenge to our practices for establishing it are what enable the alien to function as an icon of postmodern anxieties."[123] In the early 1970s,

Robert Spencer Carr resurrected the story from a quarter century earlier of a supposed flying saucer crash in Roswell, New Mexico, and claimed that the military had recovered an alien body. Over the years, the legend and associated claims grew—the alien had been necropsied, filmed—the narrative entering the mainstream. (Carr's son refuted the story in the pages of CSICOP's magazine *Skeptical Inquirer*.)[124] At the same time, more people came forward claiming that they, like Adamski, had contacted aliens—and these aliens did not merely converse with people but abducted them, performed invasive experiments.[125] Although these stories could be taken as expressions of a confident, if credulous, realism, they can also be understood as anxiety about truth. Medieval witch hunters, notes historian Walter Stephens, were driven not by assertive faith in the supernatural but by a fear that Christianity might be wrong in some important way; in eliciting confessions, inquisitors sought to reassure themselves.[126] The Roswell crash, the alien necropsy, the contactees—all of these sought to similarly prove truth by extracting it from bodies.[127]

A Conspiracy of Conspiracies

> I can feel the heat closing in, feel them out there making their
> moves.[128]
>
> WILLIAM BURROUGHS, *Naked Lunch*

First published in 1959, *Naked Lunch* circulated in Paris, alongside Henry Miller's books and others deemed too illicit for the American market. THEM was Burroughs's constant bugaboo, the antagonist of his books and his life, the conspiracy against which he fought, and the reason he sought allies on the margins, allies like Fort. THEM kept *Naked Lunch* out of America.

Mainstream religion was one of THEM, one of the forces of control. Against it, Burroughs enlisted crystals and telepathy and Reichian Orgone accumulators and spirits and demons and magic and curses and the occult and coincidences. "I do believe in the magical universe, where nothing happens unless one wills it to happen, and what we see is not one god but many gods in power and conflict," he said.[129] He turned to Scientology, achieving the rank of Clear. Burroughs was coy about his Fortean debt, but he moved through Fortean networks, influenced by science fiction, an exemplar of the New Wave, associate of the Beats, connected to Black Mountain College.[130] He was familiar with Fort, fascinated even, according to his friend the poet Terry Wilson. Supposedly, he cribbed the aphorism "history is fiction" from *Wild Talents*, inserted it into his novel *Nova Express*, picked up other bits of Fortean wisdom too.[131]

He mimicked Fort's collagistic style, as well. Went a passage in another novel, *The Soft Machine*:

> Calling partisans of all nations—Cut word lines—Shift linguals—Vibrate tourists—Free doorways—Word falling—Photo falling—Break through in Grey Room.[132]

He developed a writing method, the cut-up technique, based on a game used by Dadaists and surrealists. Take a text. Cut it into pieces. Rearrange those into something new. The result often sounded Fortean, in style and subject. Hence, he transformed articles from the Paris *Herald-Tribune* into the prose poem "Viruses Were by Accident?"

> (Reservoir of rabies and other virus? discovered in *Brown* fat of vampire bats and their well known and easily chosen human constituent.) . . .
>
> *Brown attempted to make such a deal with plants and animals over thousands of years . . .*
>
> Unusual beings dormant in cancer feel towards the day already overpopulated with hungry cows. . . .
>
> Viruses were by accident?[133]

Although rooted in surrealism, the cut-up technique belongs to postmodernism. In its quest to illuminate the consciousness of individuals, modernism had exhausted the stock of literary techniques, leaving to postmodernists empty gestures and a distrust of individuality.[134] Scholar David Harvey notes that postmodernism can be characterized by its attention to surfaces and reproduction as opposed to modernism's concern with depth and original production.[135] One example is postmodernism's interest in pastiche. Pastiche, like parody, mimics what came before it but, unlike parody, has no satirical intent—often, no concern for meaning at all. Burrough's cut-up technique, less concerned with depth of meaning than the arrangement of words, fit the postmodern temper. His cut-ups looked like, and sometimes read like, Fort's jazzy aphorisms but were emptied of the metaphysical import Fort's writing conveyed, instead undermining the creation of meaning and the notion of authorship.[136]

Even as he explored the limits of the cut-up technique, Burroughs was folded into a conspiracy of conspiracies, a paean to Fort's god Decomposition restyled as Eris, goddess of disorder. In 1975, Robert Anton Wilson and Robert Shea published *The Illuminatus Trilogy*. While working at *Playboy*, Wilson and Shea started discussing Discordianism, a mock religion invented in the early 1960s based on the principle of chaos, a religion that parodied religion.[137] Discordianism was different from the magical religions created from stories published in *Weird Tales*. Those assumed that there was a truth, though their mechanisms for accessing it were irreverent.

Discordianism dispensed with any concern for truth. Combining this interest with ideas taken from Burroughs, from the counterculture, from surrealism, and from Fort, Wilson and Shea crafted a postmodern narrative that built on the avant-garde experiments of Joyce and Pound but transcended them. *The Illuminatus Trilogy* was nonlinear, the perspective continually shifting, the story complex and contradictory.[138] They structured the narrative around a panoply of conspiracies, some of their own invention, some poached from the Fortean community and its offshoots. Wilson and Shea continually deconstructed the very ideas they presented, proposing and disposing in equal measure.[139] More so than Fort and DeCasseres, their skepticism was hungry, devouring identity and belief, making truth an impossibility. The point was to break down human categories of thought and reveal that underneath "reality" was a seething disorder. Theirs was, perhaps, the most destructive Fortean innovation to arise after Thayer's death—if also an endlessly creative one—but by no means the only twist on Forteanism.[140]

The Paranoid Style

The November 1964 issue of *Harper's* magazine carried an essay by the country's preeminent historian, Richard Hofstadter, "The Paranoid Style in American Politics." Based on a talk he'd given at Oxford University a year prior about a neglected strand of the nation's history, the essay took aim at the angry presidential campaign of Barry Goldwater. "I call it the paranoid style simply because no other word adequately evokes the sense of heated exaggeration, suspiciousness, and conspiratorial fantasy that I have in mind." Hofstadter was quick to distance the term from any clinical connotations, but not from negative ones. "Of course this term is pejorative, and it is meant to be; the paranoid style has a greater affinity for bad causes than good."[141] The burden of the essay was to trace this style—briefly—through American history, describing the young nation's introduction, in 1797, to the Illuminati and the conspiracies that attached to them: the fear of Masons, anti-Catholic bigotry—a constant right-wing sense of persecution. Those who practice the paranoid style, what today are called conspiracy theorists, are twice besieged. "We are all sufferers from history," Hofstadter writes, "but the paranoid is a double sufferer, since he is afflicted not only by the real world, with the rest of us, but by his fantasies as well."[142] This aggrievement, combined with the sense that the world is under the command of the most nefarious of forces, justifies the persecuted in using any means necessary to resist the powers that be. Hofstadter's piece is the best-known and (with Frederick Turner's "Significance of the Frontier in American History") most influential writing by an American historian ever.

And certainly, evidence of his claims swirled about Hofstadter. Only recently had Joseph McCarthy lost influence. There was the Goldwater campaign, with its John Birch supporters. The same year the *Harper's* essay was published, Louis Pauwels and Jacques Bergier's *Morning of the Magicians* was released. Even as he gave his lecture, the world moved under Hofstadter's feet. John F. Kennedy was assassinated the very next day, launching a thousand paranoid conspiracies, some of which persist to this day. Conspiracism changed, as did its sibling Forteanism, retaining some of its prior characteristics, developing some new tricks.

"The higher paranoid scholarship is nothing if not coherent—in fact the paranoid mind is far more coherent than the real world," Hofstadter concedes. "The entire right-wing movement of our time is a parade of experts, study groups, monographs, footnotes, and bibliographies."[143] Fort can be made to fit this model, and many Forteans as well. Fort did have an appreciation for facts. Why else spend all those days in libraries reading newspapers, magazines, academic journals, collecting tens of thousands of notes and citations? Why else lard his four books with repeated reference to his sources? *Doubt*, R. DeWitt Miller, Vincent Gaddis—these Fortean descendants followed suit, gathering notes, systematizing them, weighting their writings with pointers to the underlying facts. They worried over trivial inconsistencies, tried to make coherent every last irregularity, even if the stories they spun were outrageous. But Fort was skeptical of facts too, willing to undermine and dismiss them. As an "intermediatist," Fort had insisted that nothing was real and nothing was unreal, which gave him the freedom to play with the categories of true and false: to see true things damned and false things made heavenly. His ludic treatment of facts was done with a wink, coyly, leaving the reader always slightly confused: exactly how much did Fort value the facts he wrote about? Pauwels and Bergier elaborated on this ambiguous relationship, pushing Forteanism away from playfulness into more dangerous realms; they were pioneers, and the trail they blazed was followed by others. Even as Pauwels and Bergier insisted that they were dedicated truth seekers, *The Morning of the Magicians* unapologetically weaved together arguments from science, pseudoscience, mysticism, science fiction, and fantasy. When a fact was not available, bare assertion substituted just fine, as did reference to a story from the pulp magazines—and all this in the service of proving what was undeniably true. Not play, then, but serious matters. Any single work, certainly, can fail, and the world is full of bad books filled with illogical arguments. But *The Morning of the Magicians* was not exceptional, rather a sign of what was to come, presage to Michell and Rickard's phenomenalism and so much of the Forteanism of the 1970s. Over the next half century and more, this tendency to blend the factual and the fictional to make supposedly true arguments, to eschew any pretense of citing sources and argue from assertion, to shrug off Fortean playfulness in favor of earnestness, became pronounced, not just in the cultic or the occult, but in American political culture—American culture—more generally.

The reasons for this change are many. Some are ironic. Hofstadter was right to accentuate the pathology of conspiratorial thought, but there have been, as well, real conspiracies, reasons to take seriously the idea that small cabals of the craven and perfidious act in secret to carry out large, illicit schemes—cover-ups by big business, by the government, by religious institutions—and these instances have fueled other conspiracy theories while also saving them from the need for evidence. Everyone already knows there is corruption.[144] Why bother with supporting facts when pronouncement works just as well? Experts' involvement in politics further undermined the notion of neutral arbiters of truth. Experts came to be seen not as dispensers of wisdom from above the fray but just another variety of politician.[145] No more could a Hofstadter clinically define some groups as paranoid without being attacked as partisan, part of a conspiracy, paranoid himself. The same fate befell scientists, doctors, anyone who claimed to speak for the truth. More obscure changes in American culture also eroded the distinction between the real and the unreal.[146] The confidence in facts that Hofstadter observed has gone, replaced by an anxious sense that truth is too quicksilver to be pinned by anything as insubstantial as a citation.

Political scientists Russell Muirhead and Nancy L. Rosenblum have it that in the twenty-first century America entered a fundamentally new era of conspiracizing, one that dispensed with theory altogether. "Instead we have innuendo . . . or bare assertion," they write. "What validates the new conspiracism is not evidence but repetition."[147] Paranoid fables unspool on social media and cable networks, unslowed by fact-checking, absorbing the false and the rumored into capacious maws. The second decade of the new millennium brought another angry presidential campaign, this one victorious, which carried the new conspiracism into the White House. From the bully pulpit came evidence-free claims of the worst kind of chicanery; on social media, these fantasies intersected with QAnon, or Q, the dominant right-wing conspiracy of the era, a ramshackle, centerless prophecy of the imminent clash between the forces of good and a vast conspiracy of evildoers—a worldwide cabal of moneyed elites, political centrists, and liberals involved in pedophilia and other horrendous acts. "For QAnon," notes the journalist Adrienne LaFrance, "every contradiction can be explained away; no form of argument can prevail against it."[148] As with other late forms of conspiracy, QAnon effortlessly blended the fanciful and fantastic with the stuff of everyday life. There were now conspiracies about lizard-people operating in the halls of Congress; the masses being controlled through a combination of vaccines, nanotechnology, and cell phones; mass school shootings being staged.

Fort and Forteans played their part in the creation of this world; they are not blameless. They eroded the distinctions between truth and falsity, undermined the authority of experts and expertise. They launched a thousand conspiracies into the national consciousness. But conspiracism had changed since *The Book of the Damned*, as had science fiction, the avant-garde, and UFOlogy. Fort's playfulness had been replaced by Thayer's acerbic nihilism, which became omnipresent and decoupled from any need to compile evidence or craft arguments. "From attacking malevolent individuals, conspiracists move on to assaulting institutions," write Muirhead and Rosenblum. "Conspiracism corrodes the foundations of democracy."[149] The paranoid style exhausted society, and itself. When a conspiracy can be reduced to a single phrase—"Fake news!"—a single word—"Rigged!"—a single letter even—Q—where else can it go?

Caitlín R. Kiernan, *To Charles Fort, with Love*

"By the damned, I mean the excluded," Charles Fort wrote in the second sentence of his first book, at the dawn of Fortean modernity.[150] "His writing could not have come at a more opportune time; if Charles Fort had not existed, it would have been necessary to invent him," says the scholar of hidden histories Colin Dickey. "The twentieth century was a Fortean time."[151] A few years into the twenty-first century, the fiction author and paleontologist Caitlín R. Kiernan contemplated that line from Fort's *Book of the Damned.* "I would mark that as one of the single most powerful sentences in 20th-Century literature," they averred.[152] Kiernan was in the process of building a new Forteanism from the remnants of the past, returning to Fort, to Lovecraft and other writers of the weird, to modernists like Joyce, to the postmodernist Burroughs, while rejecting the offerings of the *Fortean Times,* the new conspiracists, those who would blur fact and fiction; theirs was something different.

Born in Dublin, Ireland, in 1964, Kiernan moved to Alabama when they were a child. Raised on Poe and Lovecraft, on Bram Stoker, that scribbler of vampire tales, and on a host of other authors of the weird—likely it was during a childhood of devotedly reading such literature that they encountered Charles Fort. In 2006, they listed Fort as their second-favorite writer of nonfiction—unlike all those Forteans in the 1930s who could not muster enough enthusiasm to list him among their most-liked authors.[153] Kiernan also developed an early interest in paleontology, the two twining about each other, informing a singular theme: that the universe was indifferent to the concerns of humanity. "Our smallness and insignificance in the universe at large. In all *possible* universes. Within the concept of infinity. No one and nothing cares for us. No one's watching out for us."[154] In their twenties, Kiernan changed their name and transitioned to female, coming later to think of themselves as genderfluid and not transgender; they preferred plural pronouns.[155] Kiernan attended the University of Alabama at Birmingham and the University of Colorado at Boulder, where they studied paleontology.[156]

Through the 1990s, Kiernan continued their paleontological research; they also started writing fiction, building on their influences. For a time, they lived, like Lovecraft, in Rhode Island, before returning to Alabama. Within two decades, they became one of the most highly regarded weird writers of the twenty-first century. In 2012, the horror writer Peter Straub placed them in "the new vanguard, still being formed, of our best and most artful authors of the gothic and fantastic."[157] Lovecraft scholar S.

T. Joshi opined that, with Ramsey Campbell, Kiernan was the best writer of weird tales in the past half century—that is, since the early 1960s.[158] Kiernan has written comics, chapbooks, novels, and short stories.

In his book *The Unidentified*, Dickey contrasts Fort with cranks: "What Charles Fort offered was a means of looking at . . . strange events that militated against ideology seeping in." By contrast, "The crank view of the world swaps out the splendor and dynamism of everything around us for pat, one-size-fits-all explanations that flatter our perceptions and prejudices. It chases order and reassurance at the expense of wonder and variety, and it too easily defaults to paranoia and conspiracy as a defense against being wrong."[159] Dickey too easily separates Fort from the cranks, the conspiracists, and the paranoid. Neither Fort nor Forteanism was ever unsullied. From its earliest moments, Fortean modernity blended the Fortean with the conspiratorial, the New Age, and the fundamentalist. But as it proceeded through the twentieth century and into the twenty-first, the conspiratorial had become particularly conspicuous, necessitating a reaction from those, like Dickey, who did not want to throw out the Fortean baby with the paranormal bathwater. *Fortean Times* tried to thread the same needle. The magazine approached the world with Fort's twinkling eye and sense of humor, willing to indulge outrageous hypotheses. But at the end of the day, its editorial position was to retreat to the safe harbor of reason, science, and likelihood; the weird was a nice place to visit, but no one would want to live there.

Kiernan, too, rejects the conspiratorial, but as firmly refuses to foreclose the possibility of the weird. *Fortean Times*, they say, was "a silly rag."[160] Theirs is not horror fiction, they insist. They feel no compulsion toward story: "Ulysses should have freed writers from plot." Rather, Kiernan focuses on atmosphere, mood, language, theme. "That's the stuff that fascinates me." The horrific contributes to the mood, the atmosphere, but "never predominates."[161] Horror is one emotion among many present in any work and can be alloyed with other sentiments. Kiernan is Fortean in foregrounding awe: "I need a world filled with wonder, with awe, with awful things. I couldn't exist in a world devoid of marvels, even if the marvels are terrible marvels. Even if they frighten me to consider them. What would be the point of a world like that, a humdrum world of known quantities and everyday expectations?"[162] The horrible and the awesome—often one and the same in their stories.

The key, though, is that the stories resist easy interpretations. There's no attempt to play with the bizarre only to return, at the end, to the rational, nor to stitch everything up into some coherent conspiracy. Kiernan's goal is to leave the reader alone with a feeling of uncanniness. There is a rigor in the method that was often not present in Fort or Lovecraft themselves; Fort offered his expressions, although he often went on to doubt them. Kiernan withholds even these, expression and doubt: "The 'reveal' is something I do my best to avoid. . . . What is weird fiction but a journey into the unknown, and if you make the unknown known, why bother? If you want to know what's going on, read Agatha Christie. Or a science textbook. 'What happened?' is absolute anathema to weird fiction."[163] Rather, they've wanted to push the boundaries as far as they can go, to force the reader into weirder and weirder situations, more and more disorienting. The Fortean world is a world without bearings—the laws may be mundane and multifarious, but they are also confounding, beyond any human's ability to fully comprehend. The mundane world itself offers moments so inexplicable that they threaten to expose everything everyone thinks they know as a sham.

A clear fictional statement on this matter is the short story "Standing Water." Written in September 2000, it was included in 2005's *To Charles Fort, with Love*, a collection of thirteen of Kiernan's stories that display their Forteanism, blending it with Lovecraft, leavening the horrible with the awesome. They wrote "Standing Water," Kiernan says, to "to explore a 'wrong thing.'" This error in existence was trivial but, in Lovecraftian fashion, hints at eldritch horrors, impossible geometries. "The cosmic as seen through the lens of the mundane."[164] In this case, the mundane is not a Super-Sargasso Sea, swirling in the heavens above the earth, nor weird rains of meat or blood or frogs but only a puddle of water—not unlike the swamp of frogs in the middle of Liverpool that fascinated Eric Frank Russell, lo, those many years before. This mud puddle is in an alley, that most humdrum of urban spaces, yet nothing else about the puddle makes sense. It appears when there's been no rain, where there is no leak, and is deep enough to swallow, without a burp, an entire broom. And then, without a warning, it disappears.

The puddle's passage is a Fortean event, one meant to send shivers down the spine and leave the reader awestruck, horrified, but not wiser. There are elements of Forteanisms past: the unexplained event, the suggestion of cosmic forces within the quotidian; even the genre is an old-fashioned Fortean one, the short story—not a television script, a website, a pseudodocumentary. But the story is something new, too. Kiernan hints that the puddle is infernal but offers no explanation, either mundane or mystical. The authorial voice suggests nothing conspiratorial nor fundamentalist nor New Age. There is no ironically raised eyebrow, no playing *as if* the thing were true. Nor is there an attempt to blur the factual and the fictional. Kiernan follows the Lovecraftian path of evoking horror by placing before the reader something that cannot be, something at odds with reality, but without the archaic language and without the cosmic pomposity. There is no Fortean philosophical and scientific tomfoolery, merely a Fortean event, and humanity stripped bare of its defenses.

Glimmering, too, in the story are future Forteanisms, when the world will be—as it was, as it is, as it ever will be—enchanted. Fort's philosophy, like most things, will pass in time; it will take its turn among the damned things. But as long as there are humans, there will remain phenomena that humans cannot explain, and these *things*—to borrow Ivan Sanderson's vague but difficult-to-resist noun—will continue to prod the human imagination in new directions, be they envisioning new structures of reality, new relationships of power, or something else altogether. In the meantime, one waits to see where Forteanism goes, how Kiernan and their generation think to new worlds.

Notes

Chapter 1

1. Damon Knight, *Charles Fort: Prophet of the Unexplained* (New York: Doubleday, 1970), 180–83; Jim Steinmeyer, *Charles Fort: The Man Who Invented the Supernatural* (New York: Penguin, 2008), 266–72.

2. Jeffrey J. Kripal, *Authors of the Impossible: The Paranormal and the Sacred* (Chicago: University of Chicago Press, 2010), 9, 125.

3. Solomon F. Bloom, "The Spirits Move Him," *Brooklyn Daily Eagle*, 29 May 1932, 58.

4. Knight, *Charles Fort*; Colin Bennett, *Politics of Imagination: The Life, Work, and Ideas of Charles Fort* (Manchester, UK: Critical Vision, 2002); Steinmeyer, *Charles Fort*; Kripal, *Authors of the Impossible*; Colin Dickey, *The Unidentified: Mythical Monsters, Alien Encounters, and Our Obsession with the Unexplained* (New York: Viking, 2020).

5. But see Deborah Dixon, "A Benevolent and Sceptical Inquiry: Exploring 'Fortean Geographies' with the Mothman," *Cultural Geographies* 14 (2007): 189–210; Jack Hunter, ed., *Damned Facts: Fortean Essays on Religion, Folklore and the Paranormal* (Paphos, Cyprus: Aporetic Press, 2016).

6. Christopher Partridge, *The Re-enchantment of the West: Alternative Spiritualities, Sacralization, Popular Culture, and Occulture*, vol. 1 (London: T & T Clark International, 2004), 68.

7. Henri Ellenberger, *The Discovery of The Unconscious: The History and Evolution of Dynamic Psychiatry* (New York: Basic Books, 1970); Anthony Giddens, *Modernity and Self-Identity: Self and Society in the Late Modern Age* (Cambridge, UK: Polity Press, 1991); James C. Scott, *Seeing Like a State: How Certain Schemes to Improve the Human Condition Have Failed* (New Haven, CT: Yale University Press, 1998).

8. Cathy Gere, *Knossos and the Prophets of Modernism* (Chicago: University of Chicago Press, 2009).

9. Michael Saler, "Modernity and Enchantment: A Historiographic Review," *American Historical Review* 111 (2006): 692–716.

10. Friedrich Nietzsche, *The Gay Science*, trans. Walter Kaufmann (New York: Random House, 1974), 181.

11. Max Weber, "Science as a Vocation," in *From Max Weber: Essays in Sociology*, ed. H. H. Gerth and C. Wright Mills (New York: Oxford University Press, 1946), 129–56.

12. Rodney Stark and Roger Finke, *Acts of Faith: Explaining the Human Side of Religion* (Berkeley: University of California Press, 2000).

13. Jason Ā. Josephson-Storm, *The Myth of Disenchantment: Magic, Modernity, and the Birth of the Human Sciences* (Chicago: University of Chicago Press, 2017).

14. Charles Taylor, *A Secular Age* (Cambridge, MA: Harvard University Press, 2007).

15. Egil Asprem, *The Problem of Disenchantment: Scientific Naturalism and Esoteric Discourse 1900–1939* (Leiden: Brill, 2016).

16. Saler, "Modernity and Disenchantment," 700.

17. Janet Oppenheim, *The Other World: Spiritualism and Psychical Research in England, 1850–1914* (Cambridge: Cambridge University Press, 1985); Joscelyn Godwin, *The Theosophical Enlightenment* (Albany: State University of New York Press, 1994); Anne Harrington, *Reenchanted Science: Holism in German Culture from Wilhelm II to Hitler* (Princeton, NJ: Princeton University Press, 1996); Alex Owen, *The Place of Enchantment: British Occultism and the Culture of the Modern* (Chicago: University of Chicago Press, 2004); Corinna Treitel, *A Science for the Soul: Occultism and the Genesis of the German Modern* (Baltimore, MD: Johns Hopkins University Press, 2004); Lynn L. Sharp, *Secular Spirituality: Reincarnation and Spiritism in Nineteenth-Century France* (Plymouth, UK: Lexington Books, 2006); M. Brady Bower, *Unruly Spirits: The Science of Psychic Phenomena in Modern France* (Urbana: University of Illinois Press, 2010); Courtenay Raia, *The New Prometheans: Faith, Science, and the Supernatural Mind in the Victorian Fin de Siècle* (Chicago: University of Chicago Press, 2019).

18. Kripal, *Authors of the Impossible*, 29.

19. Daniel P. Thurs, "Tiny Tech, Transcendent Tech: Nanotechnology, Science Fiction, and the Limits of Modern Science Talk," *Science Communication* 29 (2007): 65–95.

20. T. J. Jackson Lears, *No Place of Grace: Antimodernism and the Transformation of American Culture, 1880–1920* (New York: Pantheon, 1981).

21. David E. Nye, *American Technological Sublime* (Cambridge, MA: MIT Press, 1996); Jane Bennett, *The Enchantment of Modern Life: Attachments, Crossings, and Ethics* (Princeton, NJ: Princeton University Press, 2016).

22. Michael Saler, *As If: Modern Enchantment and the Literary Prehistory of Virtual Reality* (New York: Oxford University Press, 2011).

23. Simon During, *Modern Enchantments: The Cultural Power of Secular Magic* (Cambridge, MA: Harvard University Press, 2004).

24. Simon Locke, *Re-crafting Rationalization: Enchanted Science and Mundane Mysteries* (London: Routledge, 2016).

25. Weber, "Science as a Vocation," 129–56.

26. Lorraine Daston, "When Science Went Modern," *Hedgehog Review* 18 (2016): 18–32.

27. Peter Baehr, "The 'Iron Cage' and the 'Shell as Hard as Steel': Parsons, Weber, and the *Stahlhartes Gehäuse* Metaphor in *The Protestant Ethic and the Spirit of Capitalism*," *History and Theory* 40 (2001): 153–69.

28. Philip J. Pauly, *Controlling Life: Jacques Loeb and the Engineering Ideal in Biology* (New York: Oxford University Press, 1987).

29. Linda Dalrymple Henderson, "Editor's Introduction: I. Writing Modern Art and Science—An Overview; II. Cubism, Futurism, and Ether Physics in the Early Twentieth Century," *Science in Context* 17 (2004): 423–66.

30. John Lardas Modern, *Secularism in Antebellum America* (Chicago: University of Chicago Press, 2011).

31. Steven Sutcliffe, *Children of the New Age: A History of Spiritual Practices* (London: Routledge, 2002); Catherine L. Albanese, *A Republic of Mind and Spirit: A Cultural History of American Metaphysical Religion* (New Haven, CT: Yale University Press, 2006); Partridge, *Re-enchantment*, 68.

32. Michael Barkun, *A Culture of Conspiracy: Apocalyptic Visions in Contemporary America* (Berkeley: University of California Press, 2013).

33. Kathryn S. Olmsted, *Real Enemies: Conspiracy Theories and American Democracy, World War I to 9/11* (New York: Oxford University Press, 2011).

34. For example, Victor Burr, "Wandering through the Year," *Times Leader* (Wilkes-Barre, PA), 11 August 1932, 17.

35. Bergen Evans, *The Natural History of Nonsense* (New York: Knopf, 1946), 29.

36. A. D. Bajkov, "Do Fish Fall from the Sky?" *Science* 109 (1949): 402.

37. Evans, *Natural History of Nonsense*, 24.

38. Evans, *Natural History of Nonsense*, 26.

39. "Fish Fall from Sky in Marksville," *Times* (Shreveport, LA), 24 October 1947, 1.

40. Bajkov, "Do Fish Fall?" 402.

41. W. L. McAtee, "Showers of Organic Matter," *Monthly Weather Review*, May 1917, 217–24.

42. Lola T. Dees, "Rains of Fishes," US Department of the Interior, US Fish and Wildlife Service, Bureau of Commercial Fisheries, Fishery Leaflet 513,

April 1961, 3. See Charles Fitzhugh Talman, *Meteorology: The Science of the Atmosphere* (New York: P. F. Collier & Son, 1922), 356; E. W. Gudger, "Rains of Fishes—Myth or Fact?" *Science* 103 (1946): 683–94.

43. Bergen Evans, "Concerning Rains of Fishes," *Science* 103 (1946): 713.

44. Charles Fort, "Red Lizards, Snakes and Frogs Out of the Sky," *Philadelphia Public Ledger*, 27 July 1924, in "The Correspondence of, and to, Charles Hoy Fort," ed. Mr. X, 2004, http://www.resologist.net/corfpl01.htm.

45. Fort, "Red Lizards."

46. F. Neave, "Alexander Dimitrivitch Bajkov" *Progress Reports of the Pacific Coast Stations* (Fisheries Research Board of Canada) 105 (1956): 26.

47. Kenneth Johnstone, *The Aquatic Explorers: A History of the Fisheries Research Board of Canada* (Toronto: University of Toronto Press, 1977), 134–35.

48. "Fish Fall," 1; Bajkov, "Do Fish Fall?" 402.

49. Bajkov, "Do Fish Fall?" 402.

50. Ivan T. Sanderson, *Investigating the Unexplained: A Compendium of Disquieting Mysteries of the Natural World* (Englewood Cliffs, NJ: Prentice-Hall, 1972), 247.

51. Waldo Lee McAtee to Don Bloch, 2 March 1952, Waldo Lee McAtee papers, Library of Congress.

52. Charles Fort, *The Book of the Damned* (New York: Boni & Liveright, 1919), 241.

53. Fort, *Book of the Damned*, 87.

54. Kripal, *Authors of the Impossible*, 122–31.

55. Fort, *Book of the Damned*, 252.

56. Knight, *Charles Fort*, 27.

57. Fort to Dreiser, 3 June 1916, in "Correspondence," http://www.resologist.net/corftd06.htm.

58. Charlotte Sleigh, "Writing the Scientific Self: Samuel Butler and Charles Hoy Fort," *Journal of Literature and Science* 8 (2015): 17–35, 24.

59. Fort to Dreiser, 19 July 1916, in "Correspondence," http://www.resologist.net/corftd07.htm.

60. Knight, *Charles Fort*, 39.

61. Jerome Loving, *The Last Titan: A Life of Theodore Dreiser* (Berkeley: University of California Press, 2005).

62. Knight, *Charles Fort*, 58.

63. Steinmeyer, *Charles Fort*, 135, 228.

64. Steinmeyer, *Charles Fort*, 231.

65. Fort, *Book of the Damned*, 8.

66. David M. Cornell, "Taking Monism Seriously," *Philosophical Studies* 173 (2016): 2397–2415.

67. Benjamin DeCasseres, "An American Wrestles with God," *American Mercury*, April 1929, 467–73.

68. Steinmeyer, *Charles Fort*, 136.

69. Knight, *Charles Fort*, 59.

70. John C. McCole, *Lucifer at Large* (New York: Longmans, Green and Co., 1937), 37; Louis J. Zanine, *Mechanism and Mysticism: The Influence of Science on the Thought and Work of Theodore Dreiser* (Philadelphia: University of Pennsylvania Press, 1993).

71. Theodore Dreiser, *A Hoosier Holiday* (New York: John Lane Co., 1916), 374.

72. Dreiser, *Hoosier Holiday*, 375; Steinmeyer, *Charles Fort*, 142.

73. Thedore Dreiser, *Hey Rub-A-Dub-Dub* (New York: Boni & Liveright, 1920), 60–73, 182–200.

74. Steinmeyer, *Charles Fort*, 140.

75. Steinmeyer, *Charles Fort*, 155.

76. Steinmeyer, *Charles Fort*, 161.

77. Steinmeyer, *Charles Fort*, 160–61.

78. Fort, *Book of the Damned*, 16.

79. Fort, *Book of the Damned*, 8, 18.

80. Fort, *Book of the Damned*, 17.

81. Steinmeyer, *Charles Fort*, 165–66.

82. Steinmeyer, *Charles Fort*, 139.

83. Fort, *Book of the Damned*, 156.

84. Fort, *Book of the Damned*, 117.

85. Steinmeyer, *Charles Fort*, 166.

86. Zanine, *Mechanism and Mysticism*, 127; W. A. Swanberg, *Dreiser* (New York: Charles Scribner's Sons, 1965), 241.

87. "Fall Publications of Boni & Liveright," *Publishers' Weekly*, 27 September 1919, 835.

88. *Philadelphia Inquirer*, January 15, 1920, 12.

89. F. F. Wetmore, "'The Book of the Damned' Arrays Unexplained Facts," *Daily Arkansas Gazette*, 4 April 1920, 41, 44.

90. Benjamin DeCasseres, "The Fortean Fantasy," *Thinker*, April 1931, 76.

91. *Fortean* 2 (October 1937): 5–6; Booth Tarkington to Harry Leon Wilson, 8 February 1920, Harry Leon Wilson papers, University of California, Berkeley.

92. *Bookman*, August 1920, advertising supplement (n.p.).

93. *Fortean* 3 (January 1940): 5.

94. Ben Hecht, *Gaily, Gaily: The Memoirs of a Cub Reporter in Chicago* (New York: Doubleday, 1963), 159.

95. Steinmeyer, *Charles Fort*, 180.

96. Steinmeyer, *Charles Fort*, 11.

97. Preserved Smith, "Waifs and Strays of Thought," *Nation*, 10 April 1920, 483.

98. "'The Damned' Are Data Science Has Rejected," *New York Herald*, 15 February 1920, 73.

99. Fort to Dreiser, 31 March 1916, in "Correspondence," http://www .resologist.net/corftd03.htm.

100. Steinmeyer, *Charles Fort*, 170.

101. Edmund Pearson, *Queer Books* (New York: Doubleday, Doran, 1928), 146.

102. "Insult and Research," *New York Tribune*, 24 October 1920, 78.

103. Steinmeyer, *Charles Fort*, 192.

104. Steinmeyer, *Charles Fort*, 190.

105. Steinmeyer, *Charles Fort*, 192.

106. Fort to DeCasseres, 5 July 1926, Benjamin DeCasseres papers, New York Public Library.

107. "New Lands," *Cincinnati Enquirer*, 6 January 1924, 87.

108. Fort to the *Philadelphia Public Ledger*, 7 July 1924, in "Correspondence," http://www.resologist.net/corfpl01.htm.

109. Steinmeyer, *Charles Fort*, 224.

110. Fort to DeCasseres, 6 July 1926, DeCasseres papers.

111. Fort to DeCasseres, 2 September 1930, DeCasseres papers.

112. Benjamin DeCasseres, "The Dash of Jamake," in *The Feather Duster; or, Is He Sincere?* (New York: Roycrofters, 1912), 75–81, 79; Fort to DeCasseres, 25 October 1925, DeCasseres papers.

113. Fort to DeCasseres, 29 November 1925, DeCasseres papers; Benjamin DeCasseres, "The Frame Up," in *The Works of Benjamin DeCasseres*, vol. 1 (New York: Gordon Press, 1939), 13–18.

114. Fort to DeCasseres, 25 October 1925, 29 November 1925, 2 September 1930, DeCasseres papers.

115. Benjamin DeCasseres, "Fantasia Impromptu," in *The Works of Benjamin DeCasseres*, vol. 2 (New York: Gordon Press, 1939), 22.

116. Benjamin DeCasseres, "Modernity and the Decadence," *Camera Work*, January 1912, 17–19, 19.

117. Fort to DeCasseres, 6 July 1926, DeCasseres papers.

118. Charles Fort to Theodore Dreiser, 13 August 1916, in "Correspondence," http://www.resologist.net/corftd08.htm.

119. Ruben van Lujik, *Children of Lucifer: The Origins of Modern Religious Satanism* (Oxford: Oxford University Press, 2016).

120. David Weir, *Decadence and the Making of Modernism* (Amherst: University of Massachusetts Press, 1995), 111.

121. Fort to DeCasseres, 6 July 1926, DeCasseres papers.

122. Fort to DeCasseres, 2 September 1930, DeCasseres papers.

123. Martin A. Edwards, *H. L. Mencken and the Debunkers* (Athens: University of Georgia Press, 1984), 93–113.

124. William Manchester, *Disturber of the Peace: The Life of H. L. Mencken* (New York: Collier, 1962), 61, 65.

125. Fort to DeCasseres, 4 February 1931, 6 February 1931, DeCasseres papers.

126. Steinmeyer, *Charles Fort*, 241.

127. Charles Fort, *Lo!* (New York: Claude Kendall, 1931), 15.

128. Fort, *Lo!* 31.

129. Zanine, *Mechanism and Mysticism*, 129.

130. Steinmeyer, *Charles Fort*, 241.

131. Steinmeyer, *Charles Fort*, 144.

132. *Fortean* 3 (January 1940): 5.

133. Ben Hecht, *Humpty Dumpty* (New York: Boni & Liveright, 1924), 223.

134. Miriam Allen De Ford, "Charles Fort: Enfant Terrible of Science," *Magazine of Fantasy and Science Fiction*, January 1954, 105–16, 105.

135. Miriam Allen De Ford, "Do We Survive," *Forum and Century*, October 1935, 250–55, 252; Miriam Allen De Ford, *Up-Hill All the Way: The Life of Maynard Shipley* (Yellow Springs, OH: Antioch Press, 1956), 177–78.

136. William Rose Benét, "The Phoenix Nest," *Saturday Review of Literature*, 5 September 1942, 12–13.

137. H. L. Mencken, "Nonsense as Science," *American Mercury*, December 1932, 508–10.

138. Knight, *Charles Fort*, 171–72.

139. Fort to Edmond Hamilton, 29 March 1926, Edmond Hamilton papers, Jack Williams Science Fiction Library, Eastern New Mexico State University.

140. "Other Planets Inhabited" (letter), *Indianapolis Star*, 25 March 1926, 6.

141. "Well, Well! See a Shower of Froggies Lately? Fate of World Hangs in Discovery," *Brooklyn Daily Eagle*, 17 July 1924, 6.

142. Steinmeyer, *Charles Fort*, 227.

143. *Fortean*, 1 (September 1937): 2.

144. Fort to DeCasseres, 26 August 1930, DeCasseres papers.

145. Bennett Cerf, "Try & Stop Me," *Morning Call* (Allentown, PA), 27 September 1965, 6.

146. *Philadelphia Record*, 19 November 1930; *New York World*, 12 January 1931—both in Charles Fort papers, accession 7657-a, Special Collections De-

partment, University of Virginia Library, Charlottesville; "In the Bookmarket," *Publisher's Weekly*, 17 January 1931, 302.

147. Thayer to Mr. Cope, 22 January 1931, Fort papers.

148. Alexander Woollcott, "Fair, Fat and Fortean," *McCall's*, June 1931, 8, 59, 8.

149. Dreiser to John Cowper Powys, 21 November 1930, Fort papers.

150. Harry Elmer Barnes, *Letters* (*Time* magazine supplement), 18 October 1937.

151. Edgar Lee Masters to ALR [Arthur Leonard Ross], 7 July 1937, Fortean Society papers, 1927–1952, accession 7657, Special Collections Department, University of Virginia Library, Charlottesville.

152. Fort to DeCasseres, 29 November 1930, DeCasseres papers.

153. Steinmeyer, *Charles Fort*, 240.

154. Robert H. Elias, ed., *Letters of Theodore Dreiser*, vol. 2 (Philadelphia: University of Pennsylvania, 1959), 507–8.

155. Burton Rascoe, "The Skepticism of Charles Fort," *New York Herald Tribune*, 15 February 1931, 5.

156. "What Mr. Powys Wrote When He Read Charles Fort's *Book of the Damned* the First Time," *Fortean* 6 (January 1942): 7; John Cowper Powys, *The Diary of John Cowper Powys, 1930* (London: Greymitre Books, 1987), 207.

157. Donald Pizer, ed., *Theodore Dreiser: A Picture and a Criticism of Life; New Letters*, vol. 1 (Champaign: University of Illinois Press, 2008), 148.

158. Powys, *Diary, 1930*, 196.

159. "What Mr. Powys Wrote," 7.

160. Powys, *Diary, 1930*, 196.

161. H. Allen Smith, *Low Man on a Totem Pole* (New York: Doubleday, Doran, 1945), 30–37.

162. Louis Sherwin, "The Roving Reporter: Charles Fort Becomes a Society for the Frustration of Science," *New York Evening Post*, 27 January 1931, 3.

163. "First Meeting of the Fortean Society, January 26, 1931," Damon Knight papers, Syracuse University.

164. "First Meeting of the Fortean Society," Knight papers.

165. H. Allen Smith, *To Hell in a Handbasket: The Education of a Humorist* (Garden City, NY: Doubleday, 1962), 65.

166. H. Allen Smith, "Supercheckers Latest Game for New Yorkers," *Courier-Express* (Dubois, PA), 21 February 1931, 3.

167. Smith, *To Hell in a Handbasket*, 243.

168. Jesse Walker, *The United States of Paranoia: A Conspiracy Theory* (New York: Harper Perennial, 2014).

169. Edwin Morgan, *From Glasgow to Saturn* (Manchester, UK: Carcanet Press, 1973), 35.

170. Fortean Society membership card, 1954 file, box 1, MS Morgan T, Edwin Morgan papers, University of Glasgow Special Collections.

171. Brian Taves, "'Verne's Best Friend and His Worst Enemy: I. O. Evans and the Fitzroy Edition of Jules Verne," *Verniana–Jules Verne Studies* 4 (2011–2012): 25–54.

172. Ron Westrum, "Knowledge about Sea-Serpents," *Sociological Review* 27 (1979): 293–314.

173. "No Such Sanderson," *Doubt* 18 (July 1947): 274.

174. Jonathan Betts, *Time Restored: The Harrison Timekeepers and R. T. Gould, the Man Who Knew (Almost) Everything* (New York: Oxford University Press, 2011), 249–61.

175. Steve Lohr, "Myth or Fact, Nessie Is Still Lure to Many," *New York Times*, 11 October 1987, 17.

176. H. Allen Smith, "Existence of Sea Serpent in Scotland Speculated," *Sun* (San Bernardino, CA), 11 January 1934, 4.

177. "MFS Evans Writes," *Doubt* 27 (October 1949): 419–20.

178. I. O. Evans to Russell, 22 Sept 1957, Eric Frank Russell papers, University of Liverpool.

179. "Fortean Losses," *Doubt* 23 (December 1948): 351.

180. "Parade of Pallid Data from the Day's News," *Fortean* 1 (September 1937): 4–5; "No Such Animal," *Doubt* 30 (October 1950): 36–37.

181. "No Such Sanderson," 274; Earle Ennis, "In the Tribune 25 Years Ago," *Oakland Tribune*, 14 February 1947, 19.

182. Ivan T. Sanderson, "Don't Scoff at Sea Monsters," *Saturday Evening Post*, 8 March 1947, 22–23, 84–87.

183. "No Such Sanderson," 274.

184. George M. Eberhart, *Mysterious Creatures: A Guide to Cryptozoology* (Santa Barbara, CA: ABC-CLIO, 2002), 380.

185. John Graham to Russell, 1 May 1953, Russell papers.

186. John J. Graham to British Museum of Natural History, 20 December 1954, Loch Ness Monster/John J Graham file, DF239/1/92, Natural History Museum, London.

187. Evans to Russell, 22 Sept 1957, Russell papers.

188. Katharine A. Rodger, ed., *Renaissance Man of Cannery Row: The Life and Letters of Edward F. Ricketts* (Tuscaloosa: University of Alabama Press, 2002), 183.

189. Partridge, *Re-enchantment*, 68.

190. Kripal, *Authors of the Impossible*, 125.

191. John Steinbeck, *The Log from the Sea of Cortez* (New York: Penguin, 1989), 28.

192. Samuel Hopkins, *Alexander Woollcott, His Life and His World* (Freeport, NY: Books for Libraries Press, 1946).

193. Woollcott, "Fair, Fat and Fortean," 8, 59.

194. Fort, *Lo!* 12, 90.

195. Alexander Woollcott, *While Rome Burns* (New York: Viking Press, 1935), 93.

196. Alexander Woollcott, "Shouts & Murmurs: The Triple Warning," *New Yorker*, 19 September 1931, 36.

197. Gillian Bennett and Paul Smith, eds., *Contemporary Legend: A Reader* (London: Routledge, 1996), xxviii.

198. Beatrice Kaufman and Joseph Hennessey, eds., *The Letters of Alexander Woollcott* (New York: Viking, 1944), 96.

199. "In the Bookmarket," *Publishers' Weekly*, 14 February 1931, 832.

200. Mary Dyer Lemon, "Tarkington Commenting on Stage 'Shocks,' Observes There is Not Art Unless It Is Clean Art," *Indianapolis Star*, 29 March 1931, 11.

201. Victor Gollancz, *My Dear Timothy* (New York: Simon & Schuster, 1953), 142; Waveney Girvan to Russell, 9 January 1951, Russell papers.

202. H. Allen Smith, "Book Tastes in Symposium," *Record* (Terril, IA), 28 July 1932, 3.

203. Jack Alexander, "70,000 Things That 'Science Can't Explain,'" *St. Louis Post-Dispatch*, 31 May 1931, 53, 59 (Van Doren); William Rose Benét, "The Phoenix Nest," *Saturday Review of Literature*, 1 July 1942, 17 (Darrow); Tiffany Thayer to Gene Fowler, 5 August 1946, Gene Fowler papers, University of California, Riverside (Fowler).

204. Alexander Woollcott to ALR [Arthur Leonard Ross], 4 July 1937, Fortean Society papers.

205. Martin Kamin to Sussman, 12 March 1931; "Scheduled Fortean Gathering at George Washington Hotel," Fortean Society papers.

206. Steinmeyer, *Charles Fort*, 267–68.

207. Charles Fort, *The Complete Books of Charles Fort* (New York: Dover, 1975), 1042.

208. Mary Dyer Lemon, "Wild Talents by Charles Fort Declared Something Very New," *Indianapolis Star*, 19 June 1932, 25; "Dips into Realm of the Unknown," *Detroit Free Press*, 12 June 1932, 33.

209. Laurence Stallings, "The Book of the Day," *Oakland Tribune*, 17 June 1932, 25.

210. H.B., "Strange Signs and Fiery Portents," *Louisville Courier-Journal*, 19 June 1932, 21.

211. Joseph Jastrow, "Lo! Charles Fort!" *Saturday Review of Literature*, 2 July 1932, 819.

212. Mencken, "Nonsense as Science."

213. Smith, *Low Man*, 37.

214. Smith, *Low Man*, 37.

215. Steinmeyer, *Charles Fort*, 260.

216. Fort, *Complete Books*, 877.

217. Helen Dreiser, *My Life with Dreiser* (New York: World Publishing, 1951), 223–24.

218. Steinmeyer, *Charles Fort*, 272.

219. Steinmeyer, *Charles Fort*, 273.

220. "Dreiser Speaks at Fort's Rites," *Pittsburgh Press*, 7 May 1932, 11.

221. Donald R. Murphy, "About Books: Poltergeist Girls Built the Pyramids," *Des Moines Register*, 31 July 1932, 34.

222. Swanberg, *Dreiser*, 397.

223. Steinmeyer, *Charles Fort*, 279.

224. Zanine, *Mechanism and Mysticism*, 157.

225. Linus Pauling to Ava Helen Pauling, 3 March 1932, filed under LP Safe: box #1.011, folder #11.6, Linus Pauling papers, Oregon State University; Edward Stone, "Whodunit? Moby Dick!" *Journal of Popular Culture* 8 (1974): 280–85.

226. Thayer to Richard B. Glaenzer, 9 August 1933, Richard B. Glaenzer papers, University of Virginia.

227. Howard V. Sutherland, "Knot's from a Rambler's Log," *Honolulu Advertiser*, 12 March 1933, 22.

228. "Writers Settle in Film Colony by the Hundred," *Washington Post*, 6 October 1935, A1.

229. Edwin Schallert, "Dr. Rockwell, Noted Comedian, Ensnared by Films," *Los Angeles Times*, 15 August 1933, A7; "New Tiffany Thayer Story to Be Filmed," *Hartford Courant*, 3 September 1933, A8; Hope Ridings Miller, "Daughter of Mrs. Borden Harriman Is Going Places with Her Pen," *Washington Post*, 18 September 1935, 8.

230. George Shaffer, "Hollywood Happenings," *Chicago Daily Tribune*, 20 November 1932, F8; Mollie Merrick, "Habits of Scenarists Described," *Hartford Courant*, 4 December 1932, A6; Tip Poff, "That Certain Party," *Los Angeles Times*, 28 October 1934, A1; Tip Poff, "That Certain Party," *Los Angeles Times*, 19 May 1935, A1; "Russell's Best," *Doubt* 44 (April 1954): 270–71.

231. "Writer Free Pending Drunk-Driving Quiz," *Los Angeles Times*, 28 February 1933, 16.

232. Thayer to Annie Fort, 12 March 1935 and 13 June 1935, Theodore Dreiser papers, University of Pennsylvania.

233. Steinmeyer, *Charles Fort*, 277.

234. Steinmeyer, *Charles Fort*, 279.

235. "Dreiser Speaks at Fort's Rites," 11.

236. Swanberg, *Dreiser*, 248.

237. De Ford, *Up-Hill All the Way*, 177.

238. Zanine, *Mechanism and Mysticism*, 117–18.

239. Swanberg, *Dreiser*, 245.

240. Henry Miller, *The Books in My Life* (New York: New Directions, 1952), 136.

241. Morine Krissdóttir, *Descents of Memory: A Life of John Cowper Powys* (New York: Overlook Press, 2007), 165, 187, 242–50.

242. Krissdóttir, *Descents of Memory*, 266.

243. Powys, *Diary, 1930*, 207.

244. Morine Krissdóttir and Roger Peers, eds., *The Dorset Year: The Diary of John Cowper Powys, June 1934–July 1935* (UK: Powys Press, 1998), 154; John Cowper Powys to ALR [Arthur Leonard Ross], 12 July 1937, Fortean Society papers; Thayer to Russell, 28 January 1956, FS, Russell papers; John Cowper Powys, *Autobiography* (Hamilton, NY: Colgate University Press, 1967), 523, 551–52. "FS" (Fortean Style) dates, as explained in chapter 2, use a thirteen-month calendar that designates 1931, when the Fortean Society was founded, as Year One. Dates for Thayer's correspondence in notes are a combination of the FS days and months but use conventional years for ease of chronology; note that he stopped using the dating system for a time in the mid-1950s, and dates for this correspondence are not marked FS.

245. John Cowper Powys, *The Inmates* (London: MacDonald, 1952).

246. Powys, *Diary, 1930*, 207.

Chapter 2

1. "Who Killed Earhart and Noonan?" *Fortean* 1 (September 1937): 1–2, 1.

2. "The Wilson's [*sic*] Father and Son," *Fortean* 8 (December 1943): 11–12; Tiffany Thayer, *Little Dog Lost* (New York: Julian Messner, 1938), 89, 381.

3. Thayer, *Little Dog Lost*, 381.

4. Thayer to Harry L. Wilson, 10 September 1937, box 2, E–F Miscellany file; Booth Tarkington to Harry L. Wilson, 24 October 1937, box 4, Tarkington file (5 of 5), Harry Leon Wilson papers, University of California, Berkeley.

5. Alexander Woollcott to Booth Tarkington, 7 February 1942, Booth Tarkington papers, Princeton University.

6. Robert H. Elias, ed., *Letters of Theodore Dreiser*, vol. 2 (Philadelphia: University of Pennsylvania Press, 1959), 733–34.

7. Thayer to Don Bloch, 14 March 1942, Donald Beaty Bloch papers, New York Public Library.

8. E. C. K., "The Book Nook," *Palm Beach Post-Times*, 3 October 1937, 2.

9. "How Well Do You Know Your JWT'ers?" box MN 26, J. Walter Thompson Company, Biographical Information papers, Duke University.

10. "A Midnight Marriage," *News-Leader* (Springfield, MO), 4 January 1898, 3; "Chancery Cases," *Journal-Standard* (Freeport, IL), 29 August 1908, 5; "Elmer Thayer Wants Divorce," *Journal-Standard* (Freeport, IL), 12 August 1908, 1; "Divorces Granted," *Journal-Standard* (Freeport, IL), 15 September 1908, 1.

11. Thayer to Bloch, 7 November 1952, FS, Bloch papers.

12. "'Up the Ladder,' with Freeporter in Cast, at Germania Tonight," *Journal Standard* (Freeport, IL), 27 October 1923, 6.

13. "Night in Honolulu Best Play in Years," *Greenville* (NC) *News*, 21 September 1921, 2.

14. Thayer to Ezra Pound, 14 June 1948, FS, Ezra Pound papers, Indiana University.

15. Thayer to Bloch, 21 September 1946, Bloch papers.

16. "Talking It Over at 805," *Brief Stories*, November 1922, 2.

17. Fort to Thayer, 8 June 1925, Charles Fort papers, accession 7657-a, Special Collections Department, University of Virginia Library, Charlottesville.

18. Some insight into Thayer's early life can be gleaned from a later novel: Tiffany Thayer, *Three-Sheet* (New York: Liveright, 1932).

19. "How Well Do You Know Your JWT'ers?"

20. Thayer to Russell, 28 December 1950, Eric Frank Russell papers, University of Liverpool.

21. "Warren Lee and Tanagra to Give Dance Concert," *Los Angeles Times*, 23 December 1934, A6.

22. Katherine T. Von Blon, "Studio and Theater Comings and Goings," *Los Angeles Times*, 1 July 1934, A2; Katherine T. Von Blon, "Studio and Theater Comings and Goings," *Los Angeles Times*, 8 July 1934, A2; Katherine T. Von Blon, "Young Texas Millionaire, Turned Film Producer, to Substitute Soul for Sex," *Los Angeles Times*, 22 July 1934, A1; Katherine T. Von Blon, "Tiny Theater Arts Playbox Presents Impressive Work," *Los Angeles Times*, 29 July 1934, A2.

23. Thayer to Bloch, 7 November 1952, FS, Bloch papers; Piglet and Diana to Damon Knight, n.d., Damon Knight papers, Syracuse University.

24. Hope Ridings Miller, "Daughter of Mrs. Borden Harriman Is Going Places with Her Pen," *Washington Post*, 18 September 1935, 8.

25. "Books: Jesus in California," *Time*, 25 September 1933.

26. Thayer, *Little Dog Lost*, 359.

27. "How Well Do You Know Your JWT'ers?"; J. Walter Thompson Co., *The 1940 Radio Annual* (New York: Radio Daily, 1940), 228.

28. Harry Leon Wilson, *The Wrong Twin* (New York: Doubleday, 1921), 95; Thayer, *Little Dog Lost*, 116; "Wilson's Father and Son," 11–12.

29. "Plan Revolt against All Dogmatism: Fortean Society Will Spread Ideas

and Philosophy of Charles Fort in Magazine," *Hartford Courant*, 27 March 1935, 2; "Lament for Fort," *Letters* (*Time* magazine supplement), 5 August 1935, 17–18.

30. Elias, *Letters of Theodore Dreiser*, 2:733.

31. Theodore Dreiser to ALR [Arthur Leonard Ross], 26 June 1937, 27 July 1937, 4 August 1937, Fortean Society papers, 1927–1952, accession 7657, Special Collections Department, University of Virginia Library, Charlottesville; Elias, *Letters of Theodore Dreiser*, 2:733–34, 740–41.

32. Deborah Benson Covington, *The Argus Book Shop, a Memoir* (West Cornwall, CT: Tarrydiddle Press, 1977); Thayer to Ben Abramson, 29 March 1941, 26 April 1941, 5 May 1941, 8 October 1941, folder 410, Ben Abramson papers, Beinecke Library, Yale University.

33. H. Allen Smith, *Low Man on a Totem Pole* (New York: Doubleday, Doran, 1945), 37.

34. Tiffany Thayer, "Alfred Henry Barley," *Fortean* 6 (January 1942): 3–4; James H. Holden, *A History of Horoscopic Astrology* (Tempe, AZ: American Federation of Astrology, 2006), 207; Patrick Curry, *A Confusion of Prophets: Victorian and Edwardian Astrology* (London: Collins & Brown, 1992), 149; James A. Santucci, "The Aquarian Foundation," *Communal Societies* 9 (1989): 39–61; Thayer to Russell, 14 April 1948, FS, Russell papers; Albert E. Page, *The Chief Aspects of Western Civilization's Decline*, Little Blue Book 1800 (Girard, KS: Haldeman-Julius Publications, n.d.); Helge Kragh, "The Vortex Atom: A Victorian Theory of Everything," *Centaurus* 44 (July 2002): 32–114.

35. *Fortean* 1 (October 1937): 13; Stephen Bowers, *The Vailan or Annular Theory: A Synopsis of Prof. I. N. Vail's Argument* (Ventura, CA: Observer Press, 1892); "New Life Member," *Fortean* 9 (Spring 1944): 2.

36. "Growing Pains," *Fortean* 3 (January 1940): 6–7; "Follow-up," *Doubt* 11 (Winter 1944–1945): 162; "Atlantis Letter," *Doubt* 39 (January 1953): 187–88.

37. "Earhart Echo," *Fortean* 6 (January 1942): 12–13.

38. Thayer to T. Swann Harding, 26 August 1937, T. Swann Harding papers, Library of Congress; T. Swann Harding, *The Degradation of Science* (New York: Farrar & Rinehart, 1931); T. Swann Harding, *The Joy of Ignorance* (New York: William Godwin, 1932).

39. T. Swann Harding, *Why I Am a Skeptic*, Little Blue Book 1334 (Girard, KS: Haldeman-Julius Publications, 1929); T. Swann Harding, "Limitations of Science as a System of Belief," *Open Court*, October 1930, 577–85. On Harding, see Joshua Blu Buhs, "T. Swann Harding as a Fortean," 27 January 2016, http://www.joshuablubuhs.com/blog/t-swann-harding-as-a-fortean.

40. T. Swann Harding, *The Popular Practice of Fraud* (London: Longmans, Green and Company, 1935).

41. Thayer to Harding, December 1937, Harding papers. Thayer considers the matter in an untitled discussion of R. Buckminster Fuller's *Nine Chains to the Moon*; see *Fortean* 3 (January 1940): 4.

42. T. Swann Harding, "Fictions Men Live By," *American Spectator*, August 1933, 3; Thayer to Harding, [no day] December 1937, 6 March 1938, Harding papers.

43. Thayer to Harding, 20 September 1937, Harding papers.

44. Thayer to Harding, October 1937, Harding papers.

45. Kathleen Thayer to Bloch, 26 December 1959, Bloch papers.

46. Thayer to Bloch, 7 November 1952, FS, Bloch papers; Kathleen Thayer to E Russes, 1 August 1953, Russell papers.

47. Tiffany Thayer, "A Nation of Babes in Arms," *Ken*, May 1938, 90–93; Thayer to Harding, 6 March 1938, [no day] January 1940, Harding papers.

48. "N.Y. Publisher Is Found Dead, Face Bruised: Body of Claude H. Kendall Discovered in Hotel after Party," *Washington Post*, 26 November 1937, 3; "The Publishers Plan Promotion for High Spots of Spring Lists: Henry Holt & Co.," *Publishers Weekly*, 25 January 1941, 403–4.

49. William Sloane to Thayer, 10 July 1940, 28 October 1940; Thayer to Sloane, 25 October 1940, Henry Holt and Company papers, Princeton University.

50. Sloane to Sussman, 19 November 1940, Henry Holt papers; Miriam Allen De Ford, "Charles Fort: Enfant Terrible of Science," *Magazine of Fantasy and Science Fiction*, January 1954, 105–16, 116.

51. "They Teach Minds to Kneel," *Fortean* 2 (October 1937): 4.

52. "We Have the Freedom, but Where's the Press?" *Fortean* 3 (January 1940): 1–3, 2.

53. "We Have the Freedom," 3.

54. Thayer to Pound, 14 January 1948, FS, Pound papers.

55. Thayer to Pound, 14 January 1948, FS, Pound papers.

56. "Where Credit Is Due," *Doubt* 43 (February 1954): 256–58, 258.

57. Benjamin DeCasseres, "An American Wrestles with God," *American Mercury*, April 1929, 467–73, 472.

58. "Circus Day Is Over," *Fortean* 6 (January 1942): 1–2.

59. Woollcott to Tarkington, 7 February 1942, Tarkington papers; Woollcott to Hecht, 23 February 1942, Ben Hecht papers, Newberry Library, Chicago.

60. Stern to Tarkington, 27 April 1942, Tarkington papers.

61. Thayer to Bloch, 24 July 1943, Bloch papers.

62. Thayer to Tarkington, 24 July 1943, Tarkington papers.

63. Sussman to Thayer, 23 July 1943, Xeroxes, articles, etc. file, box 3, Knight papers.

64. Thayer to Sussman, 31 July 1943, Knight papers.

65. J. Edgar Hoover to Assistant Attorney General Tom Clar, 9 December 1943, Tiffany Thayer file, Federal Bureau of Investigation (FBI) archives, Washington, DC.

66. "Astrology and the Common Cold," *Fortean* 7 (June 1943): 10.

67. "Wild Plum Speaks," *Fortean* 10 (Summer 1944): 137–38; Alva Johnston, "The Magic Lie Detector," *Saturday Evening Post*, 15 April 1944, 9, 29; Ken Alder, *The Lie Detectors: The History of an American Obsession* (New York: Free Press, 2007).

68. "What Made the Coal Lumps Jitterbug in the Scuttle at the Wild Plum Schoolhouse?" *American* (Odessa, TX), 14 April 1944, 1.

69. US Census, 1940: Township 137, Range 92, Stark, North Dakota; roll m-t0627-03014, page 2A, enumeration district 45-17.

70. "Wild Plum School Mystery Deepens," *Nonpareil* (Council Bluffs, IA), 15 April 1944, 2.

71. Ancestry.com, "U.S., Find a Grave Index, 1600s–Current" (online database).

72. "Science: Witchery in North Dakota," *Time*, 24 April 1944.

73. "Wild Plum Speaks," 137–38.

74. "What Made the Coal Lumps Jitterbug?" 1.

75. "Wild Plum Speaks," 137–38.

76. "Solved: The Leaping Coal Mystery of Rural Schoolhouse," *Mercury* (Manhattan, KS), 19 April 1944, 6.

77. "Teacher Near Sighted, Pupils Playful—Hence Great Mystery of Wild Plum's 'Dancing Coal,'" *News-Record* (Neenah, WI), 18 April 1944, 1.

78. "Nelson Bond Writes," *Doubt* 11 (Winter 1944): 164.

79. "Wild Plum Speaks," 137–38.

80. John L. Ingham, *Into Your Tent: The Life, Work and Family Background of Eric Frank Russell* (UK: Plantech, 2010).

81. Susan Jacoby, *Freethinkers: A History of American Secularism* (New York: Metropolitan Books, 2004); Catherine L. Albanese, *A Republic of Mind and Spirit: A Cultural History of American Metaphysical Religion* (New Haven, CT: Yale University Press, 2006); Charles Taylor, *A Secular Age* (Cambridge, MA: Harvard University Press, 2007).

82. Willard Gatewood, *Controversy in the Twenties: Fundamentalism, Modernism, and Evolution* (Nashville, TN: Vanderbilt University Press, 1969); David J. Hoeveler, *The New Humanism: A Critique of Modern America, 1900–1940* (Charlottesville: University Press of Virginia, 1977); Donald H. Meyer, "Secular Transcendence: The American Religious Humanists," *American Quarterly* 34 (1982): 524–42.

83. Richard van der Riet Woolley, "Interplanetary Travel," *Nature* (sup-

plement), 14 March 1936, 442; Oliver Dunnett, "The British Interplanetary Society and Cultures of Outer Space, 1930–1970," PhD diss., University of Nottingham, 2011.

84. P. E. Cleator, ed., *Letters from Baltimore: The Mencken-Cleator Correspondence* (Madison, NJ: Fairleigh Dickinson University Press, 1982), 201, 204, 210.

85. Arthur C. Clarke, *Greetings, Carbon-Based Bipeds! Collected Essays, 1934–1998* (New York: Macmillan, 2001), 114–19.

86. Neil McAleer, *Arthur C. Clarke: The Authorized Biography* (Chicago: Contemporary Books, 1992), 28–38.

87. Mike Ashley, *The Time Machines: The Story of the Science-Fiction Pulp Magazines from the Beginning to 1950* (Liverpool: Liverpool University Press, 2001), 125.

88. C. A. Brandt, "In the Realm of Books," *Amazing Stories*, July 1931, 379.

89. Eric Frank Russell, "Eric Frank Russell on Charles Fort—The Bronx Jeer," *Fantasy Review* 2, no.10 (August–September 1948): 2–4, 3.

90. Sam Moskowitz, *Strange Horizons: The Spectrum of Science Fiction* (New York: Scribner, 1976), 244.

91. Donald Snively to Howard Snively, 28 September 1936, 8 November 1937, box 3, folder 8, Howard Snively papers, Schenectady (NY) Museum of Innovation and Science.

92. Ingham, *Into Your Tent*, 137.

93. Eric Frank Russell, "Fort—The Colossus," *Spaceways* 8 (October 1939), 4–6, 4; box 244, folder 3, Bruce Pelz papers, University of California, Riverside.

94. C. S. Youd, "Broadside!" *The Fantast*, November 1941, 14–17, 15, https://fanac.org/fanzines/Fantast/.

95. Campbell to Russell, 17 May 1950, Russell papers.

96. Eric Frank Russell, letter, *Unknown*, September 1940, 162.

97. "Chronological Notes—1941–1942," "Harry Turner's Footnotes to Fandom," 2014, http://www.htspweb.co.uk/fandf/romart/het/footnotes/fanac2.htm.

98. Eric Frank Russell, letter, *Unknown*, January 1940, 129; "Russell Writes," *Doubt* 25 (Summer 1949): 383.

99. Youd, "Broadside!" 15.

100. Moskowitz, *Strange Horizons*, 244; Ingham, *Into Your Tent*, 92, 275.

101. Ingham, *Into Your Tent*, 130.

102. William Rose Benét, "The Phoenix Nest," *Saturday Review of Literature*, 12 September 1942, 18.

103. Ingham, *Into Your Tent*, 139.

104. Ingham, *Into Your Tent*, 130.

105. Moskowitz, *Strange Horizons*, 244.

106. Ingham, *Into Your Tent*, 137, 140.

107. Russell, "Bronx Jeer," 4.

108. Joe Patrizio, "Extracts from Bill Temple's Diaries, 1935–42," *Relapse* 16, 12–17, 17, https://efanzines.com/Prolapse/Relapse16.pdf.

109. Eric Frank Russell, "Over the Border," *Unknown*, September 1939, 129–40.

110. Ron Holmes, "Gleanings," *Futurian War Digest* 8 (May 1941): 4.

111. Eric Frank Russell, "Spontaneous Frogation," *Unknown*, July 1940, 83.

112. Youd to Russell, 26 February 1947, Russell papers; Harry Turner, "Remembering Eric Frank Russell 1905–1978," "Harry Turner's Footnotes to Fandom," 2011, http://www.htspweb.co.uk/fandf/romart/het/footnotes/efr.htm; Ingham, *Into Your Tent*, 163.

113. "The Reaper Reaps," *Doubt* 53 (February 1957): 414–15.

114. "New Motto," *Doubt* 17 (Winter 1946): 251; "Our Manneken," *Doubt* 18 (July 1947): 269.

115. Kathleen Thayer to Bloch, 26 December 1959, Bloch papers; Thayer to Pound, 18 May 1946, Pound papers.

116. *Doubt* 14 (Spring 1946): 202.

117. "Where Honor is Due," *Doubt* 49 (August 1955): 349–51; Thayer to Helen Dreiser, 18 May 1946, Theodore Dreiser papers, University of Pennsylvania.

118. "Our Cover," *Doubt* 20 (March 1948): 298.

119. Julian Parr to Russell, 17 April 1947, 19 December 1948, Russell papers; "Saucerzines," *Doubt* 47 (January 1955): 324; "Interplanetaries Convene," *Doubt* 53 (February 1957): 421; Guy Powell to Russell, 1 April 1958, Russell papers.

120. William Graebner, *The Age of Doubt: American Thought and Culture in the 1940s* (Long Grove, IL: Waveland Press, 1991), 17–18.

121. Joshua Blu Buhs, "Alfred H. Barley as a Fortean," 3 February 2014, http://www.joshublubuhs.com/blog/alfred-h-barley-as-a-fortean.

122. For a consideration of alternative physics, see Margaret Wertheim, *Physics on the Fringe: Smoke Rings, Circlons, and Alternative Theories of Everything* (New York: Bloomsbury, 2011).

123. Helge Kragh, "Resisting the Bohr Atom: The Early British Opposition," *Physics in Perspective* 13 (2011): 4–35, 15–17.

124. Thomas H. Graydon, *New Laws for Natural Phenomena* (Boston: Christopher Publishing House, 1938); "Graydon's Table," *Doubt* 12 (Summer 1945): 177; "FU," *Doubt* 21 (June 1948): 319–20; Martin H. Gardner, *Fads and Fallacies in the Name of Science* (New York: Dover, 1957), 85.

125. Frederick Hammett to Harding, 26 December 1937, Harding papers; "Don't Laugh," *Fortean* 6 (January 1942): 12; Frederick S. Hammett, "Integra-

tion in Science Teaching," *Scientific Monthly*, May 1946, 429–32; T. Swann Harding, "It Rains Fish," *USDA*, 8 July 1946, 3–4, 4; "That's Our Boy," *Doubt* 33 (October 1951): 88; "We Lose Fred" *Doubt* (July 1953): 213–14.; "New Door Opened in Cancer Studies; Hope for Ultimate Control of Malignant Growths Seen in Hammett's Findings," *New York Times*, 5 December 1937, 7–8.

126. "Fortean Prescribed 'Penicillin' 60 Years Ago," *Fortean* 9 (Spring 1944): 5.

127. "No Such Orgone," *Doubt* 53 (February 1957): 415.

128. "Polio and Tonsils," *Doubt* 30 (October 1950): 38–40.

129. "Germ Theory Deloused," *Doubt* 36 (April 1952): 135; Nell Rogers and Guy Rogers, *The Medical Mischief, You Say! Degerminating the Germ Theory* (Pomeroy, WA: Health Research Books, 1996), 2.

130. "Office Notes," *Canadian Theosophist*, 15 February 1945, 370.

131. Charmaine Ortega Getz, "Maurice Doreal and His Brotherhood of the White Temple Awaited the Apocalypse in Colorado," *Westword*, 28 October 2015, https://www.westword.com/news/maurice-doreal-and-his-brotherhood -of-the-white-temple-awaited-the-apocalypse-in-colorado-7285649.

132. Jim Dennon, *The Oahspe Story* (Seaside, OR: n.p., 1965); Julia Keleher, "The Land of Shalam: Utopia in New Mexico," *New Mexico Historical Review* 19 (1944): 123–34.

133. Joscelyn Godwin, *The Theosophical Enlightenment* (Albany: State University of New York Press, 1994); Wouter J. Hanegraaff, "The Theosophical Imagination," *Correspondences* 6 (2017): 1–37; Mary Lutyens, *Krishnamurti: The Years of Fulfillment* (New York: Farrar, Straus, Giroux, 1983).

134. Robert V. Hine, *California's Utopian Colonies* (San Marino, CA: Huntington Library, 1953); Rosabeth Moss Kanter, *Commitment and Community: Communes and Utopias in Sociological Perspective* (Cambridge, MA: Harvard University Press, 1972).

135. Marie Halun Bloch, *Journey to Parnassus: A Memoir of Donald Beaty Bloch* (Northfield, MN: Carleton College, 1986).

136. W. L. McAtee to Bloch, 24 October 1951, Waldo Lee McAtee papers, Library of Congress.

137. Thayer to Bloch, 2 August 1946, FS; 14 January 1948 [1949?], FS, Bloch papers.

138. "HFFS Bloch," *Doubt* 18 (July 1947): 273.

139. Thayer to Bloch, 28 November 1943, 21 May 1947, FS, 21 June 1947, FS, Bloch papers.

140. Iktomi, *America Needs Indians* (Denver: Bradford-Robinson, Printers, 1937).

141. Thayer to Bloch, 2 August 1946, FS, 28 Fort 1947, FS, 28 February 1948, FS, Bloch papers.

142. Thayer to Bloch, 14 Fort 1946, FS, Bloch papers.

143. *Doubt* 15 (Summer 1946): 221.

144. Thayer to Bloch, 28 August 1946, FS, Bloch papers.

145. Thayer to Bloch, 21 January 1947, FS, 21 March 1947, FS, 14 June 1952, FS, Bloch papers.

146. Paula L. Wagoner, "The Search for an Honest Man: Iktomi Hcala as an Ethnographical and Humanistic Conundrum," in *Transforming Ethnohistories: Narrative, Meaning, and Community*, ed. Sebastian Felix Braun (Norman: University of Oklahoma Press, 2013), 170–80.

147. Thayer to Helen Dreiser, 1 June 1946, Dreiser papers; Thayer to Russell, 28 March 1949, FS, 28 Fort 1953, FS, Russell papers.

148. "More Monsters," *Doubt* 16 (Fall 1946): 236–37; "The Fortean University," *Doubt* 20 (March 1948): 299–301.

149. Steven J. Taylor, *Acts of Conscience: World War II, Mental Institutions, and Religious Objectors* (Syracuse, NY: Syracuse University Press, 2009).

150. "Teacher Arrested Folk School," *Leaf-Chronicle* (Clarksville, TN), 18 March 1943, 2; "Harry Leon Wilson Posts Bond on Draft Evasion," *Sentinel* (Santa Cruz, CA), 13 July 1943, 8; "Former Folk School Teacher Convicted," *Tennessean*, 22 October 1943, 20; John M. Glen, *Highlander: No Ordinary School 1932–1962* (Lexington, KY: University Press of Kentucky, 1988), 91; "Wilson's Father and Son," 1943, 11–12.

151. "Now Is the Time," *Doubt* 28 (April 1950): 9–10.

152. *Doubt* 54 (June 1957): 430.

153. "Man in Jail," *Doubt* 21 (June 1948): 319; "Now Is the Time," 9–10.

154. *Fortean* 9 (Spring 1944): 4.

155. Jesse Douglass to Thayer, n.d. (ca. December 1943), 1 April 1942, April 1942, in author's possession.

156. "Tumbril" 11, 1948, box 76, folder 7, Pelz papers.

157. "Poisoned at Its Well," *Fortean* 10 (Autumn 1944): 148; "Help Patchen Please," *Doubt* 3 (October 1951): 90–91.

158. Henry Geiger, "Toward a Golden Age," *Manas*, 16 July 1952, 1–4; Joshua Blu Buhs, "Henry Geiger as a Fortean," 10 April 2017, http://www.joshuablubuhs.com/blog/henry-geiger-as-a-fortean.

159. "The New Minority," *Pacifica Views*, 30 March 1945, 1–2, Peace Collection, Swarthmore University; Andrew Cornell, *Unruly Equality: U.S. Anarchism in the Twentieth Century* (Berkeley: University of California Press, 2016).

160. "Falls," *Doubt* 41 (July 1953): 221–22.

161. James Blish, "Scientific Method and Political Action," *Politics*, November 1946, 358–59.

162. Thayer to Bloch, 28 January 1945, FS, Bloch papers; "Mencken's Gods," *Doubt* 17 (1946): 257.

163. "Fortean University," 299–301.

164. "Replies," *Doubt* 28 (April 1950): 11.

165. "Chapter Two," *Doubt* 21 (June 1948): 320; "Chapter Three, Chicago," *Doubt* 24 (April 1949): 363; "Chapter Four, Dallas," *Doubt* 24 (April 1949): 364; "Chapter Six," *Doubt* 26 (October 1949): 394.

166. Thayer to Russell, 14 Fort 1948, FS, Russell papers.

167. Thayer to Bloch, 2 January 1945, FS; "The Fortean Society Is the Red Cross of the Human Mind," n.d., Bloch papers.

168. Thayer to Bloch, 14 Fort 1946 FS, Bloch papers.

169. Thayer to Russell, 14 June 1948, FS, Russell papers.

170. Thayer to Russell, 28 February 1949, FS, Russell papers.

171. Thayer, *Little Dog Lost*, 380; "Russell's Best," *Doubt* 42 (October 1953): 241–42; Thayer to Russell, 24 May 1958, Russell papers.

172. "Lieutenant Benjamin Franklin Pinkerton Returns (A Fortean View of the Semantic Bomb)," *Doubt* 13 (Winter 1945): 186–87; "Bring Out Your Dead!" *Doubt* 43 (February 1954): 259–60.

173. "Those Questions," *Doubt* 34 (October 1951): 99–103, 99.

174. J. Edgar Hoover to Tom Clark, 12 January 1945; [redacted] to J. Edgar Hoover, 12 June 1947; Special Agent in Charge to J. Edgar Hoover, 1 October 1951, Thayer file, FBI archives.

175. Thayer to Russell, 28 April 1955, FS, 15 September 1958, Russell papers.

176. Thayer to Caresse Crosby, 23 January 1954, FS, Caresse Crosby papers, Southern Illinois University; Thayer to James Blish, 21 July 1956, James Blish papers, Oxford University.

177. Thayer to Pound, 21 April 1947, FS, Pound papers.

178. "Fortean Society Perpetual Peace Program," *Doubt* 13 (Winter 1945): 187.

179. "Add PPP" *Doubt* 17 (Winter 1946–1947): 256.

180. Marcia Winn, "Front Views & Profiles," *Chicago Daily Tribune*, 29 March 1947, 13; Michael Hogan, *A Cross of Iron: Harry S. Truman and the Origins of the National Security State, 1945–1954* (New York: Cambridge University Press, 1998).

181. Thayer to Bloch, 21 September 1946; Robert Spencer Carr to Bloch, 4 December 1947, Bloch papers; E. Hoffmann Price, *Book of the Dead: Friends of Yesteryear: Fictioneers and Others* (Sauk City, WI: Arkham House, 2001), 197.

182. "$50,000 Verdict 2D Big Libel Loss by Smear Press," *Chicago Tribune*, 1 February 1943, 4; Thayer to Pound, 21 April 1947, FS, Pound papers.

183. "Agencies," *Broadcasting-Telecasting: The Newsweekly of Radio and Television*, 29 November 1948, 18; Thayer to Crosby, 23 January 1954, FS, 22 May 1954, Crosby papers.

184. Thayer to Bloch, 25 July 1953, FS, Bloch papers.

185. "Hi-Spots in the Mail," *Doubt* 24 (April 1949): 370–72.

186. Art Castillo to Eric Frank Russell, 2 June 1951, FS, Russell papers.

187. "Our Cover," *Doubt* 29 (July 1950): 27.

188. Thayer to Russell, 28 February 1951, FS, Russell papers.

189. Thayer to Russell, 21 October 1950, FS, 28 February 1951, FS; Art Castillo to Russell, 3 April 1951, 2 June 1951, Russell papers.

190. "Castillo on Cerberus," *Doubt* 25 (May 1949): 384–85.

191. "Giles Grist," *Doubt* 21 (Spring–Summer 1945): 173–74; Thayer to Russell, 14 February 1948, Russell papers.

192. "Balanced Living and the Realist," *Doubt* 58 (October 1958): 21–22.

193. Wesley Mason Olds, "Wilson, Edwin Henry," in *Dictionary of Modern American Philosophers*, ed. John R. Shook (Bristol: Thoemmes Continuum, 2005), 2620–23.

194. Edwin H. Wilson to the Fortean Society, 10 August 1948, 14 Fort 1948, Edwin H. Wilson papers, Southern Illinois University.

195. Thayer to Wilson, 7 January 1950, FS, 7 February 1950, FS, E. H. Wilson papers; "$50.00 Prize," *Doubt* 24 (April 1949): 362; "Los Humanistas," *Doubt* 28 (April 1950): 4.

196. "The Fortean Society," E. H. Wilson papers; "MFS Grant Speaks," *Doubt* 34 (October 1951): 108; "St. Louis Seculars," *Doubt* 48 (April 1955): 337; "Balanced Living and the Realist," 21–22.

197. Ben Abramson to Thayer, 8 October 1941, folder 410, Abramson papers; Thayer to Pound, 3 July 1948, Pound papers.

198. Rochelle Girson, "Colossal Story of Mona Lisa Includes 3½ Million Words," *Daily Press* (Newport, RI), 24 June 1956, 48.

199. Thayer to Pound, 15 August 1952, Pound papers; Girson, "Colossal Story," 48.

200. "Neapolitan Peep Show," *Time*, 11 June 1956.

201. Thayer to Russell, 19 May 1956, Russell papers.

202. John Barkham, "Opus Giganticus," *St. Louis Post-Dispatch*, 28 June 1956, 23.

203. Robert R Kirsch, "The Book Report," *Los Angeles Times*, 8 June 1956, 41; Don Morrison, "Mona Lisa's Smile Provokes a Smirk," *Star Tribune* (Minneapolis, MN), 17 June 1956, 70.

204. "Neapolitan Peep Show."

205. "On the Social Scene," *Journal-Standard* (Freeport, IL), 15 June 1956, 6.

206. "Mr. Tiffany Thayer: There Have Been Some Changes on the Street Where You Lived," *Journal-Standard* (Freeport, IL), 2 August 1956, 4.

207. Girson, "Colossal Story," 48.

208. Thayer to Russell, 27 February 1954, FS, Russell papers.

209. Thayer to Russell, 28 December 1957, FS, Russell papers.

210. "Jesus Cards," *Doubt* 27 (Winter 1949): 419.

211. "No Jesus Cards," *Doubt* 28 (April 1950): 10–11.

212. Laura McEnany, *Civil Defense Begins at Home: Militarization Meets Everyday Life in the Fifties* (Princeton, NJ: Princeton University Press, 2020).

213. *Doubt* 33 (June 1951): 88–90.

214. "Those Questions," 99–103.

215. Thayer to Russell, 7 June 1951, FS, Russell papers; Thayer to Bloch, 15 December 1951, FS, Bloch papers.

216. Thayer to Russell, 9 September 1955, Russell papers.

217. "Castillo in Toils," *Doubt* 36 (April 1952): 134; Thayer to Russell, 5 April 1952, 5 September 1953, FS, Russell papers.

218. Thayer to W. E. Edwards, 28 March 1954, FS; W. E. Edwards to J. Edgar Hoover, 12 April 1956; Mr. Nichols to M. A. Jones, 17 April 1956; J. Edgar Hoover to Thayer, 18 April 1956; Thayer to J. Edgar Hoover, 21 April 1956; W. E. Edwards to J. Edgar Hoover, 21 April 1956, FBI archives; Thayer to Russell, 29 September 1956, Russell papers.

219. Mr. Nichols to M. A. Jones, 17 April 1956, FBI archives.

220. Mr. Nichols to M. A. Jones, 20 June 1956, FBI archives.

221. Thayer to Russell, 16 July 1955, Russell papers.

222. Volunteers Wanted," *Doubt* 56 (March 1958): 480.

223. "This Computer Age, or Leave It to Singer," *Doubt* 61 (Summer 1959): 68–71, 68.

224. Thayer to Bloch, 25 March 1956, Bloch papers; Crosby to Thayer, 3 March 1956, Crosby papers.

225. Thayer to Russell, 19 May 1956, Russell papers.

226. Thayer to Russell, 25 March 1956, Russell papers; Thayer to Crosby, 3 February 1958, FS, Crosby papers.

227. Thayer to Russell, 21 July 1956, 11 May 1957, Russell papers.

228. Thayer to Russell, 11 May 1957, Russell papers.

229. "Macomb Fire Victims Planning New Start—Don't Blame Girl," *Pantagraph* (Bloomington, IL), 4 September 1948, 1.

230. "Third Building Is Destroyed by Mysterious Fires," *News-Herald* (Franklin, PA), 20 August 1948, 1; "Rash of Fires for Illinois," *Camden News*, 1 May 1948, 10.

231. "Mystery Fires Puzzle Scientists," *San Mateo Times*, 2 August 1948, 1.

232. "Crowds Hamper Investigation of Phantom Fires at Macomb," *Pantagraph* (Bloomington, IL), 24 August 1948, 1.

233. "Seek Cause for Ghostly Blazes," *Daily Freeman* (Waukesha, WI), 21 August 1948, 3; "Mystery Fires in Farm Home Continue; Officials Baffled," *Tribune* (Kokomo, IN), 20 August 1948, 1, 13.

234. "Crowds Hamper Investigation," 1; "'Ghost Fire' Victims Will Re-

build Home," *Pharos-Tribune* (Logansport, IN), 27 August 1948, 11; "Macomb Fire Victims," 1.

235. "Mysterious Fires Burn Anew on Illinois Farm," *Nonpareil* (Council Bluffs, IA), 29 August 1948, 29.

236. Thayer to Russell, 7 March 1950, FS, Russell papers.

237. "Fire Family Shuns Lie Detector," *Independent* (Long Beach, CA), 30 August 1948, 2.

238. "Conchy Specialist," *Doubt* 14 (Spring 1946): 203.

239. "Hints Weird Fires Caused by Arsonist," *Vidette-Messenger* (Valparaiso, IN), 30 August 1948, 1.

240. "We Break Down," *Doubt* 23 (December 1948): 348–49.

241. "We Break Down," 348–49.

242. "Mental Health Expert Says Mystery Fire Child 'Normal,'" *Tribune* (Kokomo, IN), 1 September 1948, 8.

243. "Macomb Fire Victims," 1.

244. "We Break Down," 348–49.

245. "Milestones," *Pantagraph* (Bloomington, IL), 28 August 1986, 19.

246. Scott Richardson, "Woman Pleads Guilty to Forgery," *Pantagraph* (Bloomington, IL), 27 April 1989, 4.

247. "Wanet Beier Romans," *Pantagraph* (Bloomington, IL), 27 June 2006, 30.

248. "Normal Home Damaged by Mattress Fire," *Pantagraph* (Bloomington, IL), 30 March 1979, 35.

249. "Further News of the Fort Omnibus," *Futurian War Digest*, August 1941, 5.

250. "Hear Ye! Hear Ye!" *Futurian War Digest*, September 1941, 4; Ingham, *Into Your Tent*, 146, 275–77, 280.

251. Tom Villis, *Reaction and the Avant-Garde: The Revolt against Liberal Democracy in Early Twentieth-Century Britain* (London: I. B. Tauris, 2006).

252. Waveney Girvan to Russell, 15 November 1950, Russell papers.

253. Thayer to Russell, 14 July 1948, FS; Miles D. S. Kirk to Russell, 24 February 1949, 4 March 1949, 14 April 1949, Russell papers.

254. Margot Metroland, "Tiffany Thayer and the Fortean Fascists, Part 1," *Counter-Currents Publishing*, 31 March 2015, https://www.counter-currents.com/2015/03/tiffany-thayer-and-the-fortean-fascists-part-1/.

255. L. Sprague de Camp to Russell, 2 December 1951, Russell papers.

256. "Foreign Correspondents," *Fortean* 2 (October 1937): 7; "Living Genius," *Doubt* 21 (June 1948): 321.

257. Thayer to Russell, 21 July 1956, Russell papers.

258. Thayer to Russell, 14 Fort 1948, FS, Russell papers.

259. Harold Chibbett to Russell, n.d. (ca. 1948); Judith L. Gee to Russell,

11 August 1948, Russell papers; Judith L. Gee, letter, *Fate*, Winter 1949, 91; Leslie Shepard, Lewis Spence, and Nandor Fodor, eds., *Encyclopedia of Occultism and Parapsychology* (Detroit: Gale, 2008), 300.

260. Thayer to Russell, 28 January 1949, FS, 26 February 1955, FS, Russell papers.

261. Thayer to Crosby, 13 April 1957, FS, Crosby papers.

262. Adina Hoffman, *Ben Hecht: Fighting Words, Moving Pictures* (New Haven, CT: Yale University Press, 2019).

263. Hoffman, *Ben Hecht*, 179–80.

264. Thayer to Russell, 28 July 1948, FS, Russell papers.

265. Gee to Russell, 10 November 1948, Russell papers.

266. "Pro-Semitic Note," *Doubt* 23 (December 1948): 350.

267. Thayer to Russell, 28 March 1949, FS, Russell papers; "Nominations," *Doubt* 51 (January 1956): 385.

268. Thayer to Russell, 3 December 1955, Russell papers.

269. Thayer to Russell, 28 January 1956, FS, 3 February 1958, FS, Russell papers.

270. Thayer to Russell, 27 April 1957, Russell papers.

271. Thayer to Russell, 7 August 1951, FS, 25 March 1956, Russell papers.

272. Youd to Russell, 26 February 1947, Russell papers.

273. Campbell to Russell, 28 July 1950, Russell papers.

274. Blish to Russell, 8 April 1950, Russell papers.

275. Dagg to Russell, 5 March 1949, Russell papers.

276. Blish to Russell, 8 April 1950; Shroyer to Russell, 15 October 1951; de Camp to Russell, 18 June 1952, Russell papers.

277. L. Sprague de Camp, *Lost Continents: The Atlantis Theme in History, Science, and Literature* (New York: Gnome Press, 1954).

278. Shroyer to Russell, 30 April 1956, Russell papers.

279. Thayer to Russell, 8 June 1957, 7 September 1957, Russell papers.

280. Eric Frank Russell, "Stargazers," *Amazing*, January 1959, 90–100.

281. Eric Frank Russell, "A Trench and Two Holes," *New Worlds*, November 1960, 2–3, 91, 3.

282. Thayer to Russell, 3 February 1958, Russell papers.

283. Thayer to Russell, 29 March 1958, Russell papers.

284. Thayer to Russell, 24 May 1958, Russell papers.

285. Thayer to Russell, 28 January 1956, FS, Russell papers.

286. Martin Gardner, *Undiluted Hocus-Pocus: The Autobiography of Martin Gardner* (Princeton, NJ: Princeton University Press, 2015), 152.

287. Gardner, *Undiluted Hocus-Pocus*, 200; George A. Reisch, *How the Cold War Transformed Philosophy of Science: To the Icy Slopes of Logic* (Cambridge: Cambridge University Press, 2005).

288. Martin Gardner, "The Hermit Scientist," *Antioch Review*, December 1950, 447–57.

289. Miriam Allen De Ford, "Science Fiction Comes of Age," *Humanist*, January 1957, 323–36.

290. "Some Do Not," *Doubt* 51 (January 1956): 385.

291. Shroyer to Russell, 30 April 1956, Russell papers.

292. De Camp to Russell, 15 March 1952, Russell papers.

293. Gardner, *Fads and Fallacies*, 51.

294. Gardner, *Fads and Fallacies*, 49.

295. Gardner, *Fads and Fallacies*, 51.

296. Norman MacBeth, "Setting the Record Straight: A Book Review," *Journal of Anthroposophy*, Spring 1965, 14–16.

297. Thayer to Russell, 28 April 1955, FS, 25 February 1956, Russell papers.

298. "New Books," *Doubt* 42 (October 1953): 246.

299. Gardner, *Undiluted Hocus-Pocus*, 151.

300. Long John Nebel, *The Way Out World* (Englewood Cliffs, NJ: Prentice-Hall, 1961).

301. David J. Hess, *Science in the New Age: The Paranormal, Its Defenders and Debunkers, and American Culture* (Madison: University of Wisconsin Press, 1993).

302. Dana Richards, ed., *Dear Martin/Dear Marcello: Gardner and Truzzi on Skepticism* (Hackensack, NJ: World Scientific Publishing, 2017).

303. George P. Hansen, *The Trickster and the Paranormal* (n.p.: Xlibris, 2001), 292.

Chapter 3

1. E. Hoffmann Price, *Book of the Dead: Friends of Yesteryear: Fictioneers and Others* (Sauk City, WI: Arkham House, 2001), 186; "Mrs. Carr Divorces Husband on Ground He Divorced Her," *Pantagraph* (Bloomington, IL), 24 November 1935, 1.

2. Price, *Book of the Dead*, 197; Thayer to Bloch, 21 September 1946; Robert Spencer Carr to Bloch, 4 December 1947, Donald Beaty Bloch papers, New York Public Library; "Keeping Posted," *Saturday Evening Post*, 6 December 1947, 10.

3. Thomas M. Disch, *The Dreams Our Stuff Is Made Of: How Science Fiction Conquered the World* (New York: Touchstone, 1998).

4. [Samuel Mines], "A Word about the Works of—Charles Fort: The Disciple of Disbelief," *Fantastic Story*, January 1925, 9; August Derleth, "Contemporary Science Fiction," *College English*, January 1952, 187–94.

5. "Our Cover," *Doubt* 20 (March 1948): 298.

6. Robert H. Elias, ed., *Letters of Theodore Dreiser*, vol. 1 (Philadelphia: University of Pennsylvania, 1959), 197.

7. Richard Holmes, *The Age of Wonder: How the Romantic Generation Discovered the Beauty and Terror of Science* (New York: Knopf, 2009), 445.

8. John Clute, *Pardon This Intrusion: Fantastika in the World Storm* (Essex: Beccon, 2011).

9. Franklin Rosemont, ed., *Surrealism and Its Popular Accomplices* (San Francisco: City Lights, 1980), 5.

10. Jill Fell, *Alfred Jarry: An Imagination in Revolt* (Madison, NJ: Fairleigh Dickinson University Press, 2005).

11. Brian Stableford, *Narrative Strategies in Science Fiction and Other Essays on Imaginative Fiction* (Holicong, PA: Wildside Press, 2009), 83–84, 88.

12. Victoria Nelson, *Gothicka: Vampire Heroes, Human Gods, and the New Supernatural* (Cambridge, MA: Harvard University Press, 2012), xi.

13. Mike Ashley and Robert A. W. Lowndes, *The Gernsback Days: A Study of the Evolution of Modern Science Fiction from 1911 to 1936* (Holicong, PA: Wildside Press, 2004), 65.

14. Sam Moskowitz, *Strange Horizons: The Spectrum of Science Fiction* (New York: Scribner, 1976), 233–34; Earl Kemp and Luis Ortiz, eds., *Cult Magazines: A to Z: A Compendium of Culturally Obsessive and Curiously Expressive Publications* (Greenwood, DE: Nonstop Press, 2009), 85–86.

15. Charles Fort, *Lo!* (New York: Claude Kendall, 1931), 204.

16. Charles Fort, *The Complete Books of Charles Fort* (New York: Dover, 1974), 847.

17. Michael Saler, "Modernity, Disenchantment, and the Ironic Imagination," *Philosophy and Literature* 28 (2004): 137–49.

18. Miriam Allen De Ford, "Charles Fort: Enfant Terrible of Science," *Magazine of Fantasy and Science Fiction*, January 1954, 105–16, 111.

19. Price, *Book of the Dead*, 175–80.

20. S. T. Joshi, *H. P. Lovecraft: The Decline of the West* (Rockville, MD: Wildside Press, 2016), 117–32, 198.

21. Donald Snively to Howard Snively, 28 September 1936, box 3, folder 8, Howard Snively papers, Schenectady (NY) Museum of Innovation and Science.

22. Scott Connors, "Weird Tales and the Great Depression," in *The Robert Howard Reader*, ed. Darrell Schweitzer (Holicong, PA: Borgo Press, 2010), 162–78.

23. Charles Fort to Edmond Hamilton, 13 February [1931], Edmond Hamilton papers, Jack Williams Science Fiction Library, Eastern New Mexico State University; S. T. Joshi and David E. Schultz, eds., *Mysteries of Time and Spirit:*

Letters of H. P. Lovecraft and Donald Wandrei (San Francisco: Night Shade Books, 2005), 91.

24. Joshi, *H. P. Lovecraft*, 117–32, 198; James Turner, ed., *H. P. Lovecraft: Selected Letters*, vol. 5, *1934–1937* (Sauk City, WI: Arkham House Publishers, 1976), 173.

25. John D. Squires, ed., *M. P. Shiel and the Lovecraft Circle: A Collection of Primary Documents, Including Shiel's Letters to August Derleth, 1929–1946* (Kettering, OH: Vainglory Press, 2001), 103–11.

26. David E. Schultz and Scott Connors, eds., *Selected Letters of Clark Ashton Smith* (Sauk City, WI: Arkham House, 2003), 129–30, 152.

27. Clute, *Pardon This Intrusion*, 3.

28. Simon During, *Modern Enchantments: The Cultural Power of Secular Magic* (Cambridge, MA: Harvard University Press, 2004).

29. Harry Leon Wilson to Leon Wilson, 12 April 1937, 24 March 1938, box 1, outgoing letters, 1937–38 file, Harry Leon Wilson papers, University of California, Berkeley.

30. John Cheng, *Astounding Wonder: Imagining Science and Science Fiction in Interwar America* (Philadelphia: University of Pennsylvania Press, 2013).

31. Daniel P. Thurs, "Tiny Tech, Transcendent Tech: Nanotechnology, Science Fiction, and the Limits of Modern Science Talk," *Science Communication* 29 (2007): 65–95, 72.

32. David Drake, *Night and Demons* (Wake Forest, NC: Baen, 2012), n.p.

33. John W. Campbell Jr., "Book Review," *Unknown Worlds*, December 1941, 130.

34. Thomas Sheridan, "Review: He Wrote 'The Rats Tale,'" *Fantasy Review*, February–March 1947, 7–8.

35. Adam Roberts, *The History of Science Fiction* (London: Palgrave Macmillan, 2005), 18.

36. Albert I. Berger, "The Magic That Works: John W. Campbell and the American Response to Technology," *Journal of Popular Culture* 5 (1972): 867–943.

37. "'Scientifiction' Ascending," *Fortean* 7 (June 1943): 5; "Fortean Arts," *Doubt* 15 (Winter 1945): 188.

38. William Sloane to Thayer, 11 October 1940, 12 November 1940; Thayer to Sloane, 5 November 1940, Henry Holt and Company papers, Princeton University.

39. Jesse Douglass to Thayer, n.d. (ca. December 1943), 1 April 1942, April 1942, in author's possession.

40. "Sunt(r)ails," October 1941, Bruce Pelz papers, University of California, Riverside; "'Scientifiction' Ascending," 5.

41. "And Now Campbell!" *Astounding Stories*, October 1934, 38; Walter A. Carrithers, letter, *Astounding Stories*, November 1943, 176.

42. Charlotte Sleigh, "'An Outcry of Silences': Charles Hoy Fort and the Uncanny Voices of Science," in *The Silences of Science: Gaps and Pauses in the Communication of Science*, ed. Felicity Mellor and Stephen Webster (London: Routledge, 2016), 274–95, 283.

43. George Wetzel, "Natural History in Water Pipes," 5, St. John's Reformed Church, Cemetery Records, Maryland State Archives.

44. James H. Bready, "About Books and Authors," *Baltimore Sun*, 8 May 1955, F7.

45. George Wetzel, "Natural History," *Maryland Conservation*, March 1955, 19–22.

46. "New Books," *Doubt* 48 (April 1955): 341–42.

47. Wetzel, "Natural History in Water Pipes," 2.

48. Wetzel, "Natural History in Water Pipes," 2.

49. *Ember* 4, August 1946; *Ember* 19, October 1946, Pelz papers.

50. *Ember* 8, n.d., Pelz papers.

51. Wetzel, "Natural History in Water Pipes," 5.

52. George Wetzel, "The Gothic Horror and Other Stories" (n.p., 1955), ii; Thomas Olive Abbott, ed., *Collected Works of Edgar Allan Poe: Tales and Sketches, 1831–1842* (Cambridge, MA: Belknap/Harvard University Press, 1978), 576.

53. S. T. Joshi, ed., *H. P. Lovecraft: Four Decades of Criticism* (Athens: Ohio University Press, 1980), ix.

54. George T. Wetzel, "The Cthulhu Mythos: A Study," in Joshi, *Lovecraft: Four Decades of Criticism*, 79–95.

55. "Tribute to George Townsend Wetzel (1920–1938), HP Lovecraft Researcher," n.d., 2013-39-301, Chester D. Cuthbert Fonds, University of Alberta.

56. Wetzel, "Natural History in Water Pipes," 3.

57. George Wetzel, "The Mechanistic-Supernatural of Lovecraft," *Fresco* 8 (Spring 1958): 54–60.

58. "Fanews," 275, n.d., 1946, Pelz papers; "Wants Information on Michael Zittle or His Conjure Book," *Daily Mail* (Hagerstown, MD), 13 February 1954, 18.

59. Wetzel, "Gothic Horror and Other Stories," ii.

60. Wetzel, "Natural History in Water Pipes," 2.

61. George Wetzel, "Tunnels of Baltimore," *Baltimore Sun*, 13 March 1953, 14; George Wetzel, "Tunnel Clues," *Baltimore Sun*, 25 May 1953, 12.

62. Wetzel, "Natural History in Water Pipes," 6–7.

63. George Wetzel, "The Sea Serpent—Its Existence Proved," *Umbra*, May 1955, n.p.

64. Wetzel, "Natural History in Water Pipes," 1.

65. Maynard Shipley, "Charles Fort, Enfant Terrible of Science," *New York Times*, 1 March 1931, 2.

66. Perry A. Chapdelaine Sr., Tony Chapdelaine, and George Hay, eds., *The John W. Campbell Letters*, vol. 1 (Franklin, TN: AC Projects, 1985), 83–84.

67. Martha A. Bartter, *The Way to Ground Zero: The Atomic Bomb in American Science Fiction* (Westport, CT: Greenwood Press, 1988), 108, 131.

68. David Ketterer, *Imprisoned in a Tesseract: The Life and Work of James Blish* (Kent, OH: Kent State University Press, 1987).

69. Blish to Knight, 20 April 1967, Damon Knight papers, Syracuse University.

70. Martin Pearson (pseud. Donald Wollheim), "Up There," *Science Fiction Quarterly*, Summer 1942, 85–87, 137.

71. Mallory Kent (pseud Robert Lowndes), letter, *Science Fiction Quarterly*, Fall 1942, 145–46.

72. James Blish, letter, *Science Fiction Quarterly*, Spring 1943, 144–45.

73. Fort, *Complete Books*, 1030–31.

74. Frederick Hehr, letter, *Astounding*, February 1935, 157.

75. William H. Patterson Jr., *Robert A. Heinlein: In Dialogue with His Century*, vol. 1, *Learning Curve* (New York: Macmillan, 2010).

76. Alexei Panshin and Cory Panshin, *The World beyond the Hill: Science Fiction and the Quest for Transcendence* (Los Angeles: Jeremy P. Tarcher, 1989), 365.

77. Jack Williamson, *Wonder's Child: My Life in Science Fiction* (New York: Bluejay, 1984), 123–24.

78. Sandra Sizer Frankiel, *California's Spiritual Frontiers: Religious Alternatives in Anglo-Protestantism, 1850–1910* (Berkeley: University of California Press, 1988), 77–78.

79. John Carter, *Sex and Rockets: The Occult World of Jack Parsons* (Port Townsend, WA: Feral House, 2005), 79–81.

80. Panshin and Panshin, *World beyond the Hill*, 380.

81. Panshin and Panshin, *World beyond the Hill*, 368.

82. Panshin and Panshin, *World beyond the Hill*, 380.

83. Alexei Panshin, *Heinlein in Dimension: A Critical Analysis* (Chicago: Advent, 1968), 21; Panshin and Panshin, *World beyond the Hill*, 384; Robert A. Heinlein, *Grumbles from the Grave* (New York: Ballantine, 1990), 12.

84. Jack Seabrook, *Martians and Misplaced Clues: The Life and Work of Fredric Brown* (Bowling Green, OH: Bowling Green State University Popular Press, 1993); Robert Bloch, *The Man Who Collected Psychos* (New York: McFarland, 2009).

85. Seabrook, *Martians*, 175.

86. Editorial, *Thrilling Wonder Stories*, February 1951, 6.

87. Editorial, *Startling Stories*, March 1948, 6.

88. Editorial, *Fantastic Story*, March 1951, 6.

89. Sam Merwin, "Star Tracks," *Astounding*, March 1952, 144–52.

90. "Sunt(r)ails," October 1941, Pelz papers; Robert Heinlein to Forrest J. Ackerman, 28 January 1945, Heinlein Society, https://mycroft.heinleinsociety .org/HeinleinNexus/HeinleinNexus/heinleinsociety.org/html/viewtopic4990 .html?p=6035&sid=c6a2a19ce5fc5385ff9b457ba159eafd; "A. L. (The Man) Joquel," *Doubt* 25 (Summer 1949): 382.

91. Arthur Louis Joquel II, *The Challenge of Space* (Hollywood, CA: House-Warven, 1952), 190–91.

92. Jason Colavito, *The Cult of Alien Gods: H. P. Lovecraft and Extraterrestrial Pop Culture* (Buffalo, NY: Prometheus, 2005); Wouter J. Hanegraaff, "From Imagination to Reality: An Introduction to Esotericism and the Occult," in *Hilma af Klint: The Art of Seeing the Invisible*, ed. Kurt Almqvist and Louise Belfrage (Stockholm: Axel and Margaret Ax:son Johnson Foundation, 2015), 59–71.

93. Williamson, *Wonder's Child*, 128.

94. Michael Barkun, *A Culture of Conspiracy: Apocalyptic Visions in Contemporary America* (Berkeley: University of California Press, 2013), 115–21; Charmaine Ortega Getz, "Maurice Doreal and His Brotherhood of the White Temple Awaited the Apocalypse in Colorado," *Westword*, 28 October 2015, https://www.westword.com/news/maurice-doreal-and-his-brotherhood-of-the -white-temple-awaited-the-apocalypse-in-colorado-7285649.

95. "Too Late for Dreams," *Doubt* 35 (January 1952): 122; "Hi-Spots in the Mail," *Doubt* 37 (June 1952): 151–52.

96. Norman G. Markham to Bloch, 26 September 1954, Bloch papers; Joshua Blu Buhs, "Norman/David Garrett Markham as a Fortean [Edited]," 24 November 2015, http://www.joshuablubuhs.com/blog/normandavid -garrett-markham-as-a-fortean.

97. Eric Frank Russell, "Over the Border," *Unknown*, September 1939, 129–40; Markham to Russell, 15 July 1948, Eric Frank Russell papers, University of Liverpool.

98. David G. Markham, letter, *Unknown*, February 1940, 122.

99. Loren E. Gross, *UFOs: A History*, vol. 1, *July 1947–December 1948* (Fremont, CA: n.p., 1988), 21.

100. Fred Nadis, *The Man from Mars: Ray Palmer's Amazing Pulp Journey* (New York: Penguin, 2013).

101. Ray Palmer, *The Secret World: The Diary of a Lifetime Questioning the "Facts,"* vol. 1 (Amherst, WI: Amherst Press, 1975), 10.

102. Jim Dennon, *The Oahspe Story* (Seaside, OR: n.p., 1965).

103. Richard Toronto, *War over Lemuria: Richard Shaver, Ray Palmer and the Strangest Chapter of 1940s Science Fiction* (Jefferson, NC: McFarland, 2013).

104. Alan Devereux, "Mr. Shaver's Memories," *Fantasy Review*, October–November 1948, 10–11.

105. Roy Lavender, "The Cinvention Seen Dimly," 68–71, and Stan Skirvin, "Wha' Happened," 75–81, in *The Cinvention Memory Book*, ed. Don Ford and Stan Skirvin, May 1950, https://fanac.org/conpubs/Worldcon/Cinvention/.

106. Martin Gardner, *The New Age: Notes of a Fringe Watcher* (Buffalo, NY: Prometheus Books, 1988), 213.

107. Markham to Russell, 15 July 1948, Russell papers.

108. Editor, "The Observatory," *Amazing Stories*, June 1948, 6; Jim Wentworth, *Giants in the Earth: The Amazing Story of Ray Palmer, Oahspe and the Shaver Mystery* (Amherst, WI: Palmer Publications, 1973), 51; Jim Wentworth, "Charles Fort's Corroboration of the Shaver Mystery," *Search*, February 1958, 40–50.

109. Jim Wentworth, "If the Sky Ever Opened Up," *Forum*, June 1967, 12–15.

110. Nadis, *Man from Mars*, 117.

111. "Church," *American Contractor*, 19 June 1920, 65; Vincent Gaddis, *The Story of Winona Lake: A Memory and a Vision* (Winona Lake, IN: Winona Lake Christian Assembly, 1960), 105.

112. Ernst Groth (pseud. Gaddis), "Beyond the Etheric Veil," *Fate*, Spring 1948, 95; "Vincent Gaddis, 1913–1997," *Strange Magazine*, Spring 1998, 29, 58.

113. "Life Energy from Carbon?" *Amazing Stories*, September 1945, 166.

114. Vincent H. Gaddis, "Notes on Subterranean Shafts," *Amazing Stories*, June 1947, 148–51.

115. Vincent Gaddis, "Tales from Tibet," *Amazing Stories*, February 1946, 172–74; Vincent Gaddis, "The Truth about Tibet," *Amazing Stories*, July 1946, 168–70.

116. Vincent H. Gaddis, "The Shaver Mystery," *Round Robin*, May–June 1947, 6–9.

117. Vincent Gaddis, "Strange Secrets of the Sea," *Amazing Stories*, February 1947, 158–60.

118. John Keel, "The Man Who Invented Flying Saucers," *Fortean Times* 41 (Winter 1983): 52–57; Jesse Walker, *The United States of Paranoia: A Conspiracy Theory* (New York: Harper Perennial, 2014).

119. Vincent H. Gaddis, "Visitors from the Void," *Amazing Stories*, June 1947, 159–61.

120. Judith L. Gee to Russell, 13 September 1948, Russell papers; John L. Ingham, *Into Your Tent: The Life, Work and Family Background of Eric Frank Russell* (UK: Plantech, 2010), 115.

121. Thayer to Russell, 14 August 1948, FS, Russell papers.

122. "On the Fortean Arts," *Doubt* 46 (October 1954): 300.

123. Thayer to Jesse Douglass, 5 December 1942, in author's possession.

124. C. A. Brandt, "In the Realm of Books," *Amazing Stories*, July 1931, 379; Moskowitz, *Strange Horizons*, 219–20.

125. J. H. Heunigar, letter, *Astounding Stories*, August 1934, 153.

126. Richard Dodson, letter, *Astounding Stories*, October 1934, 156.

127. Edward C. Love, letter, *Astounding Stories*, November 1934, 155.

128. Isaac Asimov, ed., *Before the Golden Age: A Science Fiction Anthology of the 1930s* (New York: Doubleday, 1974), 815.

129. P. Schuyler Miller, "Spawn," *Weird Tales*, August 1939, 26–45; P. Schuyler Miller, "The Reference Library," *Astounding*, May 1953, 150.

130. Sleigh, "Outcry of Silences," 283.

131. Price, *Book of the Dead*, 196–97; "Giles Grist," *Doubt* 12 (Spring–Summer 1945): 173–74; E. Hoffmann Price, "Translators," *Doubt* 15 (Summer 1946): 226–27.

132. Richard Matheson to George Haas, 30 September 1950, courtesy of Paul Stuve.

133. George Haas to Richard Matheson, 22 October 1950, courtesy of Paul Stuve.

134. Campbell to Russell, 1 October 1952, Russell papers.

135. Blish to Russell, 8 April 1949, Russell papers.

136. "Frappe," Virginia Blish, n.d., box 118, Pelz papers.

137. Eric Frank Russell, "How High Is the Sky?" *Tumbrils* 15, 1948, 6–9, Pelz papers.

138. James Blish, *Jack of Eagles* (New York: Avon, 1952), 49.

139. Alec Nevala-Lee, *Astounding: John W. Campbell, Isaac Asimov, Robert A. Heinlein, L. Ron Hubbard, and the Golden Age of Science Fiction* (New York: HarperCollins, 2018), 241–96.

140. Nevala-Lee, *Astounding*, 252.

141. Martin Gardner, *Fads and Fallacies in the Name of Science* (New York: Dover, 1957), 265.

142. "Dianetics," *Doubt* 30 (October 1950): 43.

143. Thayer to Russell, 28 December 1950, FS, Russell papers; Thayer to Nelson Bond, 28 April 1951 FS, Charles Fort papers, accession 7657-a, Special Collections Department, University of Virginia Library, Charlottesville.

144. Campbell to Russell, 16 May 1952, Russell papers.

145. Campbell to Russell, 28 July 1950, Russell papers.

146. Campbell to Russell, 7 February 1952, Russell papers.

147. Russell Miller, *Bare-Faced Messiah: The True Story of L. Ron Hubbard* (London: Joseph, 1987).

148. Susan Raine, "Astounding History: L. Ron Hubbard's Scientology Space Opera," *Religion* 45 (2015): 66–88.

149. Campbell to Russell, 28 July 1950, Russell papers.

150. "The Chapters," *Doubt* 27 (Winter 1949): 413–14.

151. Thayer to Russell, 7 March 1950, FS, Russell papers.

152. "The Chapters," *Doubt* 25 (Summer 1949): 382–83.

153. Thayer to Russell, 14 Fort 1948, FS, Russell papers.

154. Garen Drussai, "The Tainted," *Doubt* 35 (January 1952): 116–18; "On the Fortean Arts," 300.

155. Robert Barbour Johnson to Knight, n.d. (ca. 1968), Knight papers.

156. Jenny Pegg, "Remnants of a Spiritualist's Belief," *Stanford Magazine*, September/October 2012, https://stanfordmag.org/contents/remnants-of-a-spiritualist-s-belief.

157. Clarkson Dye, "Through Solid Walls," *Fate*, 2 July 1949, 38–45; "Mrs. Robert Cross," *Fate*, 3 January 1950, 95–97; Robert Barbour Johnson, "The Problem of the Stanford 'Apports,'" *INFO Journal*, Summer 1973, 18–20.

158. George Haas to Richard Matheson, 22 October 1950, courtesy of Paul Stuve.

159. "The Chapters," 382–83.

160. Robert Barbour Johnson, "Charles Fort and a Man Named Thayer," *Rhodomagnetic Digest*, September–October 1951, 3–9.

161. Robert Barbour Johnson, "Charles Fort: His Objectives Fade in the West," *Worlds of If*, July 1952, 134–37, 151, 151.

162. Johnson, "His Objectives Fade," 134.

163. George Haas to Richard Matheson, 22 October 1950, courtesy of Paul Stuve.

164. Joseph Henry Jackson, "Bookman's Notebook," *San Francisco Chronicle*, 1 September 1943, 14.

165. Miriam Allen De Ford to Anthony Boucher, 1 September 1943, Anthony Boucher papers, Indiana University; William Rose Benét, "The Phoenix Nest," *Saturday Review of Literature*, 5 September 1942, 12; William Rose Benét, "The Phoenix Nest," *Saturday Review of Literature*, 12 September 1942, 18.

166. Boucher to De Ford, 31 August 1943, Boucher papers.

167. "House Stoning Stumps Police," *Oakland Tribune*, 29 August 1943, 7.

168. Joseph Henry Jackson, "The Bookman's Daily Notebook," *San Francisco Chronicle*, 1 May 1941, 17.

169. Benét, "Phoenix Nest," 5 September 1942, 12.

170. "House Stoning Stumps Police," 7.

171. Ambrose Bierce, *Tales of Soldiers and Civilians* (New York: Heritage Press, 1943), ix–x.

172. Jackson, "Bookman's Notebook," 1 September 1943.

173. Michael Saler, "Clap If You Believe in Sherlock Holmes: Mass Culture

and the Re-enchantment of Modernity, c. 1890–c. 1940," *Historical Journal* 46 (2003): 599–622, 604.

174. De Ford, "Charles Fort."

175. "Mystery Bombs Fall from Bridge," *Oakland Tribune*, 13 September 1943, 10.

176. Joseph Henry Jackson, "Bookman's Notebook," *Los Angeles Times*, 15 March 1954, 39; Deborah Blum, *Ghost Hunters: William James and the Search for Scientific Proof of Life after Death* (New York: Penguin, 2007).

177. Michael Saler, *As If: Modern Enchantment and the Literary Prehistory of Virtual Reality* (New York: Oxford University Press, 2011).

178. Boucher to De Ford, 3 August 1944, Boucher papers.

179. Jackson, "Bookman's Notebook," 15 March 1954, 39.

180. Marilynn S. Johnson, *The Second Gold Rush: Oakland and the East Bay in World War II* (Berkeley: University of California Press, 1996).

181. Jeffrey Marks, *Anthony Boucher: A Biobibliography* (New York: McFarland, 2008), 129.

182. Anthony Boucher, "The Chronokinesis of Jonathan Hull," *Astounding*, June 1946, 118–31; Francis M. Nevins, *The Anthony Boucher Chronicles* (n.p.: Ramble House, 2009), 300, 420–21.

183. De Ford, "Charles Fort," 116.

184. Miriam Allen De Ford, "Henry Martindale, Great Dane," *Beyond Fantasy Fiction*, March 1954, 77–89.

185. Joseph Henry Jackson, "Bookman's Notebook," *Los Angeles Times*, 15 July 1952, II.5.

186. "Hosier," *Santa Cruz Sentinel*, 10 August 1943, 1.

187. "Girl, 11, on Bicycle, is Killed in Collision on Highway to Big Basin," *Santa Cruz Sentinel*, 10 February 1945, 1.

188. Anthony Boucher, "The Tenderizers," *Magazine of Fantasy and Science Fiction*, January 1972, 77–80.

189. "Pebbles from Heaven," *PRS Journal* 3 (1943): 29.

190. Brad Steiger, *Real Ghosts, Restless Spirits, and Haunted Places* (Canton, MI: Visible Ink Press, 2003), 68–70.

191. "The Compleat and Unexpurgated Who Killed Science Fiction?" *e*I*29* 5, no. 6 (December 2006), http://efanzines.com/EK/eI29/#6.

192. Michael Saler, "Modernity and Enchantment: A Historiographic Review," *American Historical Review* 111 (2006): 692–716.

193. Kenneth C. Davis, *Two-Bit Culture: The Paperbacking of America* (New York: Houghton-Mifflin, 1984); Mike Ashley, *Transformations: The Story of the Science Fiction Magazines from 1950 to 1970* (Liverpool: Liverpool University Press, 2005), 34–40.

194. Clarke to Russell, 26 July 1948, Russell papers.

195. Malcolm Cowley, *The Literary Situation* (New York: Viking, 1954), 104–5.

196. Ashley, *Transformations*, 69–70, 189–90.

197. Adam Parfrey, *It's a Man's World: Men's Adventure Magazines, the Postwar Pulps* (Port Townsend, WA: Feral House, 2015).

198. William K. Schafer and William F. Nolan, eds., *California Sorcery: A Group Celebration* (New York: Ace, 2001).

199. Bradford Lyau, *The Anticipation Novelists of 1950s French Science Fiction: Stepchildren of Voltaire* (Jefferson, NC: McFarland, 2014).

200. Wouter J. Hanegraaff, "Fiction in the Desert of the Real: Lovecraft's Cthulhu Mythos," *Ares* 7 (2007): 85–109.

201. Sumathi Ramaswamy, *The Lost Land of Lemuria: Fabulous Geographies, Catastrophic Histories* (Berkeley: University of California Press, 2004), 53–54.

202. Peter Bishop, *The Myth of Shangri-La: Tibet, Travel Writing, and the Western Creation of Sacred Landscape* (Berkeley: University of California Press, 1989), 217.

203. Barry Shipman to Thayer, 29 June 1941; Thayer to Shipman, 19 July 1941, Barry Shipman papers, Boise State University; Thayer to Russell, 14 March 1948, 14 August 1948, Russell papers.

204. Joseph Millard, "World's Greatest Heckler," *True*, April 1946, 143.

205. Tiffany Thayer, *33 Sardonics I Can't Forget* (New York: Philosophical Library, 1946), 14.

206. Thayer to Ezra Pound, 23 March 1946, FS, 21 April 1947, FS, Ezra Pound papers, Indiana University; Liesl Olson, *Chicago Renaissance: Literature and Art in the Midwest Metropolis* (New Haven, CT: Yale University Press, 2017).

207. Tiffany Thayer, "This Reading Racket," *Writer's Digest*, January 1931, 33.

208. "On the Fortean Arts," 300.

209. Sam Moskowitz, "Lo! The Poor Forteans," *Amazing Stories*, June 1965, 46.

210. Clarke, *Greetings, Carbon-Based Bipeds!*, 116–17.

211. "All About Sputs," *Doubt* 56 (March 1958): 460–80.

212. Anthony Boucher, "Science Fiction Still Leads Science Fact," *New York Times*, 1 December 1957, 285.

213. Thayer to Russell, 30 November 1957, Russell papers.

214. "All About Sputs," 460.

215. Clarke to Russell, 18 July 1956, Russell papers.

216. Joshua Blu Buhs, *Bigfoot: The Life and Times of a Legend* (Chicago: University of Chicago Press, 2009), 90.

217. Dick Bothwell, "Scientists, Reporters Search for Answers," *Tampa*

Bay Times, 14 November 1948, 4; "Gag or Not, Thing's Tracks Excite Suwannee River Area," *Tampa Tribune*, 14 November 1948, 13.

218. "Sanderson Reports," *Doubt* 24 (April 1949): 366.

219. Ivan T. Sanderson, "Don't Scoff at Sea Monsters," *Saturday Evening Post*, 8 March 1947, 22–23, 84–87.

220. Ivan T. Sanderson, "There Could Be Dinosaurs," *Saturday Evening Post*, January 3, 1948, 17, 53–56.

221. "Hi-Spots in the Mail," *Doubt* 24 (April 1949): 370–72.

222. "More Monsters," *Doubt* 16 (Fall 1946): 236–37.

223. Ivan T. Sanderson, *Invisible Residents: The Reality of Underwater UFOs* (New York: World Publishing, 1970.)

224. Ivan Sanderson, "That Forgotten Monster: Old Three-Toes" (two parts) *Fate*, December 1967, 66–75; January 1968, 85–93.

225. Brian Regal, *Searching for Sasquatch: Crackpots, Eggheads, and Cryptozoology* (London: Palgrave-MacMillan, 2011).

226. Robert E. Kohler, *All Creatures: Naturalists, Collectors, and Biodiversity, 1850–1950* (Princeton, NJ: Princeton University Press, 2013).

227. Buhs, *Bigfoot*, 90–92.

228. Ivan T. Sanderson, "The Patterson Affair," *Pursuit*, June 1968, 8–10.

229. Richard Grigonis, "Ivan T Sanderson—Chapter 10—Charles Hoy Fort and the Founding of SITU," in *A Tribute to Ivan T. Sanderson*, http://www.richardgrigonis.com/Ch10%20Charles%20Hoy%20Fort.html.

230. Ivan Sanderson to Martin Gardner, 26 March 1968, Ivan T. Sanderson papers, American Philosophical Society.

231. Clarke to Russell, 18 July 1956, Russell papers.

232. "Florida 'Giant Penguin' Hoax Revealed," *ISC Newsletter* 7 (Winter 1988): 1–3.

233. International Society of Cryptozoology, Records, 1981–1991, SIA acc. 95-031, Smithsonian Institution Archives.

234. Saler, *As If*.

235. Hanegraaff, "Desert of the Real."

Chapter 4

1. Antony Borrow, "The Banishing," *Springtime* 2, 1958, 57–68; Antony Borrow to Russell, 22 October 1948, Eric Frank Russell papers, University of Liverpool.

2. Howard Finn, "A Conversation with Eva Tucker," *Pilgrimages: A Journal of Dorothy Richardson Studies* 7 (2015): 82–100, 83–84.

3. Borrow to Russell, 2 September 1951, Russell papers.

4. Borrow to Russell, 1 October 1948, 9 October 1948, 22 October 1948, 6 November 1948, Russell papers.

5. "Antony Borrow Writes," *Doubt* 28 (April 1950): 6.

6. Thayer to Russell, 16 February 1957, Russell papers.

7. Atkins to Russell, 6 August 1947, Russell papers.

8. John Atkins, "The Challenge of Charles Fort," *The Glass* 4 (ca. 1950): 1–4.

9. Colin Bennett, *Politics of Imagination: The Life, Work, and Ideas of Charles Fort* (Manchester, UK: Critical Vision, 2002), 19, 57.

10. Paula Rabinowitz, *American Pulp: How Paperbacks Brought Modernism to Main Street* (Princeton, NJ: Princeton University Press, 2014).

11. David Weir, *Decadence and the Making of Modernism* (Amherst: University of Massachusetts Press, 1995).

12. Marja Härmänmaa and Christopher Nissen, eds., *Decadence, Degeneration, and the End: Studies in the European Fin de Siècle* (London: Palgrave Macmillan, 2014).

13. Lorraine Daston, "When Science Went Modern," *Hedgehog Review* 18 (2016): 18–32.

14. Benjamin DeCasseres, "Modernity and the Decadence," *Camera Work*, January 1912, 17–19.

15. Benjamin DeCasseres, "Bankrupt Science," *New York Sun*, 3 January 1911, 6.

16. Benjamin DeCasseres, "The Philosophy of Hypocrisy," *International*, July 1915, 222–23.

17. David Weir, *Decadent Culture in the United States: Art and Literature against the American Grain, 1890–1926* (Albany: State University of New York Press, 2008).

18. "Castillo Writes," *Doubt* 23 (1948): 346–47; Fort to DeCasseres, 6 July 1926, 29 November 1925, Benjamin DeCasseres papers, New York Public Library.

19. Herbert N. Schneidau, "Vorticism and the Career of Ezra Pound," *Modern Philology* 65 (1968): 214–27, 215; Robert Stark, *Ezra Pound's Early Verse and Lyric Tradition* (Edinburgh: Edinburgh University Press, 2012), 7.

20. Chip Rhodes, *Structures of the Jazz Age: Mass Culture, Progressive Education, and Racial Disclosures in American Modernism* (London: Verso, 1998), 45–47.

21. Liesl Olson, *Chicago Renaissance: Literature and Art in the Midwest Metropolis* (New Haven, CT: Yale University Press, 2017).

22. William C. Wees, "Ezra Pound as a Vorticist," *Wisconsin Studies in Contemporary Literature* 6 (Winter–Spring, 1965), 56–72.

23. "London Writer to the Star Recalls Strange Events at Crawfordsville," *Indianapolis Star*, 27 July 1924, 42.

24. Fort to John T. Reid, 4 October 1925, in "The Correspondence of, and to, Charles Hoy Fort," ed. Mr. X, 2004, http://www.resologist.net/corfjr05

.htm; Miriam Allen De Ford, "Charles Fort: Enfant Terrible of Science," *Magazine of Fantasy and Science Fiction*, January 1954, 105–16, 111.

25. "The Book of the Damned," *Indianapolis News*, 25 March 1920, 12.

26. "Revolution of the Word," *Transitions*, June 1929, 13.

27. Leo Knuth, "'Finnegans Wake': A Product of the Twenties," *James Joyce Quarterly* 11 (1974): 310–22.

28. Peter Gay, *Modernism: The Lure of Heresy* (New York: Norton, 2007).

29. Evan Brier, *A Novel Marketplace: Mass Culture, the Book Trade, and Postwar American Fiction* (Philadelphia: University of Pennsylvania, 2012), 123–24; Catherine Turner, *Marketing Modernism between the Two World Wars* (Amherst: University of Massachusetts Press, 2003), 174, 205–8.

30. Michael Levenson, *A Genealogy of Modernism: A Study of English Literary Doctrine 1908–1922* (Cambridge: Cambridge University Press, 1986), 68.

31. Bruce Clark, *Dora Marsden and Early Modernism: Gender, Individualism, Science* (Ann Arbor: University of Michigan Press, 1996), 20–22.

32. Jennifer Ratner-Rosenhagen, *American Nietzsche: A History of an Icon and His Ideas* (Chicago: University of Chicago Press, 2012).

33. Benjamin DeCasseres, "The Fortean Fantasy," *Thinker*, April 1931, 71–81, 81.

34. "Ego and His Own," *Doubt* 32 (March 1951): 72.

35. "MFS Bump Writes," *Doubt* 37 (June 1952): 156.

36. Clark, *Dora Marsden*, 5.

37. Schneidau, "Vorticism," 221; A. David Moody, *Ezra Pound: Poet*, vol. 1, *The Young Genius 1885–1920* (London: Oxford University Press, 2007), 242–43.

38. Nico Israel, *Spirals: The Whirled Image in Twentieth-Century Literature and Art* (New York: Columbia University Press, 2015), 68.

39. Tim Armstrong, "Social Credit Modernism," *Critical Quarterly* 55 (2013): 50–65.

40. Moody, *Ezra Pound: Poet*, 1:227; Alec Marsh, *Money and Modernity: Pound, Williams, and the Spirit of Jefferson* (Tuscaloosa: University of Alabama Press, 2011).

41. "The Philosophy of Fort," *News Review* (London), 10 June 1948, 8.

42. Joshua Blu Buhs, "I. O. Evans as a Fortean," 16 July 2015, http://www.joshuablubuhs.com/blog/i-o-evans-as-a-fortean.

43. Thayer to Pound, 14 January 1948, FS, Ezra Pound papers, Indiana University.

44. Tom Villis, *Reaction and the Avant-Garde: The Revolt against Liberal Democracy in Early Twentieth-Century Britain* (London: I. B. Tauris, 2006).

45. Marsh, *Money and Modernity*, 68.

46. Matthew Feldman, *Ezra Pound's Fascist Propaganda, 1935–45* (London: Palgrave Macmillan, 2013), 9–33.

47. Damon Knight, *The Futurians* (New York: John Day/T. Y. Crowell, 1977), 155.

48. Margot Metroland, "Tiffany Thayer and the Fortean Fascists, Part 1," *Counter-Currents Publishing*, 31 March 2015, https://www.counter-currents .com/2015/03/ tiffany-thayer-and-the-fortean-fascists-part-1/.

49. Sonia Orwell and Ian Angus, eds., *The Collected Essays, Journalism, and Letters of George Orwell*, vol. 2 (Boston: Nonpareil Books, 2000), 180; Richard C. Thurlow, *Fascism in Britain: From Oswald Mosley's Blackshirts to the National Front, a History 1918–1998* (London: I. B. Tauris, 1998), 141, 193, 203.

50. Thayer to Russell, 21 July 1956, Russell papers.

51. "First Meeting of the Fortean Society, January 26, 1931," Damon Knight papers, Syracuse University; Thayer to Bloch, 28 January 1945, FS, Donald Beaty Bloch papers, New York Public Library.

52. "On the Fortean Arts," *Doubt* 46 (October 1954): 300; "Our Loss," *Fortean* 3 (January 1940): 4; Tiffany Thayer, *33 Sardonics I Can't Forget* (New York: Philosophical Library, 1946), 204.

53. Thayer to editor, *Saturday Review of Literature*, 25 February 1946, Pound papers.

54. "On the Fortean Arts," 300.

55. "Eugenics Is Creed, Says Cult 'Omar,'" *Oakland Tribune*, 12 March 1927, 1, 2.

56. Joshua Blu Buhs, "Lilith Lorraine as a Fortean," 10 February 2015, http://www.joshuablubuhs.com/blog/lilith-lorraine-as-a-fortean.

57. Atkins to Russell, September 1947, Russell papers.

58. Borrow to Russell, 22 October 1948, Russell papers.

59. "Different," *Doubt* 14 (Spring 1946): 208.

60. Lilith Lorraine, *Let the Patterns Break* (Rogers, AR: Avalon Press, 1947), cover blurb.

61. Lorraine, *Let the Patterns Break*, 196.

62. C. Lamarr Wright, *The Story of Avalon* (Alpine, TX: Different Press, n.d.).

63. *Avalon News* 2, no. 4, Bruce Pelz papers, University of California, Riverside.

64. Lorraine, *Let the Patterns Break*, 50.

65. *Avalon News* 2, no. 4, Pelz papers.

66. Lorraine, *Let the Patterns Break*, 190.

67. *Challenge* 1, no. 1 (Summer 1950), Pelz papers.

68. Lorraine, *Let the Patterns Break*, 190.

69. DeCasseres, "Fortean Fantasy," 74.

70. Henry Miller, *The World of Sex* (New York: Ben Abramson, 1941), 53–54.

71. James Maynard, ed., *Robert Duncan: Collected Essays and Other Prose* (Berkeley: University of California Press, 2014), 3.

72. Henry Miller, *Remember to Remember* (New York: New Directions, 1961), 287.

73. *Letters of Henry Miller and Wallace Fowlie, 1943–1972* (New York: Grove, 1975), 67; Thomas Nesbit, *Henry Miller and Religion* (New York: Routledge, 2007), 63.

74. Henry Miller, *The Books in My Life* (New York: New Directions, 1969), 190.

75. Anaïs Nin, *The Diary of Anaïs Nin*, vol. 3, *1939–1944* (New York: Houghton-Mifflin, 1971), 95.

76. Robert Dana, ed., *Against the Grain: Interviews with Maverick American Publishers* (Iowa City: University of Iowa Press, 2009), 195; Deborah Benson Covington, *The Argus Book Shop, a Memoir* (West Cornwall, CT: Tarrydiddle Press, 1977), 35.

77. Bern Porter, ed., *The Happy Rock: A Book about Henry Miller* (Berkeley, CA: Bern Porter, 1945).

78. Henry Miller, *Big Sur and the Oranges of Hieronymus Bosch* (New York: New Directions, 1957), 167.

79. Joshua Blu Buhs, "'One Measure a Circle, Beginning Anywhere': Henry Miller and the Fortean Fantasy," *Nexus* 11 (2016): 145–68.

80. Henry Miller, *Nights of Love and Laughter* (New York: New American Library, 1955), 10.

81. M. C. Bradbrook, *Malcolm Lowry: His Art and Early Life: A Study in Transformation* (London: Cambridge University Press, 1975), 118.

82. Sherrill E. Grace, ed., *Sursum Corda! The Collected Letters of Malcolm Lowry, 1926–1946*, vol. 1 (London: Jonathan Cape, 1995), 315.

83. Sherrill E. Grace, *The Voyage That Never Ends: Malcolm Lowry's Fiction* (Vancouver: University of British Columbia Press, 1982).

84. Ruth Clydesdale, "Wild Talent: Malcolm Lowry, Fort and Magic," *Fortean Times*, June 2015, 40–43, 43; Sherrill E. Grace, "The Creative Process: An Introduction to Time and Space in Malcolm Lowry's Fiction," *Studies in Canadian Literature* 2 (1977): 61–68; Malcolm Lowry, *In Ballast to the White Sea* (Ottawa, ON: University of Ottawa Press, 2014), e-edition.

85. Clydesdale, "Wild Talent," 42.

86. Sherrill E. Grace, ed., *Swinging the Maelstrom: New Perspectives on Malcolm Lowry* (Montreal: McGill–Queen's University Press, 1992), 191.

87. Malcolm Lowry, *October Ferry to Gabriola: A Novel* (New York: Open Road Media, 2012), e-edition.

88. Grace, *Swinging the Maelstrom*, 205.

89. Steven Moore, *William Gaddis* (New York: Bloomsbury, 2015), 12.

90. Moore, *William Gaddis*, 12.

91. Matthew Wilkens, "Nothing as He Thought It Would Be: William Gaddis and American Postwar Fiction," *Contemporary Literature* 51 (2010): 596–628.

92. Ezra Pound, *Make It New* (New York: Faber & Faber, 1934).

93. Peter William Koenig, "Recognizing Gaddis' *Recognitions*," *Contemporary Literature* 16 (1975): 61–72.

94. William Gaddis, *The Recognitions* (New York: NYRB Classics, 2020), 81, 87.

95. Gaddis, *Recognitions*, 92; Teddy Hamstra, "Painting the Post-Secular: The Sacred as the After in William Gaddis' *The Recognitions*," MA thesis, University of Colorado, Boulder, 2019.

96. Rodger Cunningham, "When You See Yourself: Gnostic Motifs and Their Transformation in *The Recognitions*," *Soundings* 71 (1988): 619–37.

97. Margaret Cohen, *Profane Illumination: Walter Benjamin and the Paris of Surrealist Revolution* (Berkeley: University of California Press, 1995).

98. Ruth Brandon, *Surreal Lives: The Surrealists, 1917–1945* (New York: Grove, 2000).

99. T. J. Demos, "Duchamp's Labyrinth: 'First Papers of Surrealism,' 1942," *October* 97 (2001): 91–119.

100. Robert Allerton Parker, "Explorers of the Pluriverse," *First Papers of Surrealism*, 14 October–7 November 1942 (New York: Coordinating Council of French Relief Societies, 1942), n.p.

101. Parker, "Explorers of the Pluriverse," n.p.

102. Gavin Parkinson, "We Are Property: The 'Great Invisibles' Considered alongside 'Weird' and Science Fiction in America, 1919–1943," *Space Between: Literature and Culture 1914–1945* 14 (2018), https://scalar.usc.edu/works/the-space-between-literature-and-culture-1914-1945/vol14_2018_parkinson.

103. Robert Allerton Parker, "Such Pulp as Dreams are Made On," *VVV* 2–3 (1943): 62–66.

104. Philip Lamantia, "Radio Voices: A Child's Bed of Sirens," in *Popular Culture in America*, ed. Paul Buhle (Minneapolis: University of Minnesota Press, 1987), 139–49.

105. Clarke Ashton Smith plaque dedication 1985, carton 17, folder 38, Philip Lamantia papers, University of California, Berkeley.

106. Neeli Cherkovski, *Whitman's Wild Children: Portraits of Twelve Poets* (Venice, CA: Lapis Press, 1989), 125, 132.

107. Parker, "Explorers of the Pluriverse," n.p.

108. Garret Caples, Andrew Joron, Nancy Joyce Peters, eds., *The Collected*

Poems of Philip Lamantia (Berkeley: University of California Press, 2013), xxx; Nesbit, *Miller and Religion*, 63.

109. David Stephen Calonne, *The Spiritual Imagination of the Beats* (Cambridge: Cambridge University Press, 2017), 23.

110. Philip Lamantia, 5 January 1949, carton 5, folder 23, Lamantia papers.

111. Autobiographical vignette, n.d. (ca. 1961), folder 14, carton 5, Lamantia papers.

112. Garret Caples, ed., *Tau; with Journey to the End by John Hoffman* (San Francisco: City Lights, 2008), 21, 30.

113. Margaret Parton, *Laughter on the Hill: A San Francisco Interlude* (New York: McGraw Hill, 1945), 170–71; "Science Fiction Meeting to Draw Writers, Fans," *Oakland Tribune*, 27 June 1956, 38.

114. Susan Landauer, "Painting under the Shadow: California Modernism and the Second World War," in *On the Edge of America: California Modernist Art, 1910–1950*, ed. Paul J. Karlstrom (Berkeley: University of California Press, 1996), 41–68, 56.

115. Linda Hamalian, *A Life of Kenneth Rexroth* (New York: Norton, 1992).

116. Sam Hamill and Bradford Morrow, eds., *The Complete Poems of Kenneth Rexroth* (Port Townsend, WA: Copper Canyon Press, 2003), 212.

117. Michael Davidson, *The San Francisco Renaissance: Poetics and Community at Mid-Century* (Cambridge: Cambridge University Press, 1989); Lewis Ellingham and Kevin Killian, *Poet Be Like God: Jack Spicer and the San Francisco Renaissance* (Middletown, CT: Wesleyan University Press, 1998).

118. Lisa Jarnot, *Robert Duncan, the Ambassador from Venus: A Biography* (Berkeley: University of California Press, 2012), 106.

119. Joshua Blu Buhs, "George Leite as a Fortean," 23 July 2015, http://www.joshuablubuhs.com/blog/george-leite-as-a-fortean; Warren L. d'Azevedo, "Rebel Destinies: Remembering Herskovits," PAS Working Papers no. 15, Northwestern University, 2009.

120. Christopher Turner, *Adventures in the Orgasmatron: How the Sexual Revolution Came to America* (New York: Farrar, Straus and Giroux, 2011), 276.

121. Judson Crews, *Henry Miller and My Big Sur Days: Vignettes from Memory* (El Paso, TX: Vergin Press, 1992), 19; Judson Crews, *The Brave Wild Coast: A Year with Henry Miller* (Los Angeles: Dumont Press, 1997), 90.

122. Arthur Hoyle, *The Unknown Henry Miller: A Seeker in Big Sur* (New York: Arcade Publishing, 2014), 78.

123. *Letters of Henry Miller and Wallace Fowlie*, 71.

124. Andrew Cornell, *Unruly Equality: U.S. Anarchism in the Twentieth Century* (Berkeley: University of California Press, 2016), 188.

125. Philip Lamantia, 3 December 1945, 5 January 1949, carton 5, folder 23, Lamantia papers.

126. Richard Cándida Smith, *Utopia and Dissent: Art, Poetry, and Politics in California* (Berkeley: University of California Press, 1996), 64.

127. Mildred Edie Brady, "The New Cult of Sex and Anarchy," *Harper's* (April 1947): 312–32; Clint Mosher, "Emma Goldman Inspired Carmel Hate Cult Chief," *San Francisco Examiner*, 5 May 1947, 5.

128. Lee Bartlett, ed., *Kenneth Rexroth and James Laughlin: Selected Letters* (New York: Norton, 1991), 53.

129. Lamantia to Leite, n.d. (ca. 1944), carton 1, folder 23, Lamantia papers.

130. Ian S. MacNiven, ed., *Durrell-Miller Letters, 1935–1980* (New York: New Directions, 1988), 217.

131. Lee H. Watkins, "Notes on George Leite, Editor of Circle: Written in 1945," Special Collections, University of California, Davis; Lee Bartlett, *William Everson: The Life of Brother Antoninus* (New York: New Directions, 1988), 97.

132. *Circle* 10 (1948): 1; Harriet S. Blake, "The Leaves Fall in the Bay Area: Regarding Bern Porter and Four Little Magazines," *Colby Quarterly* 9 (1970): 85–104, 90.

133. *Publisher's Weekly*, 15 October 1949, 1756.

134. Stuart D. Hobbs, *The End of the American Avant-Garde* (New York: New York University Press, 1997), 80.

135. Anaïs Nin, *The Diary of Anaïs Nin*, vol. 4, *1944–1947* (New York: Harcourt Brace Jovanovich, 1972), 214–15.

136. Caples et al., *Poems of Philip Lamantia*, xxxiv.

137. d'Azevedo, "Rebel Destinies."

138. Caples et al., *Poems of Philip Lamantia*, xxxvii.

139. Philip Lamantia, 30 August 1961, carton 5, folder 14, Autobiographical prose, poems 1960, undated, Lamantia papers.

140. Hoyle, *Unknown Henry Miller*, 220.

141. Moore, *William Gaddis*, 3.

142. Katharine A. Rodger, ed., *Renaissance Man of Cannery Row: The Life and Letters of Edward F. Ricketts* (Tuscaloosa: University of Alabama Press, 2002), 183.

143. Kenneth Rexroth, *An Autobiographical Novel* (New York: Doubleday, 1966), 10–12.

144. Stephen Schwartz, *From West to East: California and the Making of the American Mind* (New York: Free Press, 1998), 239.

145. Rexroth, *Autobiographical Novel*, 110.

146. "Henry Miller Joins," *Doubt* 13 (Winter 1945): 195; "The Chapters," *Doubt* 25 (Summer 1949): 382–83.

147. Daliel Leite to author, 2 September 2016.

148. Thayer to Bloch, 14 August 1946, FS, Bloch papers.

149. *Circle* 5 (1945): n.p.

150. *Circle* 4 (1944): n.p.

151. *Circle* 1 (1944): n.p.

152. *Circle* 4 (1944): n.p.; Gioia Woods, "'An International, Dissident, Insurgent Ferment': Lawrence Ferlinghetti and the Left Coast," in *Left in the West: Literature, Culture, and Progressive Politics in the American West*, ed. Gioia Woods (Reno: University of Nevada Press, 2018), 90–108, 101–2.

153. "The Chapters," *Doubt* 27 (October 1949): 413–14.

154. "No Such Orgone," *Doubt* 53 (February 1957): 415.

155. *Fortean* 9 (Spring 1944): 4–5; "Now Is the Time," *Doubt* 28 (April 1950): 9–10; "Conchy Book Ready," *Doubt* 29 (July 1950): 20.

156. Gavin Parkinson, "Surrealism and Everyday Magic in the 1950s: Between the Paranormal and 'Fantastic Realism,'" *Papers of Surrealism* 11 (2015): 1–22, https://research.manchester.ac.uk/en/projects/the-ahrb-centre-for-studies-of-surrealism-and-its-legacies.

157. "Saucerzines," *Doubt* 47 (January 1955): 324.

158. Bradford Lyau, *The Anticipation Novelists of 1950s French Science Fiction: Stepchildren of Voltaire* (Jefferson, NC: McFarland, 2014).

159. "Fort in French," *Doubt* 52 (May 1956): 400.

160. Parkinson, "Surrealism and Everyday Magic."

161. Jean-Paul Sartre, *What Is Literature?* trans. Bernard Frechtman (New York: Philosophical Library, 1950), 132–39.

162. Gavin Parkinson, *Futures of Surrealism: Myth, Science Fiction, and Fantastic Art in France, 1936–1969* (New Haven, CT: Yale University Press, 2015), 139.

163. Rexroth, *Autobiographical Novel*, 97.

164. Parkinson, *Futures of Surrealism*, 127.

165. Louis Pauwels and Jacques Bergier, *The Morning of the Magicians: Secret Societies, Conspiracies, and Vanished Civilizations*, trans. Rollo Myers (Rochester, VT: Destiny Books, 2008), 119.

166. Pauwels and Bergier, *Morning of the Magicians*, 117.

167. Charles Fort, *The Book of the Damned* (New York: Boni & Liveright, 1919), 119, 131.

168. Pauwels and Bergier, *Morning of the Magicians*, 28.

169. Pauwels and Bergier, *Morning of the Magicians*, xxvi.

170. Parkinson, *Futures of Surrealism*, 139.

171. Gérard Duzoi, *History of the Surrealist Movement*, trans. Alison Anderson (Chicago: University of Chicago Press, 2002), 625, 762.

172. Parkinson, "Surrealism and Everyday Magic."

173. John Warne Monroe, *Laboratories of Faith: Mesmerism, Spiritism, and Occultism in Modern France* (Ithaca, NY: Cornell University Press, 2008), 258–59.

174. Mircea Eliade, *Occultism, Witchcraft, and Cultural Fashions: Essays in Comparative Religions* (Chicago: University of Chicago Press, 1976), 8, 10.

175. Anastasia Aukeman, *Welcome to Painterland: Bruce Conner and the Rat Bastard Protective 57 Association* (Berkeley: University of California Press, 2016), 18.

176. Carolyn Burke, *Becoming Modern: The Life of Mina Loy* (New York: McMillan, 1996), 399; Herman Mhire, *A Century of Vision: Louisiana Photography* (New Orleans: Louisiana State Museum, 1986), 36.

177. "First Prize," *Doubt* 57 (July 1958): 2–4. Likely the artist was Sylvi Edith Moray, who also went by S. E. Mackey and S. E. Laurila. She lived in the San Francisco Bay Area and was associated with the CO Joseph Moray.

178. Richard Cándida Smith, "The Elusive Quest of the Moderns," in *On the Edge of America: California Modernist Art, 1900–1950*, ed. Paul J. Karlstrom (Berkeley: University of California Press, 1996), 21–40, 34.

179. Wees, "Pound as a Vorticist," 65.

180. Schneidau, "Vorticism," 216; Wees, "Pound as a Vorticist," 58.

181. Schneidau, "Vorticism," 216.

182. William C. Lipke and Bernard W. Rozran, "Ezra Pound and Vorticism: A Polite Blast," *Wisconsin Studies in Contemporary Literature* 7 (Summer 1966): 201–10, 201.

183. Thayer to editor, *Saturday Review of Literature*, 25 February 1946, Pound papers.

184. Roxana Preda, "Social Credit in America: A View from Ezra Pound's Economic Correspondence, 1933–1940," *Paideuma* 34 (2005): 201–27, 222.

185. Preda, "Social Credit," 218.

186. John Tytell, *Ezra Pound: The Solitary Volcano* (New York: Anchor Press, 1987), 266.

187. Richard Sieburth, "Introduction," in Ezra Pound, *The Pisan Cantos* (New York: New Directions, 2003), ix–xliii.

188. A. David Moody, *Ezra Pound: Poet*, vol. 3, *The Tragic Years 1939–1972* (New York: Oxford University Press, 2015), 117–370.

189. Alec Marsh, *John Kasper and Ezra Pound: Saving the Republic* (London: Bloomsbury Publishing, 2015).

190. "Pound of Flesh," *Doubt* 12 (Spring 1945): 170.

191. Thayer to Julien Cornell, 2 October 1946, Thayer file, Julien Cornell papers, Swarthmore University; Thayer to Dorothy Pound, 10 August 1946, Pound papers; Thayer to Russell, 21 January 1947, FS, Russell papers; Thayer to Bloch, 28 August 1947, FS, Bloch papers.

192. "Pound of Flesh," 170.

193. Thayer to Pound, 23 March 1946; 21 April 1947, FS, Pound papers; Thayer to Russell, 21 January 1947, FS, Russell papers; Thayer, *33 Sardonics.*

194. Thayer to Pound, 12 January 1946; 9 February 1946; 18 February 1946, Pound papers.

195. Thayer to Pound, 21 February 1946, FS; Thayer to Dorothy Pound, 2 October 1946, Pound papers. Mary de Rachewiltz, *Ezra Pound, Father and Teacher: Discretions* (New York: New Directions, 2005), 289; "Foreign Missions," *Doubt* 38 (October 1952): 164–68.

196. Thayer to Pound, n.d., Pound papers.

197. Newsletters Received by Ezra Pound at St. Elizabeth's Hospital," Hamilton College; Ellen Cardona, "Pound's Anti-Semitism at St. Elizabeth's: 1945–1958," *Flashpøint*, Spring 2007, https://www.flashpointmag.com/card .htm.

198. Thayer to Pound, 4 November 1946, Pound papers.

199. Thayer to Pound, 14 February 1948, FS, Pound papers.

200. Elmer Gertz, *The Odyssey of a Barbarian: The Biography of George Sylvester Viereck* (Buffalo, NY: Prometheus Books, 1978).

201. Thayer to Pound, 14 February 1948, FS; 7 February 1950, FS, Pound papers.

202. Donald Hall, "Ezra Pound, The Art of Poetry No. 5," *Paris Review* 28 (Summer-Fall 1962): 22–51, 43.

203. Thayer to Pound, 14 January 1948, FS, Pound papers.

204. Thayer to Pound, 5 August 1946, FS, Pound papers.

205. Thayer to Pound, 14 June 1948, FS, Pound papers.

206. Hall, "Ezra Pound," 42.

207. Ezra Pound, *The Pisan Cantos* (New York: New Directions, 1948), 92.

208. "Ezra Doubts," *Doubt* 25 (May 1949): 388.

209. Pound, *Pisan Cantos*, 24.

210. Anthony Woodward, *Ezra Pound and "The Pisan Cantos"* (London: Routledge, 1980), 65.

211. Tytell, *Solitary Volcano*, 302.

212. Frances Stonor Saunders, *The Cultural Cold War: The CIA and the World of Arts and Letters* (New York: New Press, 2013); Michael L. Krenn, *Fall-out Shelters for the Human Spirit: American Art and the Cold War* (Raleigh: University of North Carolina Press, 2005).

213. John Atkins, "The Academic Take-Over," *Mediterranean Review*, Fall 1971, 31–35.

214. Tytell, *Solitary Volcano*, 303.

215. Alan Filreis, *Counter-Revolution of the Word: The Conservative Attack on Modern Poetry, 1945–1960* (Chapel Hill: University of North Carolina Books, 2012).

216. *Avalon News* 2, no. 4, Pelz papers.

217. James Blish, "Rituals on Ezra Pound," *Sewanee Review* 58 (1950): 185–226.

218. *Tumbrils* 1, March 1945, Pelz papers.

219. "Never Kick a Lady," *Doubt* 27 (Winter 1949): 414–15.

220. Thayer to Russell, 5 November 1949, FS, Russell papers.

221. Eric Frank Russell, "In Defense of Pound," *Doubt* 27 (Winter 1949): 415.

222. Philip Lamantia, 30 August 1961, carton 5, folder 14, Autobiographical prose, poems 1960, undated, Lamantia papers.

223. Filreis, *Counter-Revolution*, 319.

224. Moody, *Ezra Pound: Poet*, 3:469.

225. Don Foster, *Author Unknown: On the Trail of Anonymous* (New York: Henry Holt, 2000), 203–20.

226. "The New Scholiasts," *Doubt* 61 (Summer 1959): 75–76.

227. Anne Conover, *Caresse Crosby: From Black Sun to Rocca Sinibalda* (Lincoln, NE: iUniverse, 2001), 139.

228. Linda Hamalian, *The Cramoisy Queen: A Life of Caresse Crosby* (Carbondale: Southern Illinois University Press, 2005), 168.

229. Charles Olson, *The Maximus Poems* (Berkeley: University of California Press, 1983), 32.

230. Ekbert Faas, *Robert Creeley: A Biography* (Kingston, ON: McGill–Queen's University Press, 2001), 84–85.

231. George F. Butterick, "Charles Olson and the Postmodern Advance," *Iowa Review* 11 (1980): 4–27; W. T. Lhamon Jr., *Deliberate Speed: The Origins of a Cultural Style in the American 1950s* (Cambridge, MA: Harvard University Press, 2002).

232. Michael Szalay, *New Deal Modernism: American Literature and the Invention of the Welfare State* (Durham, NC: Duke University Press, 2000), 259.

233. Paul Christensen, *Charles Olson: Call Him Ishmael* (Austin: University of Texas Press, 2014), 46.

234. Helen Molesworth, *Leap Before You Look: Black Mountain College, 1933–1957* (New Haven, CT: Yale University Press, 2015).

235. Thayer to Crosby, 7 March 1950, FS, Caresse Crosby papers, Southern Illinois University; "Run of the Mill," *Doubt* 32 (March 1951): 71–72.

236. Thayer to Russell, 7 March 1953, FS, Russell papers; Thayer to Crosby, 22 May 1954, FS, Crosby papers.

237. Crosby to Thayer, 16 September 1953, FS; 3 May 1956, FS, Crosby papers.

238. Thayer to Crosby, 13 April 1957, FS, Crosby papers.

239. "Our Cover," *Doubt* 41 (July 1953): 210–11.

240. Crosby to Thayer, 16 January 1953, Crosby papers.

241. Thayer to Crosby, 6 November 1954, FS, Crosby papers.

242. Dana, *Against the Grain*, 195; Damon Knight, *Charles Fort: Prophet of the Unexplained* (New York: Doubleday, 1970), xiii–xvi; Terry Wilson, *Perilous Passage: The Nervous System and the Universe in Other Words* (San Francisco: Synergetic Press, 2012).

243. "Pre-Colon Lingo," *Doubt* 18 (July 1947): 271–73.

244. Ralph Maud, *Charles Olson's Reading: A Biography* (Carbondale: Southern Illinois University Press, 1996), 147, 366; George F. Butterick, *A Guide to the Maximus Poems of Charles Olson* (Berkeley: University of California Press, 1981).

245. Donald Allen and Benjamin Friedlander, eds., *Collected Prose: Charles Olson* (Berkeley: University of California Press, 1997), 240.

246. David Harvey, *The Condition of Postmodernity: An Enquiry into the Origins of Cultural Change* (Cambridge: Blackwell, 1989), 39–65.

247. Michel Foucault, *The Order of Things: An Archaeology of the Human Sciences* (New York: Vintage, 1994).

248. Butterick, "Olson and the Postmodern Advance," 22.

249. Harvey, *Condition of Postmodernity*, 52.

250. Thayer to Crosby, 23 January 1954, FS, 22 May 1954, Crosby papers.

251. Thayer to Crosby, 28 January 1955, Crosby papers.

252. Crosby to Thayer, 24 February 1957, Crosby papers.

253. Conover, *Caresse Crosby*, 198.

Chapter 5

1. Waveney Girvan to Russell, 8 November 1950, Eric Frank Russell papers, University of Liverpool.

2. Ian Waveney Girvan, KV 2/1235, National Archives, Kew.

3. George Greenfield, *A Smattering of Monsters: A Kind of Memoir* (Columbia, SC: Camden House, 1995), 103–4.

4. Girvan to Russell, 15 November 1950, Russell papers.

5. John A. Keel, "The Flying Saucer Subculture," *Journal of Popular Culture* 8 (1975): 871–96.

6. Charles Fitzhugh Talman, *The Realm of the Air* (Indianapolis, IN: Bobbs-Merrill, 1931).

7. Will A. Page, "The Air Serpent," in *Science Fiction by Gaslight: A History and Anthology of Science Fiction in the Popular Magazines, 1891–1911*, ed. Sam Moskowitz (Cleveland, OH: World Publishing, 1968), 167–77; Bertram Atkey, "The Strange Case of Alan Moraine," in *Worlds Apart: An Anthology of Interplanetary Fiction*, ed. George W. Locke (London: Cornmarket Press, 1972), 73–86; L. Frank Baum, *Sky Island* (New York: Dover, 2012).

8. Arthur Conan Doyle, "The Horror of the Heights," *Everybody's Magazine*, November 1913, 578–90.

9. Daniel Cohen, *The Great Airship Mystery: A UFO of the 1890s* (New York: Dodd, Mead, 1981).

10. Nigel Watson, ed., *The Scareship Mystery: A Survey of Phantom Airship Scares, 1909–1918* (Corby, Northamptonshire: Domra Publications, 2000).

11. Charles Fort, *New Lands* (New York: Boni & Liveright, 1923), 177.

12. Charles Fitzhugh Talman, *Meteorology: The Science of the Atmosphere* (New York: P. F. Collier & Son, 1922), 3–4.

13. Talman, *Meteorology*, 356.

14. Charles Fitzhugh Talman, "People of Mars," *Monthly Weather Review*, December 1900, 537; C. F. Talman, "Why the Weather," *Ithaca Journal*, 26 June 1931, 15; Charles B. Driscoll, "The World and All," *Lansing State Journal*, 6 July 1932, 6; Thayer to Bloch, 3 January 1942, Donald Beaty Bloch papers, New York Public Library.

15. "Hands across the Void," *Daily Eagle* (Wichita, KS), 28 January 1920, 4; "Non-Luminous Planets Menace," *Enquirer* (Battle Creek, MI), 13 February 1920, 9.

16. Advertisement, *New York Herald*, 2 February 1920, 78; Edward L. Bernays, *Biography of an Idea: Memoirs of Public Relations Counsel* (New York: Simon & Schuster, 1965), 278.

17. "Hello, Earth! Hello!" *Tomahawk* (White Earth, MN), 18 March 1920, 6.

18. Joshua Nall, *News from Mars: Mass Media and the Forging of a New Astronomy, 1860–1910* (Pittsburgh, PA: University of Pittsburgh Press, 2019), 182.

19. Charles Fort, *The Book of the Damned* (New York: Boni & Liveright, 1919), 22; Ron Westrum, "Science and Social Intelligence about Anomalies: The Case of Meteorites," *Social Studies of Science* 8 (1978): 461–93.

20. Walter H. Kerr, "The Song of the Damned," *Doubt* 18 (July 1947): 266.

21. Clark Ashton Smith, letter, *Amazing Stories*, October 1932, 670–71.

22. David Stephen Calonne, *The Spiritual Imagination of the Beats* (Cambridge: Cambridge University Press, 2017), 23.

23. C. G. Jung, *Flying Saucers: A Modern Myth of Things Seen in the Skies*, trans. R. F. C. Hull (Princeton, NJ: Princeton University Press, 1979).

24. C. G. Jung, *On the Nature of the Psyche*, trans. R. F. C. Hull (Princeton, NJ: Princeton University Press, 1979), 105.

25. Howard H. Peckham, "Flying Saucers as Folklore," *Hoosier Folklore* 9 (1950): 103–7; Rollo May, *Man's Search for Himself* (New York: Norton, 1953), 35; Erik Davis, *TechGnosis: Myth, Magic, and Mysticism in the Age of Information* (Berkeley, CA: North Atlantic Books, 2015), 241.

26. Joshua Hoeynck, "Without Mammalia Maxima, Charles Olson and Robert Duncan Apprehend a Cosmological American Poetics," in *The New American Poetry: Fifty Years Later*, ed. John R. Woznicki (Bethlehem, PA: Lehigh University, 2014), 29–58; J. P. Telotte, *Movies, Modernism, and the Science Fiction Pulps* (New York: Oxford University Press, 2019), 162–65.

27. Daniel R. White and Alvin Wang, "Through the Dark Mirror: UFOs as a Postmodern Myth?" *Ctheory*, 6 January 1999, https://journals.uvic.ca/index.php/ctheory/article/view/14622/5488.

28. J. Gordon Melton, "The Contactees: A Survey," in *The Gods Have Landed: New Religions from Other Worlds*, ed. James R. Lewis (Albany: State University of New York Press, 1995), 1–13.

29. Thomas E. Bullard, *The Myth and Mystery of UFOs* (Lawrence: University of Kansas Press, 2009).

30. Norman Markham, "Too Late for Dreams," *Doubt* 35 (January 1952): 122.

31. Kerr, "Song of the Damned," 266.

32. Loren E. Gross, *Charles Fort, the Fortean Society, and Unidentified Flying Objects* (Fremont, CA: n.p., 1976).

33. "Rocket Fanatic Advises Staking Claims on Moon," *Daily Republican* (Belvidere, IL), 22 January 1945, 2; "Or . . . 'Did You Ever Swing on a Star?'" *Maple Leaf* (Toronto), 3 February 1945, 7.

34. Claire Cox, "Seeks Government Permit to Fly Atomic Rocket to Moon," *Dispatch* (Moline, IL), 13 August 1945, 3.

35. "'Moon Rocket Trips within Next 3 Years,'" *Star* (Lincoln, NE), 27 January 1946, 4a.

36. Joshua Blu Buhs, "Robert L. Farnsworth as a Fortean," 3 December 2014, http://www.joshuablubuhs.com/blog/robert-l-farnsworth-as-a-fortean.

37. "The Men in the Moon," *Herald and Review* (Decatur, IL), 16 December 1944, 4.

38. R. L. Farnsworth, *Rockets, New Trail to Empire: Review and Bibliography* (Glen Ellyn, IL: n.p., 1945).

39. R. L. Farnsworth, "First Target in Space," *Startling Stories*, September 1948, 98–102; R. L. Farnsworth, "Rocket Target No. 2," *Startling Stories*, May 1949, 93–95; "Rocket to the Moon? Soon!" *Popular Science*, May 1945, 2.

40. Marion Zimmer Bradley, "The Last Frontier," *Startling Stories*, April 1952, 142–44.

41. Oliver Dunnett, "The British Interplanetary Society and Cultures of Outer Space, 1930–1970," PhD diss., University of Nottingham, 2011, 115–19.

42. P. E. Cleator, ed., *Letters from Baltimore: The Mencken-Cleator Correspondence* (Madison, NJ: Fairleigh Dickinson University Press, 1982), 205;

Arthur C. Clarke, *Greetings, Carbon-Based Bipeds! Collected Essays, 1934–1998* (New York: Macmillan, 2001), 26–29; Cleator, *Letters*, 210.

43. *Ember*, 4 August 1946, 24 August 1946, 3 November 1946, Bruce Pelz papers, University of California, Riverside.

44. "Where Do the Books Go?" *Saturday Review of Literature*, 22 December 1945, 2.

45. "The Rocketeers," *Doubt* 14 (Spring 1946): 204.

46. "Terror Continues," *Doubt* 29 (January 1953): 184.

47. "The Chapters," *Fortean* 24 (April 1949): 363–64.

48. Jerome Bixby, "Review of the Current Science Fiction Fan Publications," *Startling Stories*, March 1952, 141–45.

49. "Interlocking (Intellectual) Directorates," *Doubt* 17 (March 1947): 256.

50. Harry Warner Jr., *All Our Yesterdays: An Informal History of Science Fiction Fandom in the Forties* (Chicago: Advent, 1969).

51. "Moon May Have Been Victim of Atomic War," *Daily Record* (Statesville, NC), 15 July 1947, 2.

52. Claire Cox, "Says 'Saucers' Nothing to Get Excited About," *Pharos-Tribune* (Logansport, IN), 8 July 1947, 5.

53. Ted Bloecher, "Report on the UFO Wave of 1947" (Arizona: n.p., 1967).

54. Mike Dash, *Borderlands: The Ultimate Exploration of the Unknown* (Woodstock, NY: Overlook Press, 2000), 135.

55. Hal Boyle, "It's Simple: Disk Jockeys Men from Mars," *Daily Sun* (Corsicana, TX), 8 July 1947, 1.

56. "Game of Spotting 'Flying Saucers' Sweeps Country as Mystery Holds," *Albuquerque Journal*, 8 July 1947, 1–2; "Mystery Hides 'Finding of Disc,'" *Albuquerque Journal*, 8 July 1947, 1.

57. "New Mexico 'Disc' Declared Weather Balloon and Kite," *Los Angeles Examiner*, 9 July 1947, 1; Kathryn S. Olmsted, *Real Enemies: Conspiracy Theories and American Democracy, World War I to 9/11* (New York: Oxford University Press, 2011), 183–84.

58. Thayer to Russell, 7 March 1959, FS, Russell papers.

59. "Rare Book Tells of Freak Discs in Sky Long Ago," *Albuquerque Journal*, 8 July 1947, 1.

60. Cox, "Says 'Saucers' Nothing to Get Excited About," 5.

61. Marcia Winn, "Pie in the Sky?" *Chicago Daily Tribune*, 11 July 1947, 15.

62. "No Menace from Flying Saucers, Air Force Holds," *Sun-Democrat* (Paducah, KY), 30 July 1952, 1, 15.

63. *Doubt* 19 (October 1947): 287.

64. Henry H. Smith, "1947's Flying Saucers Have Infamous Ancestry," *Pantagraph* (Bloomington, IL), 12 July 1947, 2; "Flying Saucer Is Nothing

New: Just the Successor to the Flying Cigar of the Gay Nineties," *Daily News* (Neosha, MO), 14 July 1947, 2 (quotation); "Jay Franklin," *News-Tribune* (Waco, TX), 16 July 1947, 6.

65. Thayer to Russell, 21 December 1947, FS, Russell papers.

66. George Peacock to Russell, 9 January 1949, Russell papers.

67. Thayer to Russell, 7 June 1950, FS, Russell papers.

68. Curtis Peebles, *Watch the Skies! A Chronicle of the Flying Saucer Myth* (Washington, DC: Smithsonian Institution, 1994).

69. Girvan to Thayer, 7 February 1955, Russell papers; Charles Ports, "Editor Produces Good Magazine Concerned with Flying Saucers," *Daily Times-News* (Burlington, NC), 29 October 1964, 15.

70. Martin Gardner, *Fads and Fallacies in the Name of Science* (New York: Dover, 1957), 62.

71. *Doubt* 39 (January 1953): 180.

72. Fort, *Book of the Damned*, 1; quoted in Bloecher, "UFO Wave of 1947."

73. "Lots of Folk in Eastbay See Saucers," *Oakland Tribune*, 8 July 1947, 1.

74. David F. Bascom, "Moonshine," *Oakland Tribune*, 22 May 1948, 16.

75. Fredric Brown, "Pi in the Sky," *Thrilling Wonder Stories*, Winter 1945, 46–59, 66.

76. "When the World Is So Full of a Number of Things, Along Came Sailing Saucers," *Oakland Tribune*, 8 July 1947, 26.

77. Kathleen Ludwick, "Sauce in the Sky," *Oakland Tribune* 13 July 1947, 64; Kathleen Ludwick, "Core Restored," *Oakland Tribune*, 23 July 1947, 22.

78. Charles Fort, *Lo!* (New York: Claude Kendall, 1931), 190.

79. Jack Hunter, ed., *Damned Facts: Fortean Essays on Religion, Folklore and the Paranormal* (Paphos, Cyprus: Aporetic Press, 2016).

80. Ivan Sanderson, "An Introduction to UFOlogy," *Fantastic Universe*, February 1957, 27–34; Ivan Sanderson, "What Could They Be?" *Fantastic Universe*, July 1959, 66–76.

81. Ivan T. Sanderson, *Invisible Residents: The Reality of Underwater UFOs* (New York: World Publishing, 1970).

82. Bessor to Russell, 23 March 1948, Russell papers.

83. John Philip Bessor, letter, *Life*, 28 April 1952, 10.

84. "Discs Coming in Colors Now," *Evening Telegraph* (Alton, PA), 8 July 1947, 1.

85. Joshua Blu Buhs, "Morris Ketchum Jessup as a Fortean," 26 September 2016, http://www.joshuablubuhs.com/blog/morris-ketchum-jessup-as-a-fortean.

86. Morris K. Jessup, *The Case for the UFO* (New York: Citadel Press, 1955).

87. Thomas Levenson, *The Hunt for Vulcan . . . and How Albert Einstein*

Destroyed a Planet, Discovered Relativity, and Deciphered the Universe (New York: Random House, 2015).

88. Loren E. Gross, *UFOs: A History*, vol. 1, *July 1947–December 1948* (Fremont, CA: n.p., 1988), 21; Michael D. Swords, "Project Sign and the Estimate of the Situation," *Journal of UFO Studies* 7 (2000): 27–64, 39–40.

89. Hehr to Russell, November 1955, Russell papers; Joshua Blu Buhs, "Frederick G. Hehr as a Fortean," 8 September 2014, http://www .joshublubuhs.com/blog/frederick-g-hehr-as-a-fortean; Harold T. Wilkins, *Flying Saucers on the Attack* (New York: Citadel Press, 1954), 284–86.

90. "Hehr on Gold Hill," *Doubt* 14 (Spring 1946): 202.

91. Hehr to Russell, November 1955; September 1956, Russell papers.

92. John Keel, "The Man Who Invented Flying Saucers," *Fortean Times*, no. 41 (Winter 1983): 52–57.

93. Joshua Blu Buhs, "R. Dewitt Miller and Ellora Fogle Miller as Forteans," 14 July 2014, http://www.joshublubuhs.com/blog/r-dewitt-miller-and -ellora-fogle-miller-as-forteans-conclusion; R. DeWitt Miller and Geoffrey Giles, "Fort without Theories," *Fantasy Review*, August–September 1947, 15; "The Southern California Riddle," *American Mercury*, May 1956, 99–102, 99–100; Francis M. Nevins, *The Anthony Boucher Chronicles* (n.p.: Ramble House, 2009), 300, 420–21.

94. R. Dewitt Miller, "'Disc-like Forms' Seen in Sky during the Last 150 years," *Ottawa Journal*, 8 July 1947, 12.

95. R. DeWitt Miller, *You DO Take It with You* (New York: Citadel Press, 1955).

96. "Vincent Gaddis, 1913–1997," *Strange Magazine*, Spring 1998, 29, 58.

97. Vincent H. Gaddis, "When Flesh Defies Fire," *Amazing Stories*, May 1946, 160–61; Vincent H. Gaddis, "From the Mouths of Babes," *Round Robin*, October 1946, n.p.; Vincent H. Gaddis, "With Brain Destroyed—They Live and Think!" *Fate*, Summer 1948, 76–83.

98. Vincent H. Gaddis, *Mysterious Fires and Lights* (New York: Dell, 1967), 78.

99. Joshua Blu Buhs, "Newton Meade Layne as Fortean," 2 March 2014, http://www.joshublubuhs.com/blog/newton-meade-layne-as-fortean.

100. *Round Robin* 1, no. 1 (February 1945): 1–2.

101. "Meade Layne and the Principles of BSRF," *Journal of Borderland Research* 31 (March–April 1975): https://borderlandsciences.org/journal/vol/ 31/n02/Meade_Layne_Principles_of_BSRF.html.

102. N. Meade Layne, *The Ether Ship Mystery and Its Solution* (San Diego, CA: BSRA, 1950), 4.

103. N. Meade Layne, *The Coming of the Guardians* (San Diego, CA: BSRA, 1957), 1.

104. "A Word to Our Spiritualist Friends," *Round Robin* 1, no. 3 (April 1945): 14.

105. N. Meade Layne, "Fly, Lokas, Fly," *Round Robin*, July–August 1947, 3, 8.

106. Layne, *Ether Ship Mystery*, 2.

107. Layne, *Ether Ship Mystery*, 4.

108. Layne, *Ether Ship Mystery*, 4.

109. Miller, *Reincarnation*, 112–15.

110. Rog Phillips, "What Man Can Imagine: Is There an Ether Drift?" *Amazing Stories*, June 1947, 152–55; Roger Graham, "Note Concerning Ether Drift," *Round Robin*, May–June 1947, 17.

111. Vincent Gaddis, "The Coming Spiritual Era," *Round Robin* 2, no. 6 (June 1946): 18–19; Vincent Gaddis, "The Shaver Mystery," *Round Robin* 3, no. 4 (May–June 1947): 6–9; Pitirim Sorokin, *Social and Cultural Dynamics* (Boston: Porter Sargent, 1957).

112. Meade Layne, "Mark Probert, Baffling San Diego Medium" *Fate*, May 1949, 16–21; Mark Probert, *The Magic Bag* (San Diego, CA: Inner Circle Kethra E'Da Foundation, 1963), xxiv.

113. "WELCOME? KAREETA!" *Round Robin* 2, no. 10 (October 1946): 3–7.

114. Layne, "Fly, Lokas, Fly," 3.

115. Mark O'Connell, *The Close Encounters Man: How One Man Made the World Believe in UFOs* (New York: HarperCollins, 2017).

116. Abdus Salam, "Are There Some New Fundamental Particles?" *New Scientist*, 5 May 1960, 1124–26.

117. J. Allen Hynek, "The Condon Report and UFOs," *Bulletin of Atomic Scientists*, April 1969, 39–42, 41.

118. Erik Davis, *High Weirdness: Drugs, Esoterica, and Visionary Experiences in the Seventies* (Cambridge, MA: MIT Press, 2019), 5.

119. Michael Barkun, *A Culture of Conspiracy: Apocalyptic Visions in Contemporary America* (Berkeley: University of California Press, 2013).

120. Colin Bennett, *Looking for Orthon: The Story of George Adamski, the First Flying Saucer Contactee, and How He Changed the World* (New York: Cosimo Books, 2008), 53–54.

121. Brenda Denzler, *The Lure of the Edge: Scientific Passions, Religious Beliefs, and the Pursuit of UFOs* (Berkeley: University of California Press, 2001), 36–48; Catherine L. Albanese, *A Republic of Mind and Spirit: A Cultural History of American Metaphysical Religion* (New Haven, CT: Yale University Press, 2006), 496–502.

122. Peebles, *Watch the Skies!*, 126.

123. Charles Bowen, "Thinking Aloud," *Flying Saucer Review*, November-December 1969, 26–28, 26 (emphasis in original).

124. Marx Kaye, "Fortean Aspects of the Flying Disks," *Amazing Stories*, June 1948, 154–57.

125. "New Mag," *Doubt* 21 (June 1948): 317.

126. Fred Nadis, *The Man from Mars: Ray Palmer's Amazing Pulp Journey* (New York: Penguin, 2013), 117.

127. Richard Toronto, *War over Lemuria: Richard Shaver, Ray Palmer and the Strangest Chapter of 1940s Science Fiction* (Jefferson, NC: McFarland, 2013), 8, 188.

128. Toronto, *War over Lemuria*, 190.

129. Dick McDaniel, "Amherst Publisher Deals with Reality of Uncertain Things," *Post-Crescent* (Appleton, WI), 26 November 1961, 17.

130. Buhs, "Jessup as a Fortean"; Ronald D. Story, *The Encyclopedia of UFOs* (Garden City, NY: Doubleday, 1980), 210.

131. Barkun, *Culture of Conspiracy*, 80–110; Dash, *Borderlands*, 300–301.

132. *Dickhoff v. Shaughnessy*, 142 F. Supp. 535 (S.D.N.Y. 1956); Joshua Blu Buhs, "Robert Ernst Dickhoff as a Fortean," 4 October 2016, http://www .joshuablubuhs.com/blog/robert-ernst-dickhoff-as-a-fortean.

133. Joscelyn Godwin, *Arktos: The Polar Myth in Science, Symbolism, and Nazi Survival* (Kempton, IL: Adventures Unlimited Press, 1996), 93–136.

134. Robert Ernst Dickhoff, *Homecoming of the Martians* (Ghaziabad, India: Bharti Association Publications, 1958), n.p.

135. Julian Parr to Russell, 1 August 1947, 19 October 1947, Russell papers.

136. "Astrologer Wearing Blanket Tries to Address the U.N.," *St. Louis Post Dispatch*, 9 December 1948, 2.

137. "Garry Davis Cult Spreads in Europe," *Life*, 24 January 1949, 28–29.

138. Davis to Russell, 14 July 1951, Russell papers.

139. "Garry Davis Dinner," *Doubt* 29 (July 1950): 19; Thayer to Russell, 7 June 1950, FS, Russell papers.

140. "Garry Davis Cult Spreads in Europe," 29.

141. "Connecticut Christ," *Doubt* 48 (April 1955): 335.

142. John O'Hara Cosgrave, *Man: A Citizen of the Universe* (New York: Farrar, Straus & Co., 1948).

143. "Can You Journey into Another World Yet Remain in This One?" *Chicago Tribune*, 25 September 1948, 11.

144. Bartłomiej Paszylk, *The Pleasure and Pain of Cult Horror Films: An Historical Survey* (Jefferson, NC: McFarland, 2009), 53–55.

145. Don A. Stuart, "Who Goes There?" *Astounding*, August 1938, 60–97.

146. "Garry Davis Cult Spreads in Europe," 28–29.

147. Art Buchwald, "He Just Keeps Rolling Along," *Daily News* (Newport, RI), 20 December 1957, 8.

148. "News Quotes," *Daily Notes* (Canonsburg, PA), 11 January 1957, 7.

149. Crosby to Thayer, 24 February 1957, Caresse Crosby papers, Southern Illinois University.

150. Thayer to Crosby, 21 March 1957, FS, Crosby papers.

151. Thayer to Russell, 20 January 1950, FS, Russell papers; "Our Cover," *Doubt* 41 (July 1953): 210–11; Thayer to Crosby, 21 February 1957, FS, Crosby papers.

152. Thayer to Crosby, 21 March 1957, FS, Crosby papers.

153. Thayer to Crosby, 21 March 1957, FS, Crosby papers.

154. "Saucers," *Doubt* 30 (October 1950): 35; "Credits," *Doubt* 38 (October 1952), 170.

155. David L. Miller, *Introduction to Collective Behavior and Collective Action* (Long Grove, IL: Waveland Press, 2013), 187.

156. Thayer to Russell, 14 April 1953, FS, Russell papers.

157. Thayer to Bloch, 1 January 1949, FS, Bloch papers.

158. Norman Markham to Damon Knight, 13 February 1968, Damon Knight papers, Syracuse University.

159. Bessor to Russell, 23 March 1948, Russell papers.

160. Vincent Gaddis to Knight, 19 February 1968, Knight papers.

161. "New Paper," *Doubt* 12 (1945): 178; *Round Robin* 1, no. 1 (February 1945): 8.

162. "Sizzling Zinner," *Doubt* 17 (1946): 251.

163. *Round Robin* 3, no. 4 (April 1947): 10.

164. "Goblin Market Again," *Round Robin* 7, no. 1 (May–June 1951): 21.

165. Thayer to Bloch, 28 November 1948, FS, Bloch papers.

166. Thayer to Bloch, 28 November 1948, FS, Bloch papers.

167. "What Would You Do If You Found a Good Thing?" *Mystic*, December 1954, 23; "What Would You Do If You Found a Good Thing?" *Flying Saucers from Other Worlds*, June 1957, 97.

168. Thayer to Russell, 7 March 1950, FS, Russell papers.

169. Thayer to Russell, 28 November 1953, FS, Russell papers.

170. Thayer to Russell, 28 Fort 1953, FS, Russell papers.

171. Thayer to Russell, 28 April 1955, FS, Russell papers.

172. Frank Scully to Crosby, 6 October 1954, Crosby papers.

173. Girvan to Russell, 20 November 1950, Russell papers.

174. Donald Whitacre, letter, *News-Journal* (Wilmington, OH), 10 November 1950, 3.

175. "Pertinent to the Above," *Doubt* 31 (January 1951): 52.

176. J. P. Cahn, "The Flying Saucers and the Mysterious Little Men," *True*, September 1952, 17–19, 102–12; J. P. Cahn, "Flying Saucer Swindlers," *True*, August 1956, 36–37, 69–72.

177. "Speaking of Books," *Doubt* 58 (October 1958): 22–23, 23.

178. "Sizzling Zinner," 251; Thayer to Russell, 28 August 1954, FS, Russell papers.

179. Thayer to Russell, 7 August 1950, FS, Russell papers.

180. Thayer to Russell, 7 May 1950, FS, Russell papers.

181. "Run of the Mill," *Doubt* 20 (March 1948): 303–4, 304.

182. *Doubt* 36 (April 1952): 158.

183. "Saucers Etcetera," *Doubt* 40 (April 1953): 194–206, 194.

184. De Camp to Russell, 3 August 1951, Russell papers.

185. "The United States of Dreamland," *Doubt* 19 (October 1947): 282–90.

186. "MFS Eliz. Wilson Writes," *Doubt* 31 (January 1951): 50–52.

187. "Damned Martians," *Doubt* 37 (June 1952): 154.

188. "Where Credit Is Due," *Doubt* 43 (February 1954): 256–58, 258.

189. "Where Credit Is Due," 258.

190. Thayer to Russell, 21 May 1953, FS, Russell papers.

191. "All about Sputs," *Doubt* 56 (March 1958): 460–80.

192. "All about Sputs," 461.

193. Doug Skinner, "John A. Keel: A Brief Biography," https://www.johnkeel.com/?page_id=21, 9 September 2009.

194. John Keel, "I Remember Lemuria, Too," *Fate*, November 1991, 34–37, 52–56.

195. David Clarke, "A New Demonology: John Keel and the Mothman Prophecies," in *Damned Facts: Fortean Essays on Religion, Folklore and the Paranormal*, ed. Jack Hunter (Paphos, Cyprus: Aporetic Press, 2016), 54–68.

196. John Keel, *UFOs: Operation Trojan Horse* (New York: G. P. Putnam's Sons, 1970).

197. "John Keel," *Telegraph*, 10 July 2009, https://www.telegraph.co.uk/news/obituaries/science-obituaries/5797746/John-Keel.html.

198. Peter M. Rojcewicz, "The 'Men in Black' Experience and Tradition: Analogues with the Traditional Devil Hypothesis," *Journal of American Folklore* 100 (1987): 148–60; Peter M. Rojcewicz, "Between One Eye Blink and the Next: Fairies, UFOs, and Problems of Knowledge," in *The Good People: New Fairylore Essays*, ed. Peter Navráez (Lexington: University of Kentucky Press, 1997), 479–514.

199. John Keel, *Disneyland of the Gods* (Los Angeles: Amok Press, 1988), n.p.

200. Jerome Clark, "John Keel vs. UFOlogy, *Fortean Times*, March 2002, 39–42, 42.

Chapter 6

1. Thayer to Russell, 1 August 1959, Eric Frank Russell papers, University of Liverpool.

2. Thayer to Russell, 7 March 1959, FS, Russell papers.

3. "Special Attention!!!" *Doubt* 60 (April 1959): 56.

4. "This Computer Age, or Leave It to Singer," *Doubt* 61 (Summer 1959): 68–71.

5. "Third Plateau," *Doubt* 60 (April 1959): 54–56.

6. Kathleen Thayer to Russell, 26 September 1959, Russell papers.

7. Thayer to Russell, 7 March 1953, FS, Russell papers.

8. "Dear No-Name," *Doubt* 61 (Summer 1959): 73–75.

9. Anne Ruby, "The Making of an Atlantean Scholar," *Venture Inward*, July–August 1999, 22–25, 42–43.

10. "Atlantis Research," *Doubt* 31 (January 1951): 57.

11. "More Respectability," *Doubt* 28 (April 1950): 5.

12. Joshua Blu Buhs, "David Humiston Kelley as a Fortean," 6 July 2016, http://www.joshuablubuhs.com/blog/david-humiston-kelley-as-a-fortean.

13. Dennis Tedlock, *The Olson Codex: Projective Verse and the Problem of Mayan Glyphs* (Albuquerque: University of New Mexico Press, 2017), xi.

14. Peter Mathews, "In Memoriam: David Humiston Kelley," Mesoweb, 24 May 2011, http://www.mesoweb.com/reports/DavidKelley.html.

15. Christine Stansell, *American Moderns: Bohemian New York and the Creation of a New Century* (Princeton, NJ: Princeton University Press, 2000).

16. Joshua Blu Buhs, "Charles Hutchins Hapgood as a Fortean," 16 June 2017, http://www.joshuablubuhs.com/blog/charles-hutchins-hapgood-as-a-fortean.

17. Charles Hapgood, *The Path of the Pole* (Kempton, IL: Adventures Unlimited Press, 1999), xviii.

18. Michael D. Gordin, *The Pseudoscience Wars: Immanuel Velikovsky and the Birth of the Modern Fringe* (Chicago: University of Chicago Press, 2012).

19. Robert Plumb, "Engineer Says Vast Polar Ice Cap Could Tip Earth Over at Any Time," *New York Times*, 30 August 1948, 19.

20. Waldemar Julsrud, *Enigmas del Pasado* (Acámbaro, Mexico: n.p., 1947); Lowell Harmer, "Mexico Finds Give Hint of Lost World: Dinosaur Statues Point to Men Who Lived in Age of Reptiles," *Los Angeles Times*, 25 March 1951, B1; William N. Russell, "Did Man Tame the Dinosaur?" *Fate*, February–March 1952, 20–27.

21. Hapgood to Sanderson, 9 January 1958, 14 January 1960; Hehr to Hapgood, 26 January 1959, 3 September 1959, Charles Hutchins Hapgood papers, Yale University.

22. Sanderson to Hapgood, 12 July 1959, Hapgood papers.

23. Russell, "Did Man Tame the Dinosaur?" 20–27; Charles DiPeso, "The Clay Figurines of Acambaro, Guanajuato, Mexico," *American Antiquity* 18 (1953): 388–89.

24. Hapgood to Sanderson, 13 January 1957; Sanderson to Hapgood, 20 January 1957, Hapgood papers.

25. Sanderson to Hapgood, 20 January 1957, Hapgood papers.

26. Sanderson to Hapgood, 18 September 1957, Hapgood papers.

27. Hapgood to Sanderson, 23 September 1957, Hapgood papers.

28. Sanderson to Hapgood, 29 March 1958, Hapgood papers.

29. Sanderson to Hapgood, 24 March 1958, Hapgood papers.

30. Sanderson to Hapgood, 10 July 1959, Hapgood papers.

31. Hapgood to Sanderson, 29 September 1958, Hapgood papers; Charles H. Hapgood, "Earth's Shifting Crust," *Saturday Evening Post*, 10 January 1959, 64–69; Charles H. Hapgood, *Great Mysteries of the Earth* (New York: Putnam, 1960); Charles H. Hapgood, "The Mystery of the Frozen Mammoths," *Coronet*, September 1960, 76.

32. P. Schuyler Miller, "The Reference Library," *Astounding*, March 1959, 140–41.

33. Sanderson to Hapgood, 5 July 1959, Hapgood papers.

34. Hapgood to Sanderson, 16 July 1958, 10 April 1959, 21 April 1959, Hapgood papers.

35. Sanderson to Hapgood, 22 January 1960; Frederick Hehr to Hapgood, 9 February 1959, Hapgood papers.

36. Hapgood to Sanderson, 5 December 1958, Hapgood papers.

37. Charles Hapgood, *Mystery in Acambaro* (Brattleboro, VT: Griswold Offset Printing, 1973).

38. "Speaking of Books," *Doubt* 58 (October 1958): 22–23.

39. George Sterling, "The Thirst of Satan," *Beyond the Breakers and Other Poems* (San Francisco: A. M. Robertson, 1914), 51.

40. Charles Fort, *The Books of Charles Fort* (New York: Henry Holt, 1941), 211.

41. "Spiritual Vampirism by George Christian Bump," *Doubt* 26 (October 1949): 400.

42. Vincent H. Gaddis, "The Deadly Bermuda Triangle," *Argosy*, February 1964, 28–29, 116–18; Vincent H. Gaddis, *Invisible Horizons* (New York: Ace, 1965), 190.

43. Ivan T. Sanderson, *Animal Treasure* (New York: Viking, 1937), 301.

44. Fort, *Books*, 877.

45. Franklin Walker, *The Seacoast of Bohemia: An Account of Early Carmel* (San Francisco: Book Club of California, 1966).

46. Kevin Starr, *Americans and the California Dream, 1850–1915* (New York: Oxford University Press, 1988), 270; "Nora May French, Poetess, Ends Life by Taking Poison," *San Francisco Chronicle*, 15 November 1907, 1.

47. Louis J. Zanine, *Mechanism and Mysticism: The Influence of Science on the Thought and Work of Theodore Dreiser* (Philadelphia: University of Pennsylvania Press, 1993), 117–18.

48. Fort, *Books*, 877.

49. Ivan T. Sanderson, *Invisible Residents: The Reality of Underwater UFOs* (New York: World Publishing, 1970).

50. Brett Neilson, *Free Trade in the Bermuda Triangle—and Other Tales of Counterglobalization* (Minneapolis: University of Minnesota Press, 2004), 45.

51. Fort, *Books*, 883–84.

52. Benjamin DeCasseres, *Fantasia Impromptu and Finis*, ed. Kevin I. Slaughter (n.p.: Underworld Amusements, 2016), 182.

53. Neilson, *Free Trade*, 21.

54. Fort, *Books*, 895–96.

55. Neilson, *Free Trade*, 21.

56. *Bump v. Commissioner*, 1977 T.C. Memo. 109, 36 T.C.M. 491, 1977 Tax Ct. Memo LEXIS 331.

57. George Christian Bump, "Poor Quality of Postal Service," *Stanford Daily*, 7 December 1978, 4; George Christian Bump, "'Insane' Sensitivity," *Stanford Daily*, 3 January 1979, 4.

58. George Christian Bump, "Blame Pregnancy on Amoral Society," *Stanford Daily*, 4 December 1981, 4.

59. Fort, *Books*, 895.

60. Thayer to Russell, 1 March 1958, 15 September 1958, Russell papers.

61. Thayer to Russell, 10 January 1959, FS, Russell papers.

62. Thayer to Russell, 24 May 1958, Russell papers.

63. T. Swann Harding, *The Joy of Ignorance* (New York: W. Godwin, 1932), 171.

64. "Tiffany Thayer, Author, 57, Dies," *Bridgeport Post* (Bridgeport, CT), 24 August 1959, 25.

65. "Yacht Racing Opens at Nantucket," *Herald* (Portsmouth, NH), 21 August 1959, 8; Toby Price, "Boatmen's Corner," *Herald* (Portsmouth, NH), 29 August 1959, 9.

66. "Tiffany Thayer Is Dead at 57," *New York Daily News*, 24 August 1959, 100.

67. "Tiffany Thayer Is Dead at 57," 100.

68. Kathleen Thayer to Russell, n.d. (ca. August 1959), Russell papers.

69. "Tiffany Ellsworth Thayer," 13 September 2013, https://www.findagrave.com/memorial/112268030/tiffany-ellsworth-thayer.

70. Piglet and Diana to Damon Knight, n.d., Damon Knight papers, Syracuse University.

71. Thayer to Russell, 3 February 1958, Russell papers.

72. Shroyer to Russell, 31 August 1956, Russell papers.

73. Shroyer to Russell, 21 October 1959, Russell papers.

74. I. O. Evans to Russell, 17 January 1960, Russell papers.

75. Campbell to Russell, 20 September 1959, Russell papers.

76. George Peacock to Russell, 13 August 1960, Russell papers.

77. Kathleen Thayer to Bloch, 26 December 1959, Donald Beaty Bloch papers, New York Public Library.

78. Kathleen Thayer to Russell, 13 October 1964, Russell papers.

79. Kathleen Thayer to Russell, 26 September 1963, Russell papers.

80. Kathleen Thayer to Russell, 21 July [1964], Russell papers.

81. Damon Knight, *Charles Fort: Prophet of the Unexplained* (New York: Doubleday, 1970), 195–96.

82. Swann Galleries, *The Fine Library of the Late Tiffany Thayer: Sold by Order of Mrs. Tiffany Thayer (with a few additions from others)* (New York: Swann Galleries, 1960).

83. Ronald J. Willis to Roy Lavender, 19 December 1966, Russell papers.

84. Roy Lavender to Gentlemen, 10 December [1966], Russell papers.

85. Charlotte Sleigh, "'An Outcry of Silences': Charles Hoy Fort and the Uncanny Voices of Science," in *The Silences of Science: Gaps and Pauses in the Communication of Science*, ed. Felicity Mellor and Stephen Webster (London: Routledge, 2016), 274–95.

86. Birchby to Russell, 23 September 1961, Russell papers.

87. Birchby to Russell, 22 January 1970, Russell papers.

88. Knight to Russell, 8 April 1967; Kathleen Thayer to Damon Knight, 24 October 1967, Knight papers.

89. Knight to Russell, 2 February 1968, Knight papers.

90. Leslie Frewin, *The Late Mrs. Dorothy Parker* (New York: Macmillan, 1986), 121; F. Scott Fitzgerald, *My Lost City: Personal Essays, 1920–1940*, ed. James L. West III (Cambridge: Cambridge University Press, 2005), 148.

91. Arthur M. Schlesinger Jr., *A Life in the Twentieth Century: Innocent Beginnings, 1917–1950* (Boston: Houghton Mifflin Company, 2000), 85.

92. Roy Lavender to Russell, n.d., Russell papers.

93. Thayer to Russell, 28 December 1957, FS, Russell papers.

94. Thayer to Russell, 3 February 1958, Russell papers.

95. Adam Roberts, *The History of Science Fiction* (London: Palgrave Macmillan, 2005), 230–33.

96. Lionel Fanthorpe, *Time Echo* (London: John Spencer, 1959); Stanisław Lem, *The Investigation*, trans. Adele Milch (New York: Harcourt Brace Jovanovich, 1986).

97. Richard Cándida Smith, "The Elusive Quest of the Moderns," in *On the Edge of America: California Modernist Art, 1900–1950*, ed. Paul J. Karlstrom (Berkeley: University of California Press, 1996), 21–40.

98. Kevin Dann, *Across the Great Divide: The Naturalist Myth in America* (New Brunswick, NJ: Rutgers University Press, 2000), 218.

99. "Vincent Gaddis, 1913–1997," *Strange Magazine*, Spring 1998, 29, 58;

Jeffrey J. Kripal, *Authors of the Impossible: The Paranormal and the Sacred* (Chicago: University of Chicago Press, 2010), 9, 123.

100. Delores Ballard, "'Grindstone Philosopher' Advocates Variety," *Jackson* (TN) *Sun*, 23 March 1972, 1, 26.

101. Paul J. Willis to Bloch, 14 November 1966, Bloch papers; Paul J. Willis to Roy Lavender, 29 January 1967, Carlos Roy Lavender papers, Temple University; Norman G. Markham to Knight, 13 February 1968, Knight papers.

102. Ron Willis to Roy Lavender, 19 December 1966; Paul Willis to Roy Lavender, 29 January 1967, Russell papers.

103. Paul Willis to Roy Lavender, 29 January 1969, Russell papers.

104. Robert Barbour Johnson, "The Case of the Stanford 'Apports,'" *INFO Journal*, February 1974, 20–22; Joshua Blu Buhs, *Bigfoot: The Life and Times of a Legend* (Chicago: University of Chicago Press, 2009), 169–83.

105. Jyotsna Sreenivasan, *Utopias in American History* (Santa Barbara, CA: ABC-CLIO, 2008), 330–34.

106. Charles H. Hapgood, *Voices of Spirit: Through the Psychic Experience of Elwood Babbitt* (New York: Delacorte Press/S. Lawrence, 1975), e-edition.

107. Knight to Russell, 15 June 1970, Russell papers.

108. Amy Hale, "John Michell, Radical Traditionalism, and the Emerging Politics of the Pagan New Right," *Pomegranate* 13 (2011): 77–97, 81.

109. Bob Rickard, "The First Forteans," *Fortean Times*, December 2013, 38–40; Bob Rickard, "The First Forteans: Harold Chibbett's Data of the Damned," *Fortean Times*, January 2014, 50–53.

110. Adam Parfrey, *It's a Man's World: Men's Adventure Magazines, the Postwar Pulps* (Port Townsend, WA: Feral House, 2015).

111. Buhs, *Bigfoot*, 153–58.

112. Richard Cándida Smith, *Utopia and Dissent: Art, Poetry, and Politics in California* (Berkeley: University of California Press, 1996).

113. Nat Freedland, *The Occult Explosion* (New York: Putnam, 1972); James Webb, *The Occult Underground* (La Salle, IL: Open Court, 1974); James Webb, *The Occult Establishment* (La Salle, IL: Open Court, 1976).

114. André Spears, "Warlords of Atlantis: Chasing the Demon of Analogy in the America(s) of Lawrence, Artaud and Olson," *Canadian Review of Comparative Literature* 28 (2001): 245–70.

115. Charles Olson, *The Maximus Poems*, ed. George F. Butterick (Berkeley: University of California Press, 1983), 180.

116. Olson, *Maximus Poems*, 452.

117. Spears, "Warlords of Atlantis," 266.

118. Spears, "Warlords of Atlantis," 265.

119. David Harvey, *The Condition of Postmodernity: An Enquiry into the Origins of Cultural Change* (Cambridge: Blackwell, 1989), 45.

120. Gerhard Hoffman, *From Modernism to Postmodernism: Concepts and Strategies of Postmodern American Fiction* (Amsterdam: Rodopi, 2005).

121. Jerome Clark and Loren Coleman, *The Unidentified: Notes Toward Solving the UFO Mystery* (New York: Warner, 1975).

122. John Michell and Robert Rickard, *Phenomena: A Book of Wonders* (London: Thames & Hudson, 1977), 1–2.

123. Jodi Dean, "The Truth Is Out There: Aliens and the Fugitivity of Postmodern Truth," *Camera Obscura* 14 (1997): 42–74, 45.

124. Timothy Spencer Carr, "Son of Originator of 'Alien Autopsy' Story Casts Doubt on Father's Credibility," *Skeptical Inquirer*, July–August 1997, 31–32.

125. Chris D. Bader, "The UFO Contact Movement from the 1950s to the Present," *Studies in Popular Culture* 17 (1995): 73–90.

126. Walter Stephens, *Demon Lovers: Witchcraft, Sex, and the Crisis of Belief* (Chicago: University of Chicago Press, 2002).

127. Elaine Scarry, *The Body in Pain: The Making and Unmaking of the World* (New York: Oxford University Press, 1985).

128. William Burroughs, *Naked Lunch* (New York: Grove Atlantic, 2007), 3.

129. Barry Miles, *Call Me Burroughs: A Life* (New York: Twelve, 2013), 2.

130. Terry Wilson, *Perilous Passage: The Nervous System and the Universe in Other Words* (San Francisco: Synergetic Press, 2012); David S. Wills, *Scientologist! William S. Burroughs and the "Weird Cult"* (Temple, PA: Beatdom Books, 2013).

131. Fort, *Books*, 863; William Burroughs, *Nova Express* (New York: Grove Atlantic, 2011), 5; "Terry Wilson: Cutting Up for Real," RealityStudio: A William S. Burroughs Community, https://realitystudio.org/interviews/terry-wilson-cutting-up-for-real/, 20 February 2012.

132. William Burroughs, *The Soft Machine* (New York: Grove Atlantic, 2011), 149.

133. "Cut-Up Poems from *Minutes to Go*," RealityStudio, https://realitystudio.org/texts/the-poetry-of-william-s-burroughs/cut-up-poems-from-minutes-to-go/, August 2010.

134. Fredric Jameson, "Postmodernism and Consumer Society," in *The Anti-Aesthetic: Essays on Postmodern Culture*, ed. Hal Foster (Seattle, WA: Bay Press, 1983), 111–25.

135. Harvey, *Condition of Postmodernity*, 58, 346.

136. Edward S. Robinson, *Shift Linguals: Cut-Up Narratives from William S. Burroughs to the Present* (Amsterdam: Rodolpi, 2011).

137. Adam Gorightly, *The Prankster and the Conspiracy: The Story of Kerry Thornley and How He Met Oswald and Inspired the Counterculture* (New York: Paraview Press, 2003).

138. Eric Wagner, *An Insider's Guide to Robert Anton Wilson* (Las Vegas, NV: New Falcon, 2004).

139. Erik Davis, *High Weirdness: Drugs, Esoterica, and Visionary Experiences in the Seventies* (Cambridge, MA: MIT Press, 2019).

140. Tanner F. Boyle, *The Fortean Influence on Science Fiction: Charles Fort and the Evolution of the Genre* (Jefferson, NC: McFarland, 2020).

141. Richard Hofstadter, "The Paranoid Style in American Politics," *Harper's*, November 1964, 77–86, 77.

142. Hofstadter, "Paranoid Style," 86.

143. Hofstadter, "Paranoid Style," 86.

144. Kathryn Olmsted, *Real Enemies: Conspiracy Theories and American Democracy, World War I to 9/11* (New York: Oxford University Press, 2011).

145. Brian Balogh, *Chain Reaction: Expert Debate and Public Participation in American Commercial Nuclear Power 1945–1975* (Cambridge: Cambridge University Press, 1991).

146. Wouter J. Hanegraaff, "Fiction in the Desert of the Real: Lovecraft's Cthulhu Mythos," *Ares* 7 (2007): 85–109.

147. Russell Muirhead and Nancy L. Rosenblum, *A Lot of People Are Saying: The New Conspiracism and the Assault on Democracy* (Princeton, NJ: Princeton University Press, 2020), 3.

148. Adrienne LaFrance, "The Prophecies of Q," *Atlantic*, June 2020, https://www.theatlantic.com/magazine/archive/2020/06/qanon-nothing-can-stop-what-is-coming/610567/.

149. Muirhead and Rosenblum, *Lot of People*, 4.

150. Charles Fort, *The Book of the Damned* (New York: Boni & Liveright, 1919), 1.

151. Dickey, *Unidentified*, 45–46.

152. Caitlín R. Kiernan, *To Charles Fort, with Love* (Burton, MI: Subterranean Press, 2005), 11.

153. Caitlín R. Kiernan, "Boredom, Idle Hands, Meme," *Dear Sweet Filthy World*, 16 April 2016, https://greygirlbeast.livejournal.com/226191.html.

154. Jeff VanderMeer, "Interview: Caitlín R. Kiernan on Weird Fiction," *Weird Fiction Review*, 12 March 2012, https://weirdfictionreview.com/2012/03/interview-Caitlín-r-kiernan-on-weird-fiction/.

155. Caitlín R. Kiernan, "Sentient Globs of Pleiomorphies," *Dear Sweet Filthy World*, 30 April 2020, https://greygirlbeast.livejournal.com/1544222.html.

156. Ian McDowell, "Caitlín R. Kiernan," *Encyclopedia of Alabama*, 15 December 2021, http://encyclopediaofalabama.org/article/h-3823.

157. Caitlín R. Kiernan, *The Drowning Girl* (New York: Ace, 2012), n.p.

158. S. T. Joshi, blog entry, 31 August 2013, http://stjoshi.org/news2013.html.

159. Dickey, *Unidentified*, 284.

160. Caitlín R. Kiernan, "At Once Menacing and Hard to See," *Dear Sweet Filthy World*, 1 October 2005, https://greygirlbeast.livejournal.com/165030 .html.

161. Caitlín R. Kiernan, "Howard Hughes and the End of November," *Dear Sweet Filthy World*, 30 November 2013, https://greygirlbeast.livejournal .com/1021886.html.

162. Jeremy L. C. Jones, "Finding the Language I Need: A Conversation with Caitlín R. Kiernan," *Clarkesworld*, June 2010, https://clarkesworld magazine.com/kiernan_interview/.

163. VanderMeer, "Interview: Caitlín R. Kiernan."

164. Kiernan, *To Charles Fort*, 81.

Select Bibliography

Archival Sources

Ben Abramson papers, Beinecke Library, Yale University
Anthony Boucher papers, Indiana University
James Blish papers, Oxford University
Donald Beaty Bloch papers, New York Public Library
Caresse Crosby papers, Southern Illinois University
Benjamin DeCasseres papers, New York Public Library
Theodore Dreiser papers, University of Pennsylvania
Charles Fort papers, University of Virginia Library
Fortean Society papers, 1927–1952, University of Virginia Library
Edmond Hamilton papers, Jack Williams Science Fiction Library, Eastern New
    Mexico State University
Charles Hutchins Hapgood papers, Yale University
T. Swann Harding papers, Library of Congress
Ben Hecht papers, Newberry Library, Chicago
Henry Holt and Company papers, Princeton University
International Society of Cryptozoology, Records, Smithsonian Institution
    Archives
Damon Knight papers, Syracuse University
Philip Lamantia papers, University of California, Berkeley
Carlos Roy Lavender papers, Temple University
Waldo Lee McAtee papers, Library of Congress
Bruce Pelz papers, University of California, Riverside
Ezra Pound papers, Indiana University
Eric Frank Russell papers, University of Liverpool
Howard Snively papers, Schenectady (NY) Museum of Innovation and Science
St. John's Reformed Church, Cemetery Records, Maryland State Archives
Booth Tarkington papers, Princeton University

Tiffany Thayer file, Federal Bureau of Investigation archives, Washington, DC
J. Walter Thompson Company, Biographical Information papers, Duke
    University
Edwin H. Wilson papers, Southern Illinois University
Harry Leon Wilson papers, University of California, Berkeley

Select Bibliography

Albanese, Catherine L. *A Republic of Mind and Spirit: A Cultural History of American Metaphysical Religion*. New Haven, CT: Yale University Press, 2006.

Ashley, Mike. *The Time Machines: The Story of the Science-Fiction Pulp Magazines from the Beginning to 1950*. Liverpool: Liverpool University Press, 2001.

———. *Transformations: The Story of the Science Fiction Magazines from 1950 to 1970*. Liverpool: Liverpool University Press, 2005.

Ashley, Mike, and Robert A. W. Lowndes. *The Gernsback Days: A Study of the Evolution of Modern Science Fiction from 1911 to 1936*. Holicong, PA: Wildside Press, 2004.

Asprem, Egil. *The Problem of Disenchantment: Scientific Naturalism and Esoteric Discourse 1900–1939*. Leiden: Brill, 2016.

Bader, Chris D. "The UFO Contact Movement from the 1950s to the Present." *Studies in Popular Culture* 17 (1995): 73–90.

Barkun, Michael. *A Culture of Conspiracy: Apocalyptic Visions in Contemporary America*. Berkeley: University of California Press, 2013.

Bennett, Colin. *Looking for Orthon: The Story of George Adamski, the First Flying Saucer Contactee, and How He Changed the World*. New York: Cosimo Books, 2008.

———. *Politics of Imagination: The Life, Work, and Ideas of Charles Fort*. Manchester, UK: Critical Vision, 2002.

Bennett, Jane. *The Enchantment of Modern Life: Attachments, Crossings, and Ethics*. Princeton, NJ: Princeton University Press, 2016.

Berger, Albert I. "The Magic That Works: John W. Campbell and the American Response to Technology." *Journal of Popular Culture* 5 (1972): 867–943.

Betts, Jonathan. *Time Restored: The Harrison Timekeepers and R. T. Gould, the Man Who Knew (Almost) Everything*. New York: Oxford University Press, 2011.

Blake, Joseph A. "Ufology: The Intellectual Development and Social Context of the Study of Unidentified Flying Objects." *Sociological Review* 27 (1979): 315–37.

Bower, M. Brady. *Unruly Spirits: The Science of Psychic Phenomena in Modern France*. Urbana: University of Illinois Press, 2010.

Boyle, Tanner F. *The Fortean Influence on Science Fiction: Charles Fort and the Evolution of the Genre*. Jefferson, NC: McFarland, 2020.

Bradbrook, M. C. *Malcolm Lowry: His Art and Early Life: A Study in Transformation*. London: Cambridge University Press, 1975.

Buhs, Joshua Blu. *Bigfoot: The Life and Times of a Legend*. Chicago: University of Chicago Press, 2009.

———. "'One Measure a Circle, Beginning Anywhere': Henry Miller and the Fortean Fantasy." *Nexus* 11 (2016): 145–68.

Bullard, Thomas E. *The Myth and Mystery of UFOs*. Lawrence: University of Kansas Press, 2009.

Butterick, George F. "Charles Olson and the Postmodern Advance." *Iowa Review* 11 (1980): 4–27.

———. *A Guide to The Maximus Poems of Charles Olson*. Berkeley: University of California Press, 1981.

Cándida Smith, Richard. "The Elusive Quest of the Moderns." In *On the Edge of America: California Modernist Art, 1900–1950*, edited by Paul J. Karlstrom, 21–40. Berkeley: University of California Press, 1996.

———. *Utopia and Dissent: Art, Poetry, and Politics in California*. Berkeley: University of California Press, 1996.

Caples, Garret, Andrew Joron, and Nancy Joyce Peters, eds. *The Collected Poems of Philip Lamantia*. Berkeley: University of California Press, 2013.

Carter, John. *Sex and Rockets: The Occult World of Jack Parsons*. Port Townsend, WA: Feral House, 2005.

Cheng, John. *Astounding Wonder: Imagining Science and Science Fiction in Interwar America*. Philadelphia: University of Pennsylvania Press, 2013.

Christensen, Paul. *Charles Olson: Call Him Ishmael*. Austin: University of Texas Press, 2014.

Clarke, David. "A New Demonology: John Keel and the Mothman Prophecies." In *Damned Facts: Fortean Essays on Religion, Folklore and the Paranormal*, edited by Jack Hunter, 54–68. Paphos, Cyprus: Aporetic Press, 2016.

Clute, John. *Pardon This Intrusion: Fantastika in the World Storm*. Essex: Beccon, 2011.

Clydesdale, Ruth. "Wild Talent: Malcolm Lowry, Fort and Magic." *Fortean Times*, June 2015, 40–43.

Cohen, Daniel. *The Great Airship Mystery: A UFO of the 1890s*. New York: Dodd, Mead, 1981.

Colavito, Jason. *The Cult of Alien Gods: H. P. Lovecraft and Extraterrestrial Pop Culture*. Buffalo, NY: Prometheus, 2005.

Cornell, Andrew. *Unruly Equality: U.S. Anarchism in the Twentieth Century.* Berkeley: University of California Press, 2016.

Curry, Patrick. *A Confusion of Prophets: Victorian and Edwardian Astrology.* London: Collins & Brown, 1992.,

Daston, Lorraine. "When Science Went Modern." *Hedgehog Review* 18 (2016): 18–32.

Davidson, Michael. *The San Francisco Renaissance: Poetics and Community at Mid-Century.* Cambridge: Cambridge University Press, 1989.

Davis, Erik. *High Weirdness: Drugs, Esoterica, and Visionary Experiences in the Seventies.* Cambridge, MA: MIT Press, 2019.

———. *TechGnosis: Myth, Magic, and Mysticism in the Age of Information.* Berkeley, CA: North Atlantic Books, 2015.

De Ford, Miriam Allen. "Charles Fort: Enfant Terrible of Science." *Magazine of Fantasy and Science Fiction*, January 1954, 105–16.

———. *Up-Hill All the Way: The Life of Maynard Shipley.* Yellow Springs, OH: Antioch Press, 1956.

Dean, Jodi. "The Truth Is Out There: Aliens and the Fugitivity of Postmodern Truth." *Camera Obscura* 14 (1997): 42–74.

DeCasseres, Benjamin. "The Fortean Fantasy." *Thinker*, April 1931, 71–81.

Demos, T. J. "Duchamp's Labyrinth: 'First Papers of Surrealism,' 1942," *October* 97 (2001): 91–119.

Dickey, Colin. *The Unidentified: Mythical Monsters, Alien Encounters, and our Obsession with the Unexplained.* New York: Viking, 2020.

Disch, Thomas M. *The Dreams Our Stuff Is Made Of: How Science Fiction Conquered the World.* New York: Touchstone, 1998.

Dixon, Deborah. "A Benevolent and Sceptical Inquiry: Exploring 'Fortean Geographies' with the Mothman." *Cultural Geographies* 14 (2007): 189–210.

Dreiser, Theodore. *Hey Rub-A-Dub-Dub.* New York: Boni & Liveright, 1920.

———. *A Hoosier Holiday.* New York: John Lane Co., 1916.

Dunnett, Oliver. "The British Interplanetary Society and Cultures of Outer Space, 1930–1970." PhD diss., University of Nottingham, 2011.

During, Simon. *Modern Enchantments: The Cultural Power of Secular Magic.* Cambridge, MA: Harvard University Press, 2004.

Edwards, Martin A. *H. L. Mencken and the Debunkers.* Athens: University of Georgia Press, 1984.

Evans, Bergen. *The Natural History of Nonsense.* New York: Knopf, 1946.

Filreis, Alan. *Counter-Revolution of the Word: The Conservative Attack on Modern Poetry, 1945–1960.* Chapel Hill: University of North Carolina Press, 2012.

Fort, Charles. *The Book of the Damned.* New York: Boni & Liveright, 1919.

———. *The Books of Charles Fort.* New York: Henry Holt, 1941.

———. *The Complete Books of Charles Fort*. New York: Dover, 1974.

———. *Lo!* New York: Claude Kendall, 1931.

———. *New Lands*. New York: Boni & Liveright, 1923.

———. *Wild Talents*. New York: Claude Kendall, 1932.

Frankiel, Sandra Sizer. *California's Spiritual Frontiers: Religious Alternatives in Anglo-Protestantism, 1850–1910*. Berkeley: University of California Press, 1988.

Gaddis, Vincent H. *Invisible Horizons*. New York: Ace, 1965.

Gaddis, William. *The Recognitions*. New York: NYRB Classics, 2020.

Gardner, Martin H. *Fads and Fallacies in the Name of Science*. New York: Dover, 1957.

———. *Undiluted Hocus-Pocus: The Autobiography of Martin Gardner*. Princeton, NJ: Princeton University Press, 2015.

Gay, Peter. *Modernism: The Lure of Heresy*. New York: Norton, 2007.

Gere, Cathy. *Knossos and the Prophets of Modernism*. Chicago: University of Chicago Press, 2009.

Godwin, Joscelyn. *The Theosophical Enlightenment*. Albany: State University of New York Press, 1994.

Gordin, Michael D. *The Pseudoscience Wars: Immanuel Velikovsky and the Birth of the Modern Fringe*. Chicago: University of Chicago Press, 2012.

Gorightly, Adam. *The Prankster and the Conspiracy: The Story of Kerry Thornley and How He Met Oswald and Inspired the Counterculture*. New York: Paraview Press, 2003.

Grace, Sherrill E. "The Creative Process: An Introduction to Time and Space in Malcolm Lowry's Fiction." *Studies in Canadian Literature* 2 (1977): 61–68.

———. *The Voyage That Never Ends: Malcolm Lowry's Fiction*. Vancouver: University of British Columbia Press, 1982.

Graebner, William. *The Age of Doubt: American Thought and Culture in the 1940s*. Long Grove, IL: Waveland Press, 1991.

Gross, Loren E. *Charles Fort, the Fortean Society, and Unidentified Flying Objects*. Fremont, CA: n.p., 1976.

Gudger, E. W. "Rains of Fishes—Myth or Fact?" *Science* 103 (1946): 683–94.

Hale, Amy. "John Michell, Radical Traditionalism, and the Emerging Politics of the Pagan New Right." *Pomegranate* 13 (2011): 77–97.

Hamalian, Linda. *The Cramoisy Queen: A Life of Caresse Crosby*. Carbondale: Southern Illinois University Press, 2005.

———. *A Life of Kenneth Rexroth*. New York: Norton, 1992.

Hanegraaff, Wouter J. "Fiction in the Desert of the Real: Lovecraft's Cthulhu Mythos." *Ares* 7 (2007): 85–109.

———. "The Theosophical Imagination." *Correspondences* 6 (2017): 1–37.

Hapgood, Charles. *Earth's Shifting Crust: A Key to Some Basic Problems of Earth Science.* New York: Pantheon, 1958.

Harding, T. Swann. *The Degradation of Science.* New York: Farrar & Rinehart, 1931.

———. *The Joy of Ignorance.* New York: W. Godwin, 1932.

———. *Why I Am a Skeptic.* Little Blue Book 1334. Girard, KS: Haldeman-Julius Publications, 1929.

Härmänmaa, Marja, and Christopher Nissen, eds. *Decadence, Degeneration, and the End: Studies in the European Fin de Siècle.* London: Palgrave Macmillan, 2014.

Harrington, Anne. *Reenchanted Science: Holism in German Culture from Wilhelm II to Hitler.* Princeton, NJ: Princeton University Press, 1996.

Harvey, David. *The Condition of Postmodernity: An Enquiry into the Origins of Cultural Change.* Cambridge: Blackwell, 1989.

Henderson, Linda Dalrymple. "Editor's Introduction: I. Writing Modern Art and Science—An Overview; II. Cubism, Futurism, and Ether Physics in the Early Twentieth Century." *Science in Context* 17 (2004): 423–66.

Hess, David J. *Science in the New Age: The Paranormal, Its Defenders and Debunkers, and American Culture.* Madison: University of Wisconsin Press, 1993.

Hobbs, Stuart D. *The End of the American Avant-Garde.* New York: New York University Press, 1997.

Hoeveler, David J. *The New Humanism: A Critique of Modern America, 1900–1940.* Charlottesville: University Press of Virginia, 1977.

Hoffman, Adina. *Ben Hecht: Fighting Words, Moving Pictures.* New Haven, CT: Yale University Press, 2019.

Hoyle, Arthur. *The Unknown Henry Miller: A Seeker in Big Sur.* New York: Arcade Publishing, 2014.

Iktomi. *America Needs Indians.* Denver: Bradford-Robinson, Printers, 1937.

Ingham, John L. *Into Your Tent: The Life, Work and Family Background of Eric Frank Russell.* UK: Plantech, 2010.

Jacoby, Susan. *Freethinkers: A History of American Secularism.* New York: Metropolitan Books, 2004.

Jameson, Fredric. "Postmodernism and Consumer Society." In *The Anti-Aesthetic: Essays on Postmodern Culture,* edited by Hal Foster, 111–25. Seattle, WA: Bay Press, 1983.

Josephson-Storm, Jason Ā. *The Myth of Disenchantment: Magic, Modernity, and the Birth of the Human Sciences.* Chicago: University of Chicago Press, 2017.

Joshi, S. T. *H. P. Lovecraft: The Decline of the West.* Rockville, MD: Wildside Press, 2016.

Jung, C. G. *Flying Saucers: A Modern Myth of Things Seen in the Skies.* Translated by R. F. C. Hull. Princeton, NJ: Princeton University Press, 1979.

Keel, John A. "The Flying Saucer Subculture." *Journal of Popular Culture* 8 (1975): 871–96.

———. "I Remember Lemuria, Too." *Fate*, November 1991, 34–37, 52–56.

———. "The Man Who Invented Flying Saucers." *Fortean Times*, no. 41 (Winter 1983): 52–57.

Ketterer, David. *Imprisoned in a Tesseract: The Life and Work of James Blish.* Kent, OH: Kent State University Press, 1987.

Kiernan, Caitlín R. *To Charles Fort, with Love.* Burton, MI: Subterranean Press, 2005.

Knight, Damon. *Charles Fort: Prophet of the Unexplained.* New York: Doubleday, 1970.

Knuth, Leo. "'Finnegans Wake': A Product of the Twenties." *James Joyce Quarterly* 11 (1974): 310–22.

Koenig, Peter William. "Recognizing Gaddis' *Recognitions*." *Contemporary Literature* 16 (1975): 61–72.

Kripal, Jeffrey J. *Authors of the Impossible: The Paranormal and the Sacred.* Chicago: University of Chicago Press, 2010.

Krissdóttir, Morine. *Descents of Memory: A Life of John Cowper Powys.* New York: Overlook Press, 2007.

Landauer, Susan. "Painting under the Shadow: California Modernism and the Second World War." In *On the Edge of America: California Modernist Art, 1910–1950,* edited by Paul J. Karlstrom, 41–68. Berkeley: University of California Press, 1996.

Lears, T. J. Jackson. *No Place of Grace: Antimodernism and the Transformation of American Culture, 1880–1920.* New York: Pantheon, 1981.

Levenson, Michael. *A Genealogy of Modernism: A Study of English Literary Doctrine 1908–1922.* Cambridge: Cambridge University Press, 1986.

Locke, Simon. *Re-crafting Rationalization: Enchanted Science and Mundane Mysteries.* London: Routledge, 2016.

Lorraine, Lilith. *Let the Patterns Break.* Rogers, AR: Avalon Press, 1947.

Loving, Jerome. *The Last Titan: A Life of Theodore Dreiser.* Berkeley: University of California Press, 2005.

Lyau, Bradford. *The Anticipation Novelists of 1950s French Science Fiction: Stepchildren of Voltaire.* Jefferson, NC: McFarland, 2014.

Manchester, William. *Disturber of the Peace: The Life of H. L. Mencken.* New York: Collier, 1962.

McAtee, W. L. "Showers of Organic Matter." *Monthly Weather Review*, May 1917, 217–24.

Meyer, Donald H. "Secular Transcendence: The American Religious Humanists." *American Quarterly* 34 (1982): 524–42.

Miles, Barry. *Call Me Burroughs: A Life*. New York: Twelve, 2013.

Miller, Henry. *Big Sur and the Oranges of Hieronymus Bosch*. New York: New Directions, 1957.

Modern, John Lardas. *Secularism in Antebellum America*. Chicago: University of Chicago Press, 2011.

Monroe, John Warne. *Laboratories of Faith: Mesmerism, Spiritism, and Occultism in Modern France*. Ithaca, NY: Cornell University Press, 2008.

Moody, A. David. *Ezra Pound: Poet*. Vol. 1, *The Young Genius 1885–1920*. London: Oxford University Press, 2007.

———. *Ezra Pound: Poet*. Vol. 3, *The Tragic Years 1939–1972*. New York: Oxford University Press, 2015.

Moore, Steven. *William Gaddis*. New York: Bloomsbury Publishing, 2015.

Moskowitz, Sam. *Strange Horizons: The Spectrum of Science Fiction*. New York: Scribner, 1976.

Muirhead, Russell, and Nancy L. Rosenblum. *A Lot of People Are Saying: The New Conspiracism and the Assault on Democracy*. Princeton, NJ: Princeton University Press, 2020.

Nadis, Fred. *The Man from Mars: Ray Palmer's Amazing Pulp Journey*. New York: Penguin, 2013.

Nall, Joshua. *News from Mars: Mass Media and the Forging of a New Astronomy, 1860–1910*. Pittsburgh, PA: University of Pittsburgh Press, 2019.

Neilson, Brett. *Free Trade in the Bermuda Triangle—and Other Tales of Counterglobalization*. Minneapolis: University of Minnesota Press, 2004.

Nelson, Victoria. *Gothicka: Vampire Heroes, Human Gods, and the New Supernatural*. Cambridge, MA: Harvard University Press, 2012.

Nesbit, Thomas. *Henry Miller and Religion*. New York: Routledge, 2007.

Nevala-Lee, Alec. *Astounding: John W. Campbell, Isaac Asimov, Robert A. Heinlein, L. Ron Hubbard, and the Golden Age of Science Fiction*. New York: HarperCollins, 2018.

Olmsted, Kathryn S. *Real Enemies: Conspiracy Theories and American Democracy, World War I to 9/11*. New York: Oxford University Press, 2011.

Olson, Charles. *The Maximus Poems*. Edited by George F. Butterick. Berkeley: University of California Press, 1983.

Olson, Liesl. *Chicago Renaissance: Literature and Art in the Midwest Metropolis*. New Haven, CT: Yale University Press, 2017.

Oppenheim, Janet. *The Other World: Spiritualism and Psychical Research in England, 1850–1914*. Cambridge: Cambridge University Press, 1985.

Owen, Alex. *The Place of Enchantment: British Occultism and the Culture of the Modern*. Chicago: University of Chicago Press, 2004.

Panshin, Alexei. *Heinlein in Dimension: A Critical Analysis*. Chicago: Advent, 1968.

Panshin, Alexei, and Cory Panshin. *The World beyond the Hill: Science Fiction and the Quest for Transcendence*. Los Angeles: Jeremy P. Tarcher, 1989.

Parfrey, Adam. *It's a Man's World: Men's Adventure Magazines, the Postwar Pulps*. Port Townsend, WA: Feral House, 2015.

Parkinson, Gavin. *Futures of Surrealism: Myth, Science Fiction, and Fantastic Art in France, 1936–1969*. New Haven, CT: Yale University Press, 2015.

———. "Surrealism and Everyday Magic in the 1950s: Between the Paranormal and 'Fantastic Realism.'" *Papers of Surrealism* 11 (2015): 1–22.

Partridge, Christopher. *The Re-enchantment of the West: Alternative Spiritualities, Sacralization, Popular Culture, and Occulture*. Vol. 1. London: T & T Clark International, 2004.

Pauwels, Louis, and Jacques Bergier. *The Morning of the Magicians: Secret Societies, Conspiracies, and Vanished Civilizations*. Translated by Rollo Myers. Rochester, VT: Destiny Books, 2008.

Pearson, Edmund. *Queer Books*. New York: Doubleday, Doran, 1928.

Peebles, Curtis. *Watch the Skies! A Chronicle of the Flying Saucer Myth*. Washington, DC: Smithsonian Institution, 1994.

Price, E. Hoffmann. *Book of the Dead: Friends of Yesteryear: Fictioneers and Others*. Sauk City, WI: Arkham House, 2001.

Rabinowitz, Paula. *American Pulp: How Paperbacks Brought Modernism to Main Street*. Princeton, NJ: Princeton University Press, 2014.

Raia, Courtenay. *The New Prometheans: Faith, Science, and the Supernatural Mind in the Victorian Fin de Siècle*. Chicago: University of Chicago Press, 2019.

Ramaswamy, Sumathi. *The Lost Land of Lemuria: Fabulous Geographies, Catastrophic Histories*. Berkeley: University of California Press, 2004.

Ratner-Rosenhagen, Jennifer. *American Nietzsche: A History of an Icon and His Ideas*. Chicago: University of Chicago Press, 2012.

Rhodes, Chip. *Structures of the Jazz Age: Mass Culture, Progressive Education, and Racial Disclosures in American Modernism*. London: Verso, 1998.

Roberts, Adam. *The History of Science Fiction*. London: Palgrave Macmillan, 2005.

Robinson, Edward S. *Shift Linguals: Cut-Up Narratives from William S. Burroughs to the Present*. Amsterdam: Rodolpi, 2011.

Rojcewicz, Peter M. "The 'Men in Black' Experience and Tradition: Analogues with the Traditional Devil Hypothesis." *Journal of American Folklore* 100 (1987): 148–60.

Rosemont, Franklin, ed. *Surrealism and Its Popular Accomplices*. San Francisco: City Lights, 1980.

Saler, Michael. *As If: Modern Enchantment and the Literary Prehistory of Virtual Reality*. New York: Oxford University Press, 2011.

———. "Clap If You Believe in Sherlock Holmes: Mass Culture and the Re-enchantment of Modernity, c. 1890–c. 1940." *Historical Journal* 46 (2003): 599–622.

———. "Modernity and Enchantment: A Historiographic Review." *American Historical Review* 111 (2006): 692–716.

———. "Modernity, Disenchantment, and the Ironic Imagination." *Philosophy and Literature* 28 (2004): 137–49.

Sanderson, Ivan T. *Investigating the Unexplained: A Compendium of Disquieting Mysteries of the Natural World*. Englewood Cliffs, NJ: Prentice-Hall, 1972.

———. *Invisible Residents: The Reality of Underwater UFOs*. New York: World Publishing, 1970.

Schneidau, Herbert N. "Vorticism and the Career of Ezra Pound." *Modern Philology* 65 (1968): 214–27.

Schwartz, Stephen. *From West to East: California and the Making of the American Mind*. New York: Free Press, 1998.

Seabrook, Jack. *Martians and Misplaced Clues: The Life and Work of Fredric Brown*. Bowling Green, OH: Bowling Green State University Popular Press, 1993.

Sharp, Lynn L. *Secular Spirituality: Reincarnation and Spiritism in Nineteenth-Century France*. Plymouth, UK: Lexington Books, 2006.

Sleigh, Charlotte. "'An Outcry of Silences': Charles Hoy Fort and the Uncanny Voices of Science." In *The Silences of Science: Gaps and Pauses in the Communication of Science*, edited by Felicity Mellor and Stephen Webster, 274–95. London: Routledge, 2016.

———. "Writing the Scientific Self: Samuel Butler and Charles Hoy Fort." *Journal of Literature and Science* 8 (2015): 17–35.

Smith, H. Allen. *Low Man on a Totem Pole*. New York: Doubleday, Doran, 1945.

Spears, André. "Warlords of Atlantis: Chasing the Demon of Analogy in the America(s) of Lawrence, Artaud and Olson." *Canadian Review of Comparative Literature* 28 (2001): 245–70.

Stableford, Brian. *Narrative Strategies in Science Fiction and Other Essays on Imaginative Fiction*. Holicong, PA: Wildside Press, 2009.

Stark, Robert. *Ezra Pound's Early Verse and Lyric Tradition*. Edinburgh: Edinburgh University Press, 2012.

Steinmeyer, Jim. *Charles Fort: The Man Who Invented the Supernatural*. New York: Penguin, 2008.

Stephens, Walter. *Demon Lovers: Witchcraft, Sex, and the Crisis of Belief*. Chicago: University of Chicago Press, 2002.

Stone, Edward. "Whodunit? Moby Dick!" *Journal of Popular Culture* 8 (1974): 280–85.

Sutcliffe, Steven. *Children of the New Age: A History of Spiritual Practices.* London: Routledge, 2002.

Swanberg, W. A. *Dreiser.* New York: Charles Scribner's Sons, 1965.

Szalay, Michael. *New Deal Modernism: American Literature and the Invention of the Welfare State.* Durham, NC: Duke University Press, 2000.

Talman, Charles Fitzhugh. *Meteorology: The Science of the Atmosphere.* New York: P. F. Collier & Son, 1922.

———. *The Realm of the Air.* Indianapolis, IN: Bobbs-Merrill, 1931.

Taylor, Charles. *A Secular Age.* Cambridge, MA: Harvard University Press, 2007.

Tedlock, Dennis. *The Olson Codex: Projective Verse and the Problem of Mayan Glyphs.* Albuquerque: University of New Mexico Press, 2017.

Thayer, Tiffany. *Little Dog Lost.* New York: Julian Messner, 1938.

———. *Thirteen Men.* New York: Claude Kendall, 1930.

———. *33 Sardonics I Can't Forget.* New York: Philosophical Library, 1946.

———. *Three-Sheet.* New York: Liveright, 1932.

Thurlow, Richard C. *Fascism in Britain: From Oswald Mosley's Blackshirts to the National Front, a History 1918–1998.* London: I. B. Tauris, 1998.

Thurs, Daniel P. "Tiny Tech, Transcendent Tech: Nanotechnology, Science Fiction, and the Limits of Modern Science Talk." *Science Communication* 29 (2007): 65–95.

Toronto, Richard. *War over Lemuria: Richard Shaver, Ray Palmer and the Strangest Chapter of 1940s Science Fiction.* Jefferson, NC: McFarland, 2013.

Treitel, Corinna. *A Science for the Soul: Occultism and the Genesis of the German Modern.* Baltimore, MD: Johns Hopkins University Press, 2004.

Turner, Catherine. *Marketing Modernism between the Two World Wars.* Amherst: University of Massachusetts Press, 2003.

Turner, Christopher. *Adventures in the Orgasmatron: How the Sexual Revolution Came to America.* New York: Farrar, Straus and Giroux, 2011.

Villis, Tom. *Reaction and the Avant-Garde: The Revolt against Liberal Democracy in Early Twentieth-Century Britain.* London: I. B. Tauris, 2006.

Walker, Jesse. *The United States of Paranoia: A Conspiracy Theory.* New York: Harper Perennial, 2014.

Warner, Harry, Jr. *All Our Yesterdays: An Informal History of Science Fiction Fandom in the Forties.* Chicago: Advent, 1969.

Weber, Max. "Science as a Vocation." In *From Max Weber: Essays in Sociology,* edited by H. H. Gerth and C. Wright Mills, 129–56. New York: Oxford University Press, 1946.

Weir, David. *Decadence and the Making of Modernism.* Amherst: University of Massachusetts Press, 1995.

———. *Decadent Culture in the United States: Art and Literature against the American Grain, 1890–1926*. Albany: State University of New York Press, 2008.

Westrum, Ron. "Knowledge about Sea-Serpents." *Sociological Review* 27 (1979): 293–314.

———. "Science and Social Intelligence about Anomalies: The Case of Meteorites." *Social Studies of Science* 8 (1978): 461–93.

Wilkens, Matthew. "Nothing as He Thought It Would Be: William Gaddis and American Postwar Fiction." *Contemporary Literature* 51 (2010): 596–628.

Wills, David S. *Scientologist! William S. Burroughs and the "Weird Cult."* Temple, PA: Beatdom Books, 2013.

Wilson, Terry. *Perilous Passage: The Nervous System and the Universe in Other Words*. San Francisco: Synergetic Press, 2012.

Zanine, Louis J. *Mechanism and Mysticism: The Influence of Science on the Thought and Work of Theodore Dreiser*. Philadelphia: University of Pennsylvania Press, 1993.

Index

Page numbers in *italics* refer to illustrations.